MW00814895

Nanobiomaterial Engineering

Pranjal Chandra • Rajiv Prakash
Editors

Nanobiomaterial Engineering

Concepts and Their Applications in Biomedicine and Diagnostics

Editors
Pranjal Chandra
School of Biochemical Engineering
Indian Institute of Technology (BHU)
Varanasi, Uttar Pradesh, India

Rajiv Prakash
School of Materials Science and Technology
Indian Institute of Technology (BHU)
Varanasi, Uttar Pradesh, India

ISBN 978-981-32-9839-2 ISBN 978-981-32-9840-8 (eBook)
https://doi.org/10.1007/978-981-32-9840-8

This Springer imprint is published by the registered company Springer Nature Singapore Pte Ltd.
The registered company address is: 152 Beach Road, #21-01/04 Gateway East, Singapore 189721, Singapore

Contents

1 Nanobiosensor-Based Diagnostic System: Transducers and Surface Materials 1
Gorachand Dutta

2 Biosensors Based on Nanomaterials: Transducers and Modified Surfaces for Diagnostics 15
Marcelo R. Romero and Matías L. Picchio

3 Carbon Quantum Dots: A Potential Candidate for Diagnostic and Therapeutic Application 49
S. Sharath Shankar, Vishnu Ramachandran, Rabina P. Raj, T. V. Sruthi, and V. B. Sameer Kumar

4 Carbon Nanomaterials for Electrochemiluminescence-Based Immunosensors: Recent Advances and Applications 71
Nura Fazira Noor Azam, Syazana Abdullah Lim, and Minhaz Uddin Ahmed

5 Green Synthesis of Colloidal Metallic Nanoparticles Using Polyelectrolytes for Biomedical Applications 91
Ana M. Herrera-González, M. Caldera-Villalobos, J. García-Serrano, and M. C. Reyes-Ángeles

6 Peroxidase-Like Activity of Metal Nanoparticles for Biomedical Applications 109
Swachhatoa Ghosh and Amit Jaiswal

7 Biomedical Applications of Lipid Membrane-Based Biosensing Devices 127
Georgia-Paraskevi Nikoleli, Marianna-Thalia Nikolelis, Spyridoula Bratakou, and Vasillios N. Psychoyios

8 Dextran-based Hydrogel Layers for Biosensors 139
Andras Saftics, Barbara Türk, Attila Sulyok, Norbert Nagy, Emil Agócs, Benjámin Kalas, Péter Petrik, Miklós Fried, Nguyen Quoc Khánh, Aurél Prósz, Katalin Kamarás, Inna Szekacs, Robert Horvath, and Sándor Kurunczi

9 **Electrochemical Nanoengineered Sensors in Infectious Disease Diagnosis** 165
Suryasnata Tripathy, Patta Supraja, and Shiv Govind Singh

10 **Nanobiotechnology Advancements in Lateral Flow Immunodiagnostics** 181
Vivek Borse and Rohit Srivastava

11 **Biological Acoustic Sensors for Microbial Cell Detection** 205
O. I. Guliy, B. D. Zaitsev, A. A. Teplykh, and I. A. Borodina

12 **Nanorobots for In Vivo Monitoring: The Future of Nano-Implantable Devices** 227
Mamta Gandhi and Preeti Nigam Joshi

13 **Nanobiomaterials in Drug Delivery: Designing Strategies and Critical Concepts for Their Potential Clinical Applications** 253
Chang Liu, Zhixiang Cui, Xin Zhang, and Shirui Mao

14 **Future Prospected of Engineered Nanobiomaterials in Human Health Care** 275
Guilherme Barroso L. de Freitas and Durinézio J. de Almeida

Editors and Contributors

About the Editors

Pranjal Chandra is currently employed as Assistant Professor at the School of Biochemical Engineering, Indian Institute of Technology (BHU), Varanasi, India. He earned his Ph.D. from Pusan National University, South Korea, and did his postdoctoral training at Technion-Israel Institute of Technology, Israel. His research focus is highly interdisciplinary, spanning a wide range in biotechnology, nanobiosensors, material engineering, nanomedicine, etc. He has designed several commercially viable biosensing prototypes that can be operated for on-site analysis for biomedical diagnostics. He has published four books on various aspects of biosensors/medical diagnostics from IET London, Springer Nature, and CRC Press, USA. He has also published over 85 journal articles in topmost journals of his research area. His work has been greatly highlighted in over 300 topmost news agencies globally including Rajya Sabha TV; DD Science; Science Trends, USA; Nature India; Vigyan Prasar; Global Medical Discovery, Canada; APBN Singapore; Business Wire; Dublin; etc. He is Recipient of many prestigious awards and fellowships, such as DST Ramanujan Fellowship (Government of India); ECRA (DST, Government of India); BK-21 and NRF Fellowship, South Korea; Technion Postdoctoral Fellowship, Israel; NMS Young Scientist Award; BRSI Young Scientist Award; Young Engineers Award 2018; RSC Highly Cited Corresponding Author Award (general chemistry); ACS/Elsevier Outstanding Reviewer Awards; etc. He is Reviewer of over 40 international journals and Expert Project Reviewer of various national/international funding agencies. He is Associate Editor of Sensors International and an Editorial Board Member of *Materials Science for Energy Technologies* by KeAi and Elsevier; *World Journal of Methodology*, USA; *Frontiers in Bioscience*, USA; and *Reports in Advances of Physical Sciences*, Singapore.

Rajiv Prakash received his M.Tech. degree in Materials Technology from Indian Institute of Technology, Banaras Hindu University IIT (BHU), Varanasi, India, and did his Ph.D. work from Tata Institute of Fundamental Research, Mumbai, India. Presently, he is Professor in the School of Materials Science and Technology, Dean (Research and Development) of the institute, and Professor-in-Charge of the Central Instrument Facility. He was also a Visiting Scientist/Professor at the University of Applied Sciences, Germany, and Kyushu Institute of Technology, Japan (under DST-JSPS). Before joining IIT (BHU), Varanasi, he has served as Scientist in the CSIR Laboratory, Lucknow, India, for more than 7 years. He has been Recipient of many awards including Young Scientist Award (Council of Science and Technology), Young Engineer Awards of India (Indian National Academy of Engineering), and Materials Society Medal Award of India. He is elected as Academician of Asia Pacific Academy of Materials in 2013. His current research interests include morphology-controlled synthesis of electronic polymers, nanocomposites, fabrication and characterization of organic electronic devices, sensors/biosensors, and nanobioengineering. Till now, he has more than 160 publications in international journals of repute and 20 patents to his credit (out of which 2 technologies transferred for commercialization). He is on the Editorial Board of several national and international journals. His interdisciplinary research work has been highly cited and featured on various platforms including nature news. He is Member of various national committees including India Vision Plan 2035 of DST-TIFAC and "IMPRINT" program of MHRD, Government of India.

Contributors

Emil Agócs Photonics Department, Centre for Energy Research, Hungarian Academy of Sciences, Budapest, Hungary

Minhaz Uddin Ahmed Biosensors and Biotechnology Laboratory, Chemical Science Programme, Faculty of Science, Universiti Brunei Darussalam, Gadong, Brunei Darussalam

Nura Fazira Noor Azam Biosensors and Biotechnology Laboratory, Chemical Science Programme, Faculty of Science, Universiti Brunei Darussalam, Gadong, Brunei Darussalam

I. A. Borodina Kotel'nikov Institute of Radio Engineering and Electronics, RAS, Saratov Branch, Saratov, Russia

Vivek Borse NanoBioSens Laboratory, Centre for Nanotechnology, Indian Institute of Technology Guwahati, Guwahati, Assam, India

Spyridoula Bratakou Laboratory of Inorganic & Analytical Chemistry, School of Chemical Engineering, Department 1, Chemical Sciences, National Technical University of Athens, Athens, Greece

M. Caldera-Villalobos Doctorado en Ciencias de los Materiales, Instituto de Ciencias Básicas e Ingeniería, Universidad Autónoma del Estado de Hidalgo, Hidalgo, Mexico

Zhixiang Cui School of Pharmacy, Shenyang Pharmaceutical University, Shenyang, China

Durinézio J. de Almeida Department of Biology, Midwestern State University (UNICENTRO), Guarapuava, Brazil

Guilherme Barroso L. de Freitas Department of Biochemistry and Pharmacology, Federal University of Piauí (UFPI), Teresina, Brazil

Gorachand Dutta School of Medical Science and Technology, Indian Institute of Technology Kharagpur, Kharagpur, West Bengal, India

Miklós Fried Photonics Department, Centre for Energy Research, Hungarian Academy of Sciences, Budapest, Hungary

Mamta Gandhi NCL Venture Center, FastSense Diagnostics Pvt. Ltd, Pune, India

J. García-Serrano Laboratorio de Polímeros, Instituto de Ciencias Básicas e Ingeniería, Universidad Autónoma del Estado de Hidalgo, Hidalgo, Mexico

Swachhatoa Ghosh School of Basic Sciences, Indian Institute of Technology Mandi, Mandi, Himachal Pradesh, India

O. I. Guliy Institute of Biochemistry and Physiology of Plants and Microorganisms, RAS, Saratov, Russia
Saratov State Vavilov Agrarian University, Saratov, Russia

Ana M. Herrera-González Laboratorio de Polímeros, Instituto de Ciencias Básicas e Ingeniería, Universidad Autónoma del Estado de Hidalgo, Hidalgo, Mexico

Robert Horvath Nanobiosensorics Laboratory, Centre for Energy Research, Hungarian Academy of Sciences, Budapest, Hungary

Amit Jaiswal School of Basic Sciences, Indian Institute of Technology Mandi, Mandi, Himachal Pradesh, India

Preeti Nigam Joshi NCL Venture Center, FastSense Diagnostics Pvt. Ltd, Pune, India

Benjámin Kalas Photonics Department, Centre for Energy Research, Hungarian Academy of Sciences, Budapest, Hungary

Katalin Kamarás Institute for Solid State Physics and Optics, Wigner Research Centre for Physics, Hungarian Academy of Sciences, Budapest, Hungary

Nguyen Quoc Khánh Microtechnology Department, Centre for Energy Research, Hungarian Academy of Sciences, Budapest, Hungary

V. B. Sameer Kumar Department of Biochemistry & Molecular Biology, Central University of Kerala, Kasaragod, Kerala, India

Sándor Kurunczi Nanobiosensorics Laboratory, Centre for Energy Research, Hungarian Academy of Sciences, Budapest, Hungary

Syazana Abdullah Lim School of Applied Sciences and Mathematics, Universiti Teknologi Brunei, Gadong, Brunei Darussalam

Chang Liu School of Pharmacy, Shenyang Pharmaceutical University, Shenyang, China

Shirui Mao School of Pharmacy, Shenyang Pharmaceutical University, Shenyang, China

Norbert Nagy Photonics Department, Centre for Energy Research, Hungarian Academy of Sciences, Budapest, Hungary

Georgia-Paraskevi Nikoleli Laboratory of Inorganic & Analytical Chemistry, School of Chemical Engineering, Department 1, Chemical Sciences, National Technical University of Athens, Athens, Greece

Marianna-Thalia Nikolelis Laboratory of Inorganic & Analytical Chemistry, School of Chemical Engineering, Department 1, Chemical Sciences, National Technical University of Athens, Athens, Greece

Péter Petrik Photonics Department, Centre for Energy Research, Hungarian Academy of Sciences, Budapest, Hungary

Matías L. Picchio Facultad de Ciencias Químicas, Departamento de Química Orgánica (FCQ-UNC), Universidad Nacional de Córdoba, Córdoba, Argentina
Facultad Regional Villa María, Universidad Tecnológica Nacional, Córdoba, Argentina

Aurél Prósz Nanobiosensorics Laboratory, Centre for Energy Research, Hungarian Academy of Sciences, Budapest, Hungary

Vasillios N. Psychoyios Laboratory of Inorganic & Analytical Chemistry, School of Chemical Engineering, Department 1, Chemical Sciences, National Technical University of Athens, Athens, Greece

Rabina P. Raj Department of Biochemistry & Molecular Biology, Central University of Kerala, Kasaragod, Kerala, India

Vishnu Ramachandran Department of Biochemistry & Molecular Biology, Central University of Kerala, Kasaragod, Kerala, India

M. C. Reyes-Ángeles Maestría en Ciencias de los Materiales, Instituto de Ciencias Básicas e Ingeniería, Universidad Autónoma del Estado de Hidalgo, Hidalgo, Mexico

Marcelo R. Romero Facultad de Ciencias Químicas, Departamento de Química Orgánica (FCQ-UNC), Universidad Nacional de Córdoba, Córdoba, Argentina
Consejo Nacional de Investigaciones Científicas y Técnicas (CONICET), Instituto de Investigación y Desarrollo en Ingeniería de Procesos y Química Aplicada (IPQA), Córdoba, Argentina

Andras Saftics Nanobiosensorics Laboratory, Centre for Energy Research, Hungarian Academy of Sciences, Budapest, Hungary

S. Sharath Shankar Department of Biochemistry & Molecular Biology, Central University of Kerala, Kasaragod, Kerala, India

Shiv Govind Singh Department of Electrical Engineering, Indian Institute of Technology Hyderabad, Hyderabad, India

Rohit Srivastava NanoBios Laboratory, Department of Biosciences and Bioengineering, Indian Institute of Technology Bombay, Mumbai, Maharashtra, India

T. V. Sruthi Department of Biochemistry & Molecular Biology, Central University of Kerala, Kasaragod, Kerala, India

Attila Sulyok Thin Film Physics Department, Centre for Energy Research, Hungarian Academy of Sciences, Budapest, Hungary

Patta Supraja Department of Electrical Engineering, Indian Institute of Technology Hyderabad, Hyderabad, India

Inna Szekacs Nanobiosensorics Laboratory, Centre for Energy Research, Hungarian Academy of Sciences, Budapest, Hungary

A. A. Teplykh Kotel'nikov Institute of Radio Engineering and Electronics, RAS, Saratov Branch, Saratov, Russia

Suryasnata Tripathy Department of Electrical Engineering, Indian Institute of Technology Hyderabad, Hyderabad, India

Barbara Türk Nanobiosensorics Laboratory, Centre for Energy Research, Hungarian Academy of Sciences, Budapest, Hungary

B. D. Zaitsev Kotel'nikov Institute of Radio Engineering and Electronics, RAS, Saratov Branch, Saratov, Russia

Xin Zhang School of Pharmacy, Shenyang Pharmaceutical University, Shenyang, China

Nanobiosensor-Based Diagnostic System: Transducers and Surface Materials

1

Gorachand Dutta

Abstract

There are increasing needs for the development of simple, cost-effective, portable, integrated biosensors that can be operated outside the laboratory by untrained personnel. The main challenges of point-of-care testing require to implement complex biosensing methods into low-cost technologies. Point-of-care testing is known as medical diagnostic process which is conducted to the near patient and does not need any well-trained personnel. The diagnosis technology should be affordable and disposable to provide the benefits to the large part of the population in developing countries. In this chapter, different nanobiosensors for medical diagnosis using several surface modification strategies of transducers were discussed. A unique redox cycling technology was presented to amplify the signal-to-background ratios for ultrasensitive biomarker detection which are suitable for point-of-site detection such as medical diagnostics, biological research, environmental monitoring, and food analysis. Also, some advanced nanobiosensing technologies including printed circuit board (PCB) were described on the commercial arena for next-generation point-of-care testing.

Keywords

Nanobiosensors · Transducers · Surface functionalization · Point-of-care · Lab-on-a-chip

---

G. Dutta (✉)
School of Medical Science and Technology, Indian Institute of Technology Kharagpur, Kharagpur, West Bengal, India
e-mail: g.dutta@smst.iitkgp.ac.in

P. Chandra, R. Prakash (eds.), *Nanobiomaterial Engineering*,
https://doi.org/10.1007/978-981-32-9840-8_1

1.1 Introduction

Biosensors are detection techniques which utilize one or more biologic recognition elements for specific detection of certain target analyte of interest (Gruhl et al. 2013; Hwang et al. 2018). Major applications for biosensors are clinical diagnosis and disease prevention (Akhtar et al. 2018; Lee et al. 2019a; Li et al. 2019; Wang et al. 2019; Zhang et al. 2019). There has been intense demand of HIV/AIDS, tuberculosis, and vector-borne disease biomarker biosensor (David et al. 2007) for portable devices for rapid and sensitive detection. Various groups and companies in the USA (Whitesides Research Group at Harvard University, Cortez Diagnostics, Inc. California), the UK (Liverpool School of Tropical Medicine and University of Southampton), and France (James P. Di Santo group) have been working on the development of biosensors for HIV/AIDS, tuberculosis, and vector-borne diseases at the point of care. A mobile device named Qpoc Handled Laboratory was designed by the British company QuantuMDx that provides HIV, tuberculosis, and malaria results in 10 min. There has been an increased attempt toward the technological development of biosensing devices for the past about 10 years. In this context, the research on technical feasibility and concept proving in the area of biomolecular electronic devices, technological development of some biosensors and laboratory-level technological development of some biosensors and related biomaterials, was developed. Biosensor technology offers several benefits over conventional diagnostic analysis including simplicity of use, specificity for the target analyte, and capability for continuous monitoring and multiplexing (Chandra et al. 2011; Dutta 2017; He et al. 2018).

A transducer is any device used to convert energy from one form to another (Chandra et al. 2011). A bio-recognition layer and a physicochemical transducer are the two intimately coupled parts in biosensor transducer (Aashish et al. 2018). The biochemical energy converts to an electronic or optical signal. The sensing surface which provides a solid support for the immobilization of the capture biomolecules (i.e., antibody), as well as electron transfer process from the biological/chemical reaction, plays a crucial role for ultrasensitive biomarker detection for POCT. Therefore, selecting an appropriate transducer along with proper surface modifications is an important step to build a highly sensitive biosensor.

Suitable solid surface selection and novel surface chemistry are the most important challenges for developing a viable lab-on-a-chip biosensor (González-Gaitán et al. 2017). There is an increasing demand for a multidisciplinary approach to design and manufacturing of micro-/nanodevices that could be applicable for ultrasensitive biosensing (Khan et al. 2019). Different surface materials were widely used in biosensor to obtain low and reproducible background signal (Zhu et al. 2019). The major requirements for selecting the sensing surface materials depend on (a) biocompatibility with the biological element; (b) nonexistence of diffusion barriers, (c) stability factors with temperature, pH, and ionic strength; (d) specificity and sensitivity of the analyte; and (e) cost-effectiveness for on-site diagnosis.

Over the years, there are increasing needs for the development of a simple, cost-effective, portable, integrated biosensors that can be operated outside the laboratory by untrained personnel (Eltzov and Marks 2016; Siddiqui et al. 2018; Yang et al. 2016). In recent years, many chip-based biosensors have been reported, and so far, the technology for electrochemical amplification on a chip has always relied on expensive micro- and nanofabrication technologies such as optical and e-beam lithography (Cinel et al. 2012). This approach has several drawbacks even when leading to reliable results in research laboratories. Most of the clinical analysis is carried out in centralized laboratories where high-technology equipment is available and trained personnel perform the assays under almost ideal conditions. However, a large part of the population in developing countries does not have access to state-of-the-art diagnostic methods. It is very important to perform highly sensitive detection of biomarkers with printed and flexible electronics for pint-of-site application. With the rise of printed electronics and roll-2-roll technologies, tools have been developed (Liddle and Gallatin 2011) that could potentially make diagnostics available to a much wider population.

A small instrument can offer very stable voltage/current source and detector that are always unnoticed in electrochemical biosensors (Gu et al. 2019; Nze et al. 2019). Therefore, a combination of new signal amplification technology (i.e., redox cycling) and electrochemical detection can play an important role in the development of ultrasensitive and reproducible biosensors for point-of-care testing. Point-of-care testing (POCT) of biomarkers in clinical samples is of great importance for rapid and cost-effective diagnosis (Akanda et al. 2014; Akanda and Ju 2018; Singh et al. 2013; Xiang et al. 2018). However, up to now it is extremely challenging to develop a POCT technique retaining both simplicity and very high ultrasensitivity and simplicity.

This chapter focuses on the different nanobiosensors for medical diagnosis using several surface modification strategies of transducers. Different transducer effects will be discussed for ultrasensitive biosensing. Also, some advanced nanobiosensing technologies will be explored on the commercial arena for next-generation point-of-care testing.

1.2 Different Transducers for Biosensing with Advanced Surface Materials

Over the past two decades, various transducing systems have been applied for highly sensitive biomarker detection. Gold electrodes were widely used in nanobiosensing for its unique redox property, and the extraordinary affinity of thiol compounds for its surface makes these electrodes very suitable for point-of-care immunodiagnostics (Lee et al. 2019b). Thiolated antibody could be immobilized promptly making the sensor fabrication steps easier for biomarker detection (Wang et al. 2017). To develop a washing-free immunosensing technique without any label, Dutta et al. discovered a rapid measurement of protein biomarkers in whole blood samples (Fig. 1.1) using gold transducer and thiolated capture antibody (Dutta and Lillehoj 2018). Using this nanobiosensor, *Pf*HRP2, a malaria biomarker, was quantified from 100 ng/mL to 100 μg/mL in whole blood samples. This method does not

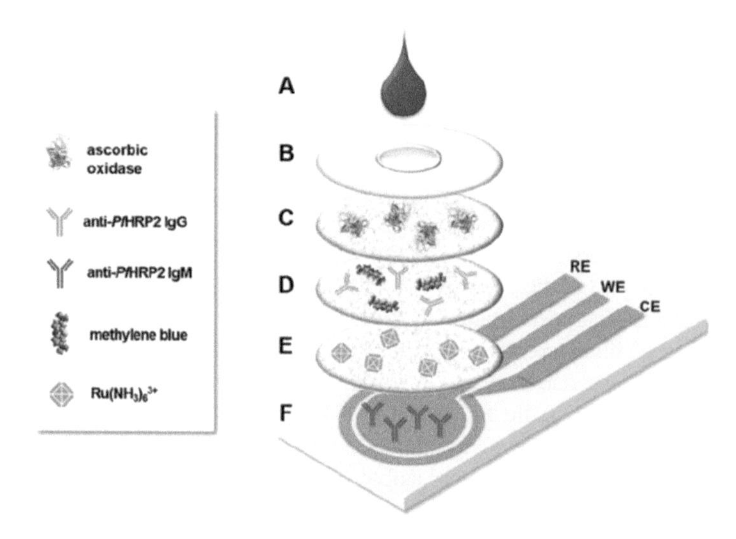

Fig. 1.1 An exploded view of a nanobiosensor using thiolated capture probes and gold transducer. (Reprinted with permission from Dutta and Lillehoj 2018. Copyright (2018) with permission from Nature)

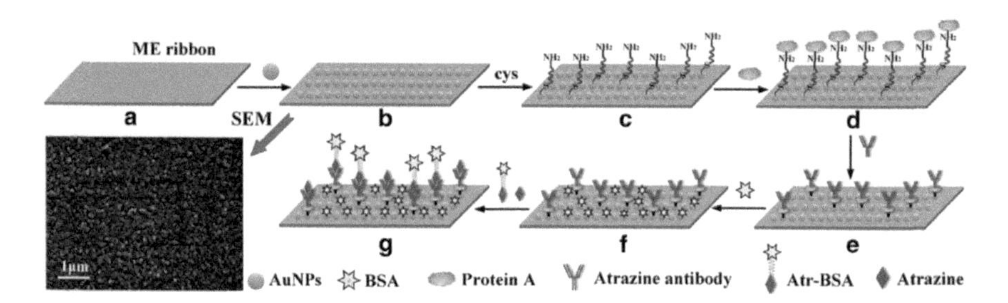

Fig. 1.2 Schematic of the functionalization procedures of the competitive immunoassay (**a**) ME ribbon; (**b**) the AuNP immobilization; (**c**) the SAM formation; (**d**) the protein formation; (**e**) the antibody immobilization; (**f**) BSA coating; and (**g**) atrazine and Atr–BSA competitive mechanism. (Reprinted with permission from Sang et al. 2018. Copyright (2018) with permission from Springer)

require any sample processing, labeling, or washing. Because of excellent stability and very good reproducibility, the device was well suited for point-of-care testing in developing countries.

Sang et al. reported magnetoelastic (ME) nanobiosensor, based on ME materials and gold nanoparticles (AuNPs) (Fig. 1.2), for highly sensitive detection of atrazine employing the competitive immunoassay (Sang et al. 2018). The biosensing results indicated that the ME nanobiosensor displayed strong specificity and stability toward atrazine with detection limit 1 ng/mL. This report also specified a novel convenient method for rapid, selective, and highly sensitive detection of atrazine which has implications for its applications in water quality monitoring and other environmental detection fields. The competitive biosensing scheme was established by oriented immobilization of atrazine antibody to protein A covalently modified on the AuNP-coated ME material surface, followed by the competitive reaction of atrazine–albumin conjugate (Atr–BSA) and atrazine with the atrazine antibody.

Proa-Coronado et al. reported a reduced graphene coated with platinum nanoparticle-based transducer in back-gated field effect transistor (GFET) nanobiosensors (Proa-Coronado 2018). X-ray photoelectron spectroscopy and cyclic voltammetry techniques were used to prove the adsorption-interaction of CS_2 and human serum albumin biomolecules on rGO/Pt. The local deposition of CS_2, rGO/Pt, and protein G was performed by using a commercial microplotter instrument.

A label-free chemiresistor nanobiosensor employing a SWCN chemiresistor transducer functionalized with antidengue NS1 monoclonal antibodies for rapid detection of the dengue nonstructural protein 1 (NS1) was described by Wasik et al. (Wasik et al. 2018) (Fig. 1.3). A wide range of NS1 was detected with high sensitivity and selectivity with the limit of detection 0.09 ng/mL.

A surface-enhanced Raman scattering (SERS) sensor was reported based on the Ag nanorice@Raman label@SiO2 sandwich nanoparticles that are coupled to a periodic Au triangle nanoarray via the linkage of hepatitis B virus (HBV) DNA (Li et al. 2013) (Fig. 1.4). This nanobiosensor is expected to result in the spatially enhanced electromagnetic (EM) field of the quasiperiodic array, leading to ultrasensitive SERS detection. In the sandwich nanoparticles, malachite green isothiocyanate (MGITC) molecules are chosen as the Raman labels that are embedded between the Ag nanorice core and the SiO_2 shell. The detection limit was 50 aM.

A new nanostructured SERS-electrochemical nanobiosensor was developed for screening chemotherapeutic drugs and to aid in the assessment of DNA modification/damage caused by these drugs (Ilkhani et al. 2016). The self-assembled monolayer protected gold-disk electrode (AuDE) was coated with a reduced graphene

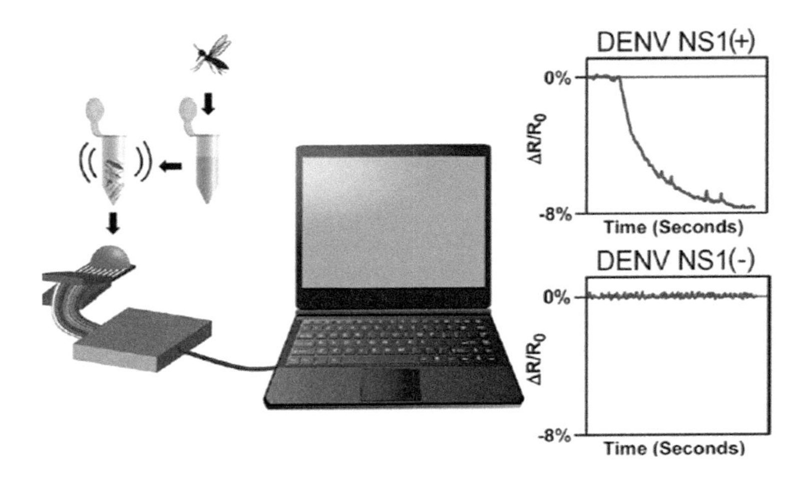

Fig. 1.3 The schematic of a label-free nanobiosensor. (Reprinted with permission from Wasik et al. 2018. Copyright (2018) American Chemical Society)

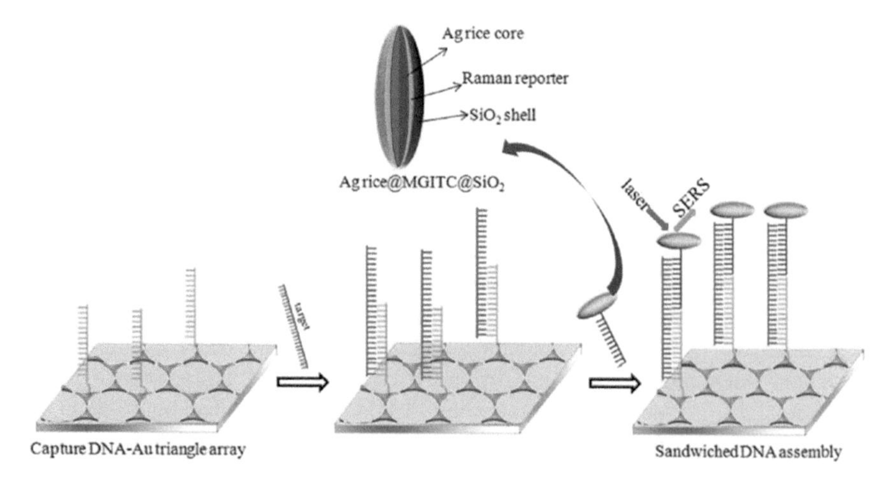

Fig. 1.4 Schematic of the sandwich structure of Ag nanorice@Raman label@SiO2 and the mechanism of SERS sensor. (Reprinted with permission from Li et al. 2013. Copyright (2013) American Chemical Society)

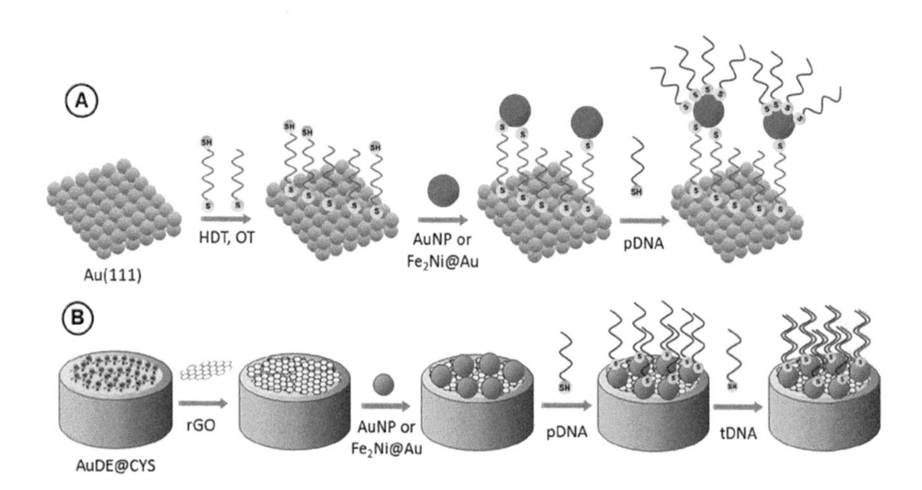

Fig. 1.5 The schematic illustration of SERS/electrochemical nanobiosensors: (**a**) the gold transducer is coated with a SAM of hexanedithiol and octanethiol, and covalently attached AuNPs or Fe_2Ni@Au NPs (magnetic) were functionalized with ssDNA probe; (**b**) SERS/electrochemical nanobiosensor with gold-disk electrode and functionalization of dsDNA. (Reprinted from Ilkhani et al. 2016. Copyright (2016), with permission from Elsevier)

oxide (rGO), decorated with plasmonic gold-coated Fe2Ni@Au magnetic nanoparticles functionalized with double-stranded DNA (dsDNA) (Fig. 1.5). The complete nanobiosensor complex was used to the action of a model chemotherapeutic drug, doxorubicin (DOX), to fit the DNA modification and its dose dependence. These

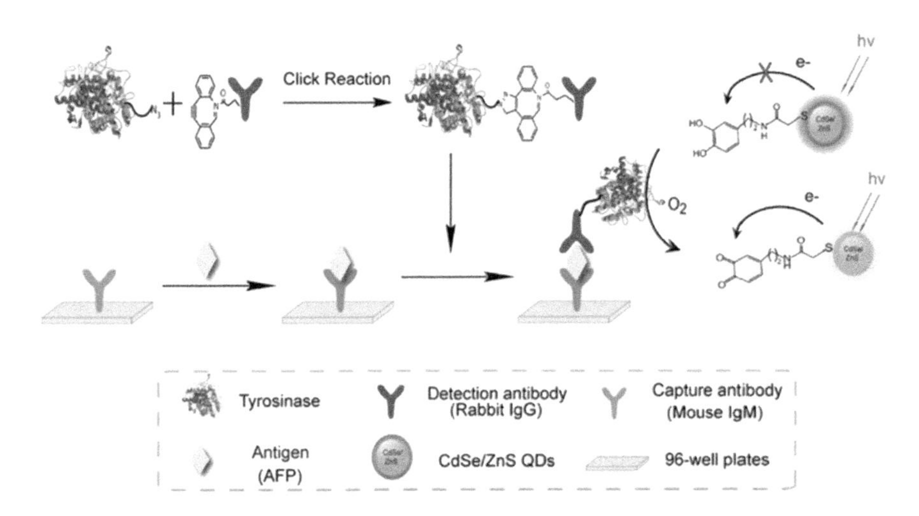

Fig. 1.6 Schematic illustration of the sandwich RMFIA for the detection of disease biomarkers. (Reprinted with permission from Zhang et al. 2016. Copyright (2016) American Chemical Society)

new biosensors are sensitive to agents that interact with DNA and facilitate the analysis of functional groups for determination of the binding mode.

A nanobiosensor for the quantification of acetylcholine (Ach) was reported on the carbon paste electrodes (CPE) modified by copper nanoparticles (Heli et al. 2009) (Fig. 1.6). Cu(III) active species was used to oxidized the Ach. Quantum dots (QDs) have many unique properties and are used in nanobiosensors such as (a) excellent brightness and photostability, (b) wide and continuous excitation spectrum, (c) narrow emission spectrum. A redox-mediated indirect fluorescence immunoassay (RMFIA) for the detection of the disease biomarker α-fetoprotein (AFP) using dopamine (DA)-functionalized CdSe/ZnS quantum dots (QDs) was reported (Fig. 1.6) (Zhang et al. 2016). The detection antibody was conjugated with tyrosinase and used as a bridge connecting the fluorescence signals of the QDs. Different concentration of the disease biomarkers was detected. The immunoassay was sensitive with the detection limit of 10 pM.

1.3 Low Electrocatalytically Active Indium Tin Oxide (ITO) Transducer for Highly Sensitive Biomarker Detection

Indium tin oxide (ITO) is a mixed composition of indium, tin, and oxygen. ITO is widely used in biosensing because of favorable platform in disease diagnosis due to their good electrical conductivity, transparency to visible wavelengths, and high surface-to-volume ratio (Jiaul Haque et al. 2015; Park et al. 2015). In many reported works (Das et al. 2006; Jiaul Haque et al. 2015; Park et al. 2015; Yang 2012), low electrocatalytic indium tin oxide (ITO) electrodes were used to obtain low and reproducible background levels, and low amounts of electroactive species were modified on ITO electrodes to obtain the rapid electrooxidation of substrate molecules. Also, in highly outer-sphere reactions (OSR-philic species), ITO electrodes reacted very slowly with highly inner-sphere reactions (ISR-philic species) like tris(2-carboxyethyl) phosphine (TCEP) (Akanda et al. 2012, 2013). Overall, the outer-sphere to inner-sphere reaction allowed a very low detection limit even in clinical samples.

ITO electrodes are very low electrocatalytically active in electrochemical detection of biomarkers in real samples with very low interfering effect such as ascorbic acid (AA) and uric acid (UA) (Dutta et al. 2014, 2015; Park et al. 2015). ITO electrodes could be used as biosensing surface by modifying with foreign materials, i.e., reduced graphene oxide (rGO), avidin, streptavidin, silicon layer ((3-aminopropyl) triethoxysilane (APTES)), and gold nanoparticles, to obtain high signal-to-noise ratios (S/N) (Akanda et al. 2012; Aziz et al. 2007; Fang et al. 2018; Singh et al. 2013). A new enzyme-free immunosensor-based ITO electrodes where a unique, competitive electrochemical scheme between MB, hydrazine, and Pt nanoparticles (NPs) was used (Dutta et al. 2017) (Fig. 1.7). This nanobiosensor offers several advantages including rapid electrokinetics, high sensitivity, and good reproducibility. Also, because of ITO electrode, this sensor offers very good detection performance even in real samples with minimal interference species effects.

An ultrasensitive immunosensor was developed based on ITO electrodes for the detection of malaria with fg/mL detection limit (Dutta and Lillehoj 2017) (Fig. 1.8). Here authors used an advanced biosensing technology called redox cycling to amplify the signal-to-background ratios. Real samples were investigated based on a unique electrochemical–chemical–chemical (ECC) redox cycling signal

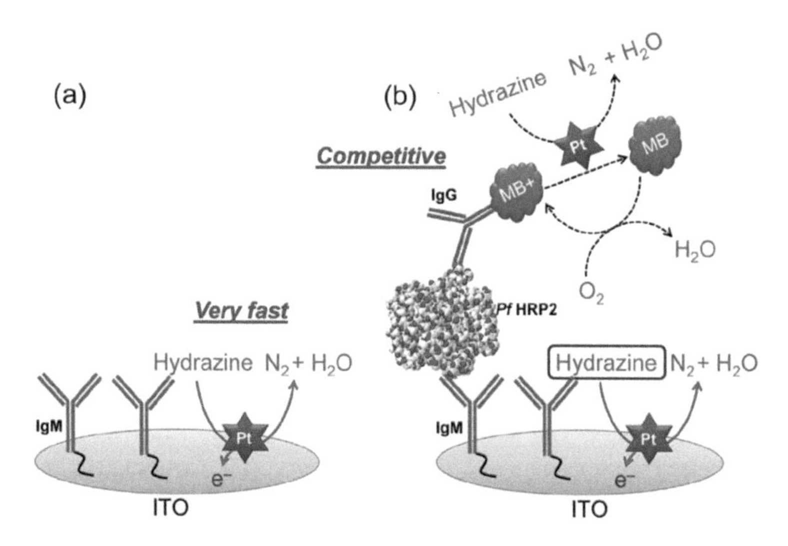

Fig. 1.7 Schematic illustration of an ITO-based nanobiosensor for highly sensitive malaria detection in the absence (**a**) and presence (**b**) of the target antigen with the MB-labeled detection antibody. (Reprinted from Dutta et al. 2017. Copyright (2017), with permission from Elsevier)

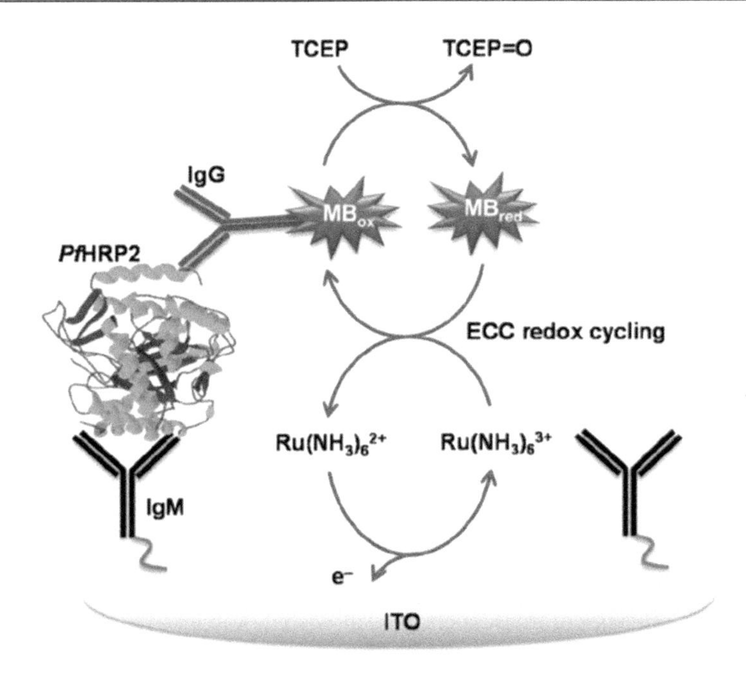

Fig. 1.8 Schematic of the electrochemical immunosensor integrating ECC redox cycling for malaria detection. (Reprinted from Dutta and Lillehoj 2017. Copyright (2017), with permission from the Royal Society of Chemistry)

amplification scheme. This scheme used methylene blue (MB) as a redox indicator which undergoes an endergonic reaction with $Ru(NH_3)_6^{3+}$ and a highly exergonic reaction with tris(2-carboxyethyl)phosphine (TCEP). Malaria biomarker was detected in human plasma and whole blood samples. The detection limit was 10 fg mL^{-1} and 18 fg mL^{-1}, respectively. This nanobiosensor exhibits excellent selectivity, very good reproducibility, and high stability making ITO-based biosensor a promising platform for point-of-care testing, especially for detecting extremely low biomarker concentrations in raw biofluids.

1.4 A Commercial Step for the Nanobiosensor

Commercialization of nanobiosensor devices for disease diagnosis is presently the "holy grail" within the micro total analysis system research community (Lin and Wang 2005; Mahato and Chandra 2019; Mahato et al. 2018). Quite a few nanobiosensors are adopted by the market and reach to the bedside diagnosis although a large variety of highly advanced chips are available and could potentially challenge our healthcare, biology, chemistry, and all related disciplines. Fortunately, the electronics industry already has at its disposal a large industrial base for the manufacturing of printed circuit boards (PCB) at extremely high volumes and at minimal production costs (Moschou and Tserepi 2017). Over the past 20 years, the rapidly increasing number of publications on lab-on-a-chip systems realized on printed circuit boards (PCB) is indicative of the future commercialization of the nanobiosensor technology and its emerging applications (Dutta et al. 2018b). Indeed, the lab-on-printed circuit board (lab-on-PCB) technology enables the seamless integration of microfluidics, sensors, and electronics and promises the commercial upscalability and standardization of microfluidics, leveraging the well-established PCB industry with standardized fabrication facilities and processes (Dutta et al. 2018a; Ghoreishizadeh et al. 2019; Moschou et al. 2015). The microfluidic devices are seamlessly integrated with PCB sensors and make the technique more useful for complex sample analysis. The amplification reaction can occur on the chip integrated with implanted heaters and will allow the system for point-of-site application (Dutta et al. 2018a; Moschou and Tserepi 2017).

1.5 Conclusions

In this chapter, different nanobiosensors for medical diagnosis using several surface modification strategies of transducers were discussed. A unique redox cycling technology was presented to amplify the signal-to-background ratios for ultrasensitive biomarker detection which is appropriate for point-of-site diagnosis such as medical application, biology-related research, environmental monitoring, and food safety. Also, some advanced nanobiosensing technologies including printed circuit board (PCB) were described on the commercial arena for next-generation point-of-care testing.

Acknowledgments Dr. Gorachand Dutta gratefully acknowledges the School of Medical Science and Technology, IIT Kharagpur, for the research support. Also Dr. Dutta thanks Prof. Peter B. Lillehoj, Dr. Despina Moschou, and Professor Haesik Yang for their guidance during his post-doctoral and PhD research work.

References

Aashish A, Sadanandhan NK, Ganesan KP, Hareesh UNS, Muthusamy S, Devaki SJ (2018) Flexible electrochemical transducer platform for neurotransmitters. ACS Omega 3(3):3489–3500

Akanda MR, Ju H (2018) Ferritin-triggered redox cycling for highly sensitive electrochemical Immunosensing of protein. Anal Chem 90(13):8028–8034

Akanda MR, Choe YL, Yang H (2012) "Outer-sphere to inner-sphere" redox cycling for ultrasensitive immunosensors. Anal Chem 84(2):1049–1055

Akanda MR, Tamilavan V, Park S, Jo K, Hyun MH, Yang H (2013) Hydroquinone diphosphate as a phosphatase substrate in enzymatic amplification combined with electrochemical–chemical–chemical redox cycling for the detection of E. coli O157:H7. Anal Chem 85(3):1631–1636

Akanda MR, Joung HA, Tamilavan V, Park S, Kim S, Hyun MH, Kim MG, Yang H (2014) An interference-free and rapid electrochemical lateral-flow immunoassay for one-step ultrasensitive detection with serum. Analyst 139(6):1420–1425

Akhtar MH, Hussain KK, Gurudatt NG, Chandra P, Shim YB (2018) Ultrasensitive dual probe immunosensor for the monitoring of nicotine induced-brain derived neurotrophic factor released from cancer cells. Biosens Bioelectron 116:108–115

Aziz MA, Park S, Jon S, Yang H (2007) Amperometric immunosensing using an indium tin oxide electrode modified with multi-walled carbon nanotube and poly(ethylene glycol)-silane copolymer. Chem Commun (Camb) 25:2610–2612

Chandra P, Noh H-B, Won M-S, Shim Y-B (2011) Detection of daunomycin using phosphatidylserine and aptamer co-immobilized on Au nanoparticles deposited conducting polymer. Biosens Bioelectron 26(11):4442–4449

Cinel NA, Butun S, Ozbay E (2012) Electron beam lithography designed silver nano-disks used as label free nano-biosensors based on localized surface plasmon resonance. Opt Express 20(3):2587–2597

Das J, Aziz MA, Yang H (2006) A nanocatalyst-based assay for proteins: DNA-free ultrasensitive electrochemical detection using catalytic reduction of p-nitrophenol by gold-nanoparticle labels. J Am Chem Soc 128(50):16022–16023

David AM, Mercado SP, Becker D, Edmundo K, Mugisha F (2007) The prevention and control of HIV/AIDS, TB and vector-borne diseases in informal settlements: challenges, opportunities and insights. J Urban Health 84(3 Suppl):i65–i74

Dutta G (2017) Electrochemical redox cycling amplification technology for point-of-care cancer diagnosis. Springer, Singapore

Dutta G, Lillehoj PB (2017) An ultrasensitive enzyme-free electrochemical immunosensor based on redox cycling amplification using methylene blue. Analyst 142(18):3492–3499

Dutta G, Lillehoj PB (2018) Wash-free, label-free immunoassay for rapid electrochemical detection of PfHRP2 in whole blood samples. Sci Rep 8(1):17129

Dutta G, Kim S, Park S, Yang H (2014) Washing-free heterogeneous Immunosensor using proximity-dependent Electron mediation between an enzyme label and an electrode. Anal Chem 86(9):4589–4595

Dutta G, Park S, Singh A, Seo J, Kim S, Yang H (2015) Low-interference washing-free electrochemical Immunosensor using Glycerol-3-phosphate dehydrogenase as an enzyme label. Anal Chem 87(7):3574–3578

Dutta G, Nagarajan S, Lapidus LJ, Lillehoj PB (2017) Enzyme-free electrochemical immunosensor based on methylene blue and the electro-oxidation of hydrazine on Pt nanoparticles. Biosens Bioelectron 92:372–377

Dutta G, Rainbow J, Zupancic U, Papamatthaiou S, Estrela P, Moschou D (2018a) Microfluidic devices for label-free DNA detection. Chemosensors 6(4):43

Dutta G, Regoutz A, Moschou D (2018b) Commercially fabricated printed circuit board sensing electrodes for biomarker electrochemical detection: the importance of electrode surface characteristics in sensor performance. PRO 2(13):741

Eltzov E, Marks RS (2016) Miniaturized flow stacked immunoassay for detecting Escherichia coli in a single step. Anal Chem 88(12):6441–6449

Fang CS, Kim KS, Ha DT, Kim MS, Yang H (2018) Washing-free electrochemical detection of amplified double-stranded DNAs using a zinc finger protein. Anal Chem 90(7):4776–4782

Ghoreishizadeh SS, Moschou D, McBay D, Gonalez-Solino C, Dutta G, Lorenzo MD, Soltan A (2019) Towards self-powered and autonomous wearable glucose sensor. In: 25th IEEE international conference on electronics, circuits and systems (ICECS), pp 701–704

González-Gaitán C, Ruiz-Rosas R, Morallón E, Cazorla-Amorós D (2017) Effects of the surface chemistry and structure of carbon nanotubes on the coating of glucose oxidase and electrochemical biosensors performance. RSC Adv 7:26867–26878

Gruhl FJ, Rapp BE, Lange K (2013) Biosensors for diagnostic applications. Adv Biochem Eng Biotechnol 133:115–148

Gu H, Hou Q, Liu Y, Cai Y, Guo Y, Xiang H, Chen S (2019) On-line regeneration of electrochemical biosensor for in vivo repetitive measurements of striatum Cu(2+) under global cerebral ischemia/reperfusion events. Biosens Bioelectron 135:111–119

He PJW, Katis IN, Eason RW, Sones CL (2018) Rapid multiplexed detection on lateral-flow devices using a laser direct-write technique. Biosensors (Basel) 8(4):pii: E97
Heli H, Hajjizadeh M, Jabbari A, Moosavi-Movahedi AA (2009) Copper nanoparticles-modified carbon paste transducer as a biosensor for determination of acetylcholine. Biosens Bioelectron 24(8):2328–2333
Hwang SG, Ha K, Guk K, Lee DK, Eom G, Song S, Kang T, Park H, Jung J, Lim E-K (2018) Rapid and simple detection of Tamiflu-resistant influenza virus: development of oseltamivir derivative-based lateral flow biosensor for point-of-care (POC) diagnostics. Sci Rep 8(1):12999
Ilkhani H, Hughes T, Li J, Zhong CJ, Hepel M (2016) Nanostructured SERS-electrochemical biosensors for testing of anticancer drug interactions with DNA. Biosens Bioelectron 80:257–264
Jiaul Haque AM, Kim J, Dutta G, Kim S, Yang H (2015) Redox cycling-amplified enzymatic ag deposition and its application in the highly sensitive detection of creatine kinase-MB. Chem Commun (Camb) 51(77):14493–14496
Khan S, Ali S, Bermak A (2019) Recent developments in printing flexible and wearable sensing electronics for healthcare applications. Sensors (Basel) 19(5):pii: E1230
Lee CW, Chang HY, Wu JK, Tseng FG (2019a) Ultra-sensitive electrochemical detection of bacteremia enabled by redox-active gold nanoparticles (raGNPs) in a nano-sieving microfluidic system (NS-MFS). Biosens Bioelectron 133:215–222
Lee EH, Lee SK, Kim MJ, Lee SW (2019b) Simple and rapid detection of bisphenol a using a gold nanoparticle-based colorimetric aptasensor. Food Chem 287:205–213
Li M, Cushing SK, Liang H, Suri S, Ma D, Wu N (2013) Plasmonic nanorice antenna on triangle nanoarray for surface-enhanced Raman scattering detection of hepatitis B virus DNA. Anal Chem 85(4):2072–2078
Li Q, Zhou D, Pan J, Liu Z, Chen J (2019) An ultrasensitive and simple fluorescence biosensor for detection of the Kras wild type by using the three-way DNA junction-driven catalyzed hairpin assembly strategy. Analyst 144(9):3088–3093
Liddle JA, Gallatin GM (2011) Lithography, metrology and nanomanufacturing. Nanoscale 3(7):2679–2688
Lin CT, Wang SM (2005) Biosensor commercialization strategy – a theoretical approach. Front Biosci 10:99–106
Mahato K, Chandra P (2019) Paper-based miniaturized immunosensor for naked eye ALP detection based on digital image colorimetry integrated with smartphone. Biosens Bioelectron 128:9–16
Mahato K, Maurya PK, Chandra P (2018) Fundamentals and commercial aspects of nanobiosensors in point-of-care clinical diagnostics. 3 Biotech 8(3):149
Moschou D, Tserepi A (2017) The lab-on-PCB approach: tackling the muTAS commercial upscaling bottleneck. Lab Chip 17(8):1388–1405
Moschou D, Trantidou T, Regoutz A, Carta D, Morgan H, Prodromakis T (2015) Surface and electrical characterization of ag/AgCl pseudo-reference electrodes manufactured with commercially available PCB technologies. Sensors (Basel) 15(8):18102–18113
Nze UC, Beeman MG, Lambert CJ, Salih G, Gale BK, Sant HJ (2019) Hydrodynamic cavitation for the rapid separation and electrochemical detection of Cryptosporidium parvum and Escherichia coli O157:H7 in ground beef. Biosens Bioelectron 135:137–144
Park S, Kim J, Ock H, Dutta G, Seo J, Shin EC, Yang H (2015) Sensitive electrochemical detection of vaccinia virus in a solution containing a high concentration of L-ascorbic acid. Analyst 140(16):5481–5487
Proa-Coronado S, Vargas-García JR, Manzo-Robledo A, Mendoza-Acevedo S, Villagómez CJ (2018) Platinum nanoparticles homogenously decorating multilayered reduced graphene oxide for electrical nanobiosensor applications. Thin Solid Films 658:54–60. Elsevier
Sang S, Guo X, Liu R, Wang J, Guo J, Zhang Y, Yuan Z, Zhang W (2018) A novel Magnetoelastic Nanobiosensor for highly sensitive detection of atrazine. Nanoscale Res Lett 13(1):414
Siddiqui MF, Kim S, Jeon H, Kim T, Joo C, Park S (2018) Miniaturized sample preparation and rapid detection of Arsenite in contaminated soil using a smartphone. Sensors (Basel) 18(3):pii: E77

Singh A, Park S, Yang H (2013) Glucose-oxidase label-based redox cycling for an incubation period-free electrochemical Immunosensor. Anal Chem 85(10):4863–4868

Wang X, Mei Z, Wang Y, Tang L (2017) Comparison of four methods for the biofunctionalization of gold nanorods by the introduction of sulfhydryl groups to antibodies. Beilstein J Nanotechnol 8:372–380

Wang Z, Yao X, Wang R, Ji Y, Yue T, Sun J, Li T, Wang J, Zhang D (2019) Label-free strip sensor based on surface positively charged nitrogen-rich carbon nanoparticles for rapid detection of Salmonella enteritidis. Biosens Bioelectron 132:360–367

Wasik D, Mulchandani A, Yates MV (2018) Point-of-use Nanobiosensor for detection of dengue virus NS1 antigen in adult Aedes aegypti: a potential tool for improved dengue surveillance. Anal Chem 90(1):679–684

Xiang H, Wang Y, Wang M, Shao Y, Jiao Y, Zhu Y (2018) A redox cycling-amplified electrochemical immunosensor for α-fetoprotein sensitive detection via polydopamine nanolabels. Nanoscale 10:13572–13580

Yang H (2012) Enzyme-based ultrasensitive electrochemical biosensors. Curr Opin Chem Biol 16(3–4):422–428

Yang M, Jeong SW, Chang SJ, Kim KH, Jang M, Kim CH, Bae NH, Sim GS, Kang T, Lee SJ, Choi BG, Lee KG (2016) Flexible and disposable sensing platforms based on newspaper. ACS Appl Mater Interfaces 8(51):34978–34984

Zhang WH, Ma W, Long YT (2016) Redox-mediated indirect fluorescence immunoassay for the detection of disease biomarkers using dopamine-functionalized quantum dots. Anal Chem 88(10):5131–5136

Zhang H, Fan M, Jiang J, Shen Q, Cai C, Shen J (2019) Sensitive electrochemical biosensor for MicroRNAs based on duplex-specific nuclease-assisted target recycling followed with gold nanoparticles and enzymatic signal amplification. Anal Chim Acta 1064:33–39

Zhu J, Ye Z, Fan X, Wang H, Wang Z, Chen B (2019) A highly sensitive biosensor based on Au NPs/rGO-PAMAM-Fc nanomaterials for detection of cholesterol. Int J Nanomedicine 14:835–849

Biosensors Based on Nanomaterials: Transducers and Modified Surfaces for Diagnostics

2

Marcelo R. Romero and Matías L. Picchio

Abstract

The use of nanoparticles has opened a new era in the development of nanobiosensors capable of achieving analytical responses that compete with the most powerful instrumental techniques. Nanobiosensors are devices that allow analytical determinations through a specific action event between an analyte of interest and a bio-recognition molecule. These recognition molecules as enzymes, antibodies, nucleic acids, and aptamers are studied in detail in this chapter. The role of nanomaterials in biosensors is described in a separate section since they play a central role, allowing the understanding of their physicochemical properties such as quantum confinement, surface plasmon resonance, magnetic properties, and the effect of area increase. In addition, a brief review is provided about some basic concepts for the integration of the sensor components and their function in sensing systems found in the literature. Subsequently, a classification is proposed to summarize its fundamental characteristics, mechanism of operation, analytical characteristics, advantages, and disadvantages. Then, the main nanobiosensor types found in the literature are detailed, and specific explanations are given, e.g., those based on the determination of electrical, piezoelectric, colorimetric, fluorescent, and chemiluminescent properties. Likewise, the functioning of recently developed nanobiosensors is discussed, such as those based on local

M. R. Romero (✉)
Facultad de Ciencias Químicas, Departamento de Química Orgánica (FCQ-UNC), Universidad Nacional de Córdoba, Córdoba, Argentina

Consejo Nacional de Investigaciones Científicas y Técnicas (CONICET), Instituto de Investigación y Desarrollo en Ingeniería de Procesos y Química Aplicada (IPQA), Córdoba, Argentina

M. L. Picchio
Facultad de Ciencias Químicas, Departamento de Química Orgánica (FCQ-UNC), Universidad Nacional de Córdoba, Córdoba, Argentina

Facultad Regional Villa María, Universidad Tecnológica Nacional, Córdoba, Argentina

P. Chandra, R. Prakash (eds.), *Nanobiomaterial Engineering*,
https://doi.org/10.1007/978-981-32-9840-8_2

surface plasmon resonance (LSPR) and surface enhancement Raman signal (SERS). Also, the applications of nanobiosensors in different fields of biomedicine and their fundamental importance to advance in the diagnosis of multiple pathologies as cancer are detailed. Finally, we discuss the state of the art and the future perspectives of scientific development.

Keywords
Nanobiosensors · Bioreceptor · Nanoparticles · SERS · FRET · LSPR · Diagnostics

2.1 Introduction

Biosensors based on nanomaterials are usually called nanobiosensors. Nanobiosensors are devices that allow analytical determinations through a specific action event between an analyte of interest and a bio-recognition molecule. In nanobiosensors, this recognition event is also amplified by the presence of a nanomaterial. The resulting event is transduced into a quantifiable signal that can be recorded by a microprocessor.

Each component of these devices, recognition molecule and nanomaterial, is fundamental for the functioning of nanobiosensors, and they will be studied in detail in this chapter. In addition, a brief review will be provided about some basic concepts for the integration of the sensor components and their function in sensing systems found in the literature. Subsequently, a principle of the main types of nanobiosensors will be shown, and a classification will be proposed to summarize its fundamental characteristics, mechanism of operation, analytical characteristics, advantages, and disadvantages. Finally, the applications of nanobiosensors in different fields of biomedicine and their fundamental importance to advance in the diagnosis of multiple pathologies are detailed (Fig. 2.1).

The following is a brief description of the main topics specifically explored in this chapter. First, we characterize the typical bio-recognition molecules present in

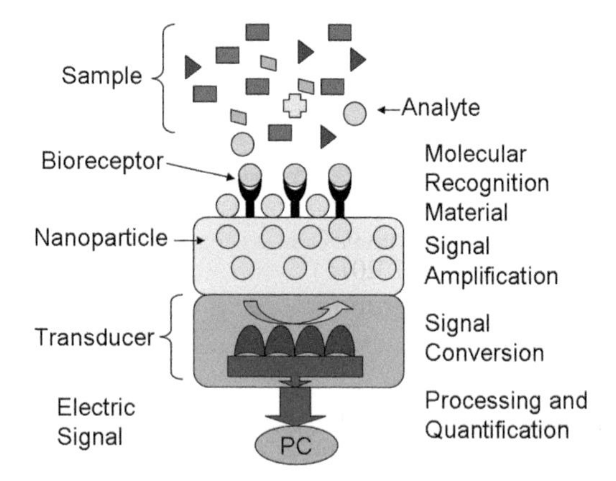

Fig. 2.1 Scheme of main nanobiosensor components

most biosensors such as enzymes, antibodies, and nucleic acids. Other more recent discoveries such as aptamers are also included. In all cases, a brief characterization is provided; yet, the analysis focuses on immobilization and the main advantages and disadvantages of its use. The role of nanomaterials in biosensors is also described in a separate section since they play a central role in this chapter, allowing the understanding of their physicochemical properties such as quantum confinement, surface plasmon resonance, magnetic properties, the effect of area increase, and how they provide improved analytical responses to biosensors. Once the effect of the nanomaterial on the functioning of nanobiosensors is explained, the characteristics of the main types found in the literature are detailed, and specific examples are given, e.g., those based on the determination of electrical properties such as electrochemical, voltammetric, and impedance biosensors. The piezoelectric, colorimetric, fluorescent, and chemiluminescent nanobiosensors are also defined. Likewise, the functioning of recently developed nanobiosensors is discussed, such as those based on local surface plasmon resonance (LSPR) and surface enhancement Raman signal (SERS). In the final section, we analyze the main applications of the diagnosis of pathologies (Chandra 2016; Chandra et al. 2017). The section addresses the scientific interest in the quantification of certain analytes and the grounds for the use of nanobiosensors in conditions in which conventional biosensors are not sufficient. Finally, we discuss the state of the art and the future perspectives of scientific development.

2.2 General Properties of *Bio-recognition Molecules*

The bio-recognition molecule is responsible for giving specificity to the sensor. The interaction is given by a chemical bond generally non-covalent and reversible between a molecule present in the sensor and the analyte of interest in the sample problem. The specificity is given by the conformation or complementary three-dimensional structure of the bio-recognition molecule and the analyte. The primary binding between the recognition molecule and the analyte must produce a chemical or physical alteration which can then be translated into a quantifiable signal. In general, the bio-recognition molecule may be the same as that used in conventional sensors, although in nanobiosensors it is usually anchored or linked to nanomaterials. The presence of nanoparticles brings about the amplification of the signal, allowing the detection of many events that previously could not be quantified with conventional biosensors either due to the low signal provided by the interaction or to a very low concentration of the analyte in the sample. Thus, the integration between the bio-recognition molecule and the nanomaterial is particularly important (Bhakta et al. 2015).

The bio-recognition molecules present in the sensors are varied; yet, they can be classified into two large groups: natural and synthetic. In general, synthetic molecules were inspired by the natural ones and show some general advantages such as greater chemical stability and lower molecular complexity, making results more predictable, although they are usually more expensive and the sensors developed from them are less profitable. The most frequently used bio-recognition molecules are detailed below.

2.2.1 Enzymes

One of the first bio-recognition molecules used for sensors are enzymes. They are proteins composed of amino acids linked through peptide bonds forming long chains folded into globular structures. The enzyme structure and conformation is highly complex and is given by the composition and arrangement of amino acids in their sequence (which amounts to several hundred), the winding process during their biosynthesis and the chemical environment in which they are found. The following figure shows typical amino acids (Fig. 2.2).

These amino acids are produced in all types of biological organisms since they are responsible for carrying out different types of biochemical reactions, acting as highly efficient, specific catalysts.

The mechanism of Michaelis-Menten explains, essentially, how enzyme (*E*) binds to substrate (*S*) to form an enzyme-substrate (ES) complex. Subsequently, product (P) is generated by the enzymatic reaction. All this is represented in the following equation:

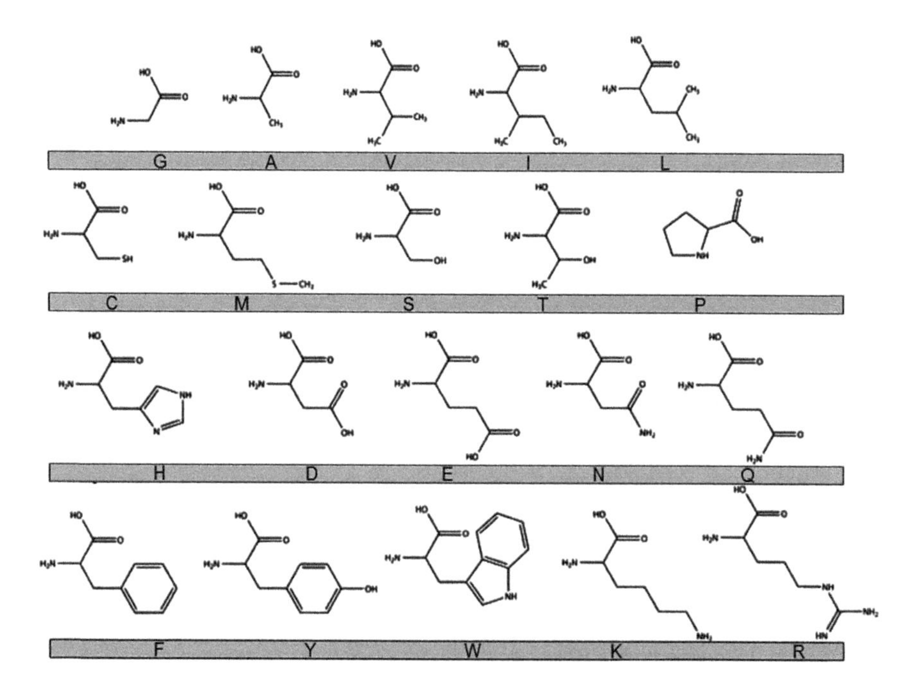

Fig. 2.2 Molecular structure of amino acids. G (glycine), A (arginine), V (valine), I (isoleucine), L (leucine), C (cysteine), M (methionine), S (serine), T (threonine), P (proline), H (histidine), D (aspartic acid), E (glutamic acid), N (asparagine), Q (glutamine), F (phenylalanine), Y (tyrosine), W (tryptophan), K (lysine), and R (arginine)

$$V_0 = V_{max}[S]/(K_m + [S]) \quad (2.1)$$

where V_0 is the speed, V_{max} is the maximum speed of catalysis, and K_m is the substrate concentration at $V_{max}/2$. There are many other mechanisms of enzymatic catalysis; however, one of the most relevant enzymes used in biosensors is ping-pong (Romero et al. 2012a, b).

They are widely used as recognition elements in sensors since they present high specificity to the analyte (substrate), binding it through molecule sites generally comprising several functional groups of amino acids that usually have hydrogen bonding interactions. Such a structure is called active or recognition site. Moreover, enzymes could hardly catalyze reactions of chemical species differing from those for the analyte of interest, being able to discriminate up to optical isomers of the same molecule, which is a great advantage in sensor applications. Among the different enzymes available, those able to catalyze the conversion of the analyte into another chemical species more easily detectable are most commonly used in sensors. A typical example is the development of lactate sensors. In this device, lactate can be degraded by the recognition molecule called lactate oxidase (LOD). The catalytic reaction produces pyruvate; yet, the relevant product which allows quantification of lactate is the electroactive species hydrogen peroxide, which can be detected electrochemically by amperometry. The following equations show chemical reactions where the enzyme catalyzes the oxidation reaction of lactate passing from an oxidized state (ox) to a reduced (red) one in the first reaction. In the second reaction, the enzyme is regenerated in the presence of oxygen. Finally, the oxidation reaction of hydrogen peroxide in the electrode, which delivers two electrons per molecule of catalyzed lactate, is seen:

$$\text{Lactate} + \text{LOD}_{ox} \rightarrow \text{pyruvate} + \text{LOD}_{red} \quad (2.2)$$

$$\text{LOD}_{red} + O_2 \rightarrow \text{LOD}_{ox} + H_2O_2 \quad (2.3)$$

$$H_2O_2 \rightarrow O_2 + 2H^+ + 2e^- \quad (2.4)$$

However, the enzyme-substrate recognition site has a conformation sustained by weak non-covalent bonds, representing a disadvantage for the intended application. Therefore, during the process of immobilization or anchoring to the sensor, conformational changes can be produced that alter or irreversibly damage both its activity and specificity. The immobilization strategy of an enzyme can be carried out by physical or chemical methods. Physical methods involve the anchoring of the enzyme through non-covalent bonds (absorption, entrapment, or encapsulation) with the advantage of minimally altering enzyme conformation; however, this methodology could lead to the release of molecules from the sensor to the medium, producing both a decrease in sensitivity and competition in catalysis of the analyte with the enzyme in solution. On the other hand, chemical anchoring involves the covalent binding to the side functional groups of the amino acid chain such as amino, carboxylic acid, sulfide, and hydroxyl sites. Common covalent agents are

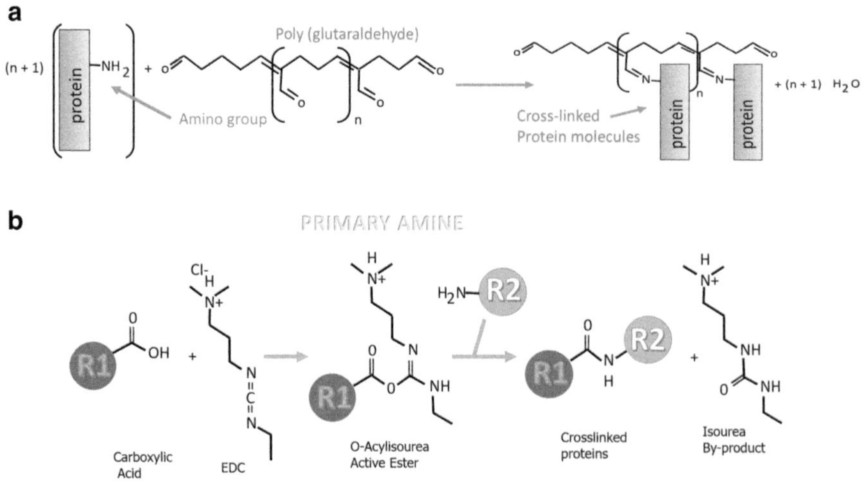

Fig. 2.3 Covalent bond of enzymes based on (a) glutaraldehyde and (b) EDC

difunctional molecules such as glutaraldehyde, able to bind amino groups, and those based on carbodiimides (EDC or DCC) alone or together with succinimides (NHS or sulfo-NHS) that bind carboxylic acids with amines. For a further description of enzyme immobilization, the book by Hermanson et al. (1992) is recommended (Fig. 2.3).

For the immobilization of enzymes, those chemical groups present in areas of the molecule away from the active site should be selected, so as to minimize the effect of anchoring on the catalytic activity. In addition, before immobilization, the number of binding sites per enzymatic molecule must be known, since it is advisable to work with low percentages in relation to the total number of amino acids in the molecule (usually <5%). The details of the three-dimensional structure and amino acid composition of enzymes can be found in an extensive database of X-ray diffraction-resolved structures at www.pdb.org. Figure 2.4 shows the three-dimensional scheme of the monomeric protein chain of the enzyme glucose oxidase (blue), peripheral carbohydrates (green), prosthetic group flavin adenine dinucleotide (FAD) in the active site (white), and possible sites of glutaraldehyde anchoring to the amino acid lysine (red). As observed, no anchorage site is close to the active site of the enzyme.

On the other hand, if the sensor is required to be stable in extreme conditions, the number of crosslinking sites should be increased, causing significant loss of activity although gaining stability, since covalent bonds stiffen the molecule. Among the coupling agents, those longer and bearing simple carbon-carbon bonds are usually chosen to promote flexibility to the molecular structure. Enzymes have been successfully combined with nanomaterials such as carbon nanotubes, achieving outstanding increase in response times due to the large increase in the surface area of the system and the decrease in mass transport times, reflected in the shortening of the response time of the sensors (Romero et al. 2017).

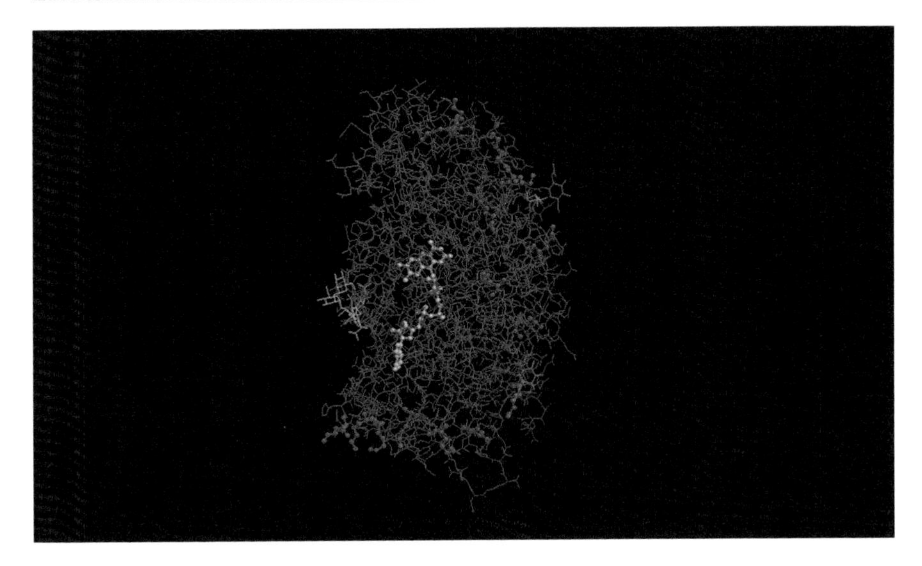

Fig. 2.4 Three-dimensional scheme of a subunit of the glucose oxidase enzyme. Available binding sites for glutaraldehyde were highlighted in red

2.2.2 Antibodies

Antibodies, like enzymes, are proteins of the glycoprotein type. They are generally produced from biosynthesis mechanisms only found in multicellular organisms such as humans and mammals. However, evidence has shown the existence of precursor and antibody-like molecules in more primitive organisms. The organisms that produce them are usually multicellular and use antibodies as specific recognition molecules to counteract the effect of the invasion of an external infectious agent. It is interesting to note that through chemical action on the remains of the infectious agent (antigen), the immune system originates fragments (epitopes) through a process called opsonization, which they later use as a template for the synthesis of specific glycoprotein molecules in immune cells called lymphocytes B. Although we can find different types of antibodies, those most largely used in the sensors field are type G immunoglobulins. These have a molecular structure of letter Y in which the lateral branches (fab) correspond to the variable sections specific to recognition by which this molecule binds to the antigen forming an antigen-antibody complex; the appropriate zone for its anchoring is a conserved and little variable fraction of the molecule (fc). This molecule can be used as a molecular recognition agent, although it is also possible to separate the fractions from each other with enzymes called pepsin or papain. The resulting fragments of fab maintain specificity and can be used separately as bio-recognition molecules in sensors (Janeway 2001) (Fig. 2.5).

For the development of sensors, two types of antibodies can be used: those of polyclonal origin, which are produced by injecting in an animal (e.g., a rabbit) the antigenic molecule, from which the sensor will be designed, and serum containing

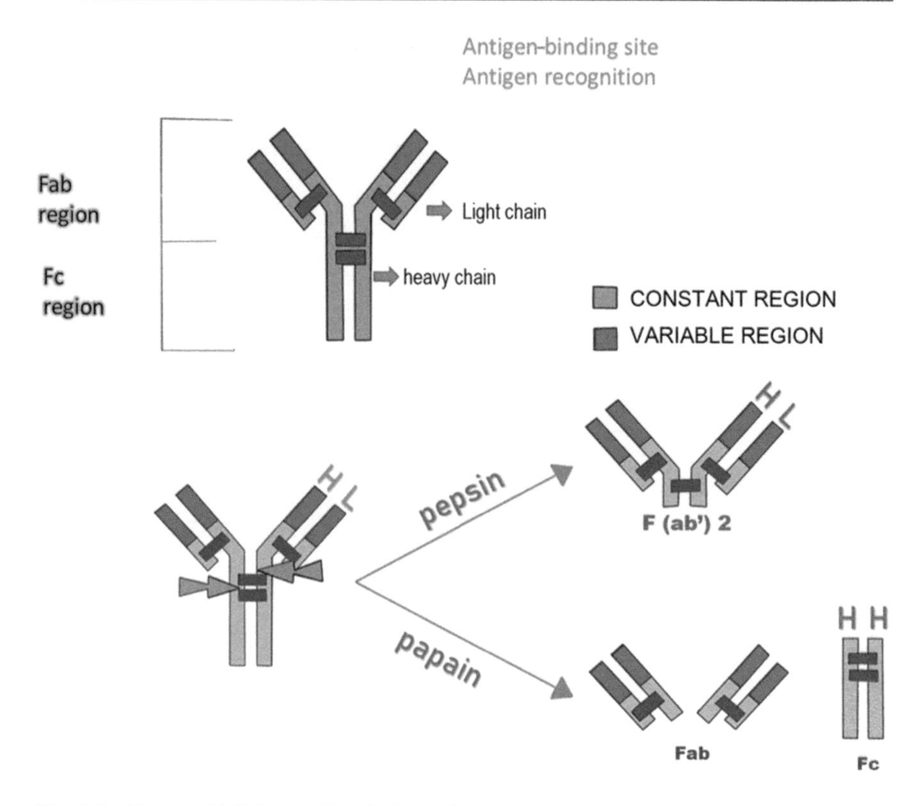

Fig. 2.5 (Top graph) Scheme of antibody regions. (Bottom graph) Cleavage of antibodies with pepsin and papain

antibodies (antiserum) which are purified. The resulting antibodies usually have different affinity constants and conformations (they can be diverse immunoglobulin types: G, A, M, etc.) and bind to the antigen by means of hardly predictable sites. However, they involve the most economical method to develop antibodies sensitive to a molecule of interest.

The second methodology is carried out through the production of monoclonal antibodies capable of recognizing a single epitope of an antigen. To do this, lymphocytes are fused with a cancerous cell line (hybridoma). Hybridomas can be selected to produce numerous monoclonal antibodies. This is the most suitable method to achieve specific antibodies of virtually identical molecular structure.

In general, all the chemical manipulations of these molecules reveal the same limitations and disadvantages as enzymes during the immobilization process, since they are easily altered, changing their affinity. However, there are alternatives to chemical binding which maintain the activity of the antibody, such as the use of specific anchor molecules including protein A or G coming from bacteria. These proteins have the ability to anchor antibodies by the fc site. The use of these proteins is often adequate for the development of sensing systems since it could only involve the optimization of the appropriate anchoring of protein A or G on the sensor surface, coupling any desired antibody and yielding sensors for different antigens. However, many authors prefer the chemical anchoring method to the fc portion, probably to overcome certain drawbacks such as molecule loss and antibody release by reversion of affinity binding with A or G, or competition for the fc binding site in case the medium problem also contains other antibodies.

In general, excellent results are obtained by anchoring antibodies to nanoparticles and studying the interaction of the biomolecule with its antigen, thus achieving high specificity and sensitivity due to the coupling with the nanomaterial. It has also been observed that incubation times may be much shorter in these configurations than with conventional methods of immunology.

2.2.3 Nucleic Acids

Nucleic acids are macromolecules responsible for the transmission of genetic information of all living beings from one generation to the next. They are composed of nucleotides that have a structure comprising phosphate, a sugar (ribose or deoxyribose), and a base (purine or pirimidic). The nucleotides are joined by phosphodiester bonds in an elongated strand conformation which may be single or double. There are two main types of nucleic acids, single-chain ribonucleic acid or RNA (whose sugar is ribose) and double-chain deoxyribonucleic acid or DNA (whose sugar is deoxyribose). The different base types that make up the DNA chain are adenine, cytosine, guanine, and thymine. Those of RNA are similar, having uracil instead of thymine. In addition each base of a chain is linked by hydrogen bonding with a base of the other chain in certain favorable configurations (complementary). Hence, adenine binds with thymine and cytosine with guanine, the latter being the most stable as it has three hydrogen bonds, whereas the other pair has only two (Alberts et al. 2014). In this way, one strand of DNA (ssDNA) can be joined with another complementary chain forming a double strand (dsDNA). This recognition event is highly specific and also thermally reversible and proportional to the length of the strand, without altering the integrity of the phosphodiester bond (Fig. 2.6).

In the field of sensors, ssDNA is preferred as a bio-recognition molecule instead of RNA, since the latter is more easily degraded by hard denaturing enzymes (ribonucleases) present in almost all biological media. In addition, ssDNA is quite stable in different chemical and thermal media. At present, there are standardized methodologies for the design of nucleotide sequences of interest and the amplification of oligonucleotides by a technique called PCR (polymerase chain reaction) that combines heating-cooling cycles to achieve hybridization and elongation of chains with a heat-resistant DNA polymerase. The event of complementary chain recognition is highly specific; hence, it is widely used for the development of sensors capable of detecting DNA or RNA sequences associated with pathologies of microorganisms, considering that after hybridization chains show a decrease in the photon absorptivity coefficient (hypochromism). While this event is sensitive, it can be amplified greatly if the nucleic acid is anchored to a nanoparticle by the effect of the biomolecular interaction on the plasmon surface.

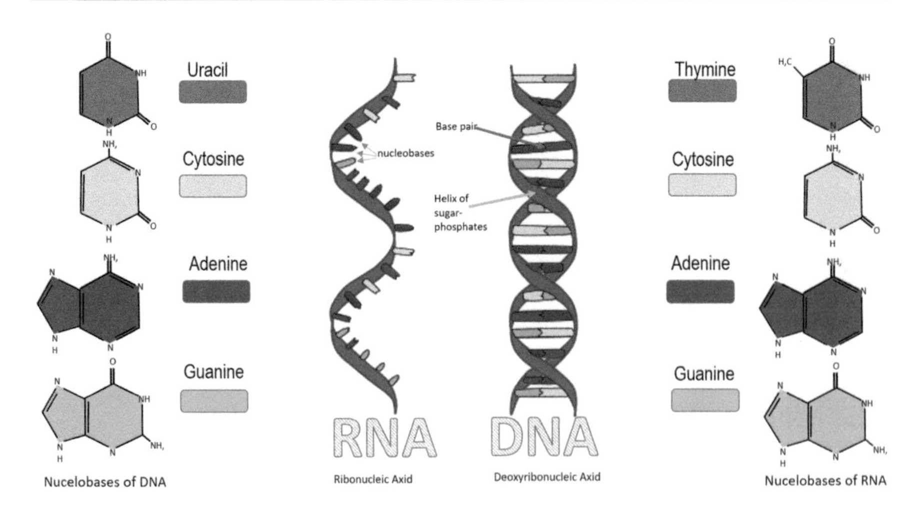

Fig. 2.6 Representation of the chemical structure of nucleotides, DNA and RNA

2.2.4 Aptamers

Aptamers are synthetic oligonucleotides or peptides designed to bind target molecules that may be nucleic acids with high specificity. They are usually created by selection from a large random pool of sequences or gene library by a method called SELEX (Sefah et al. 2010) (Fig. 2.7).

The sequences obtained show a variable central part specific to the analyte and conserved lateral sequences. In general, the reached specificity to the analyte is so good that it only requires the synthesis of oligonucleotides of around 20 bases.

Aptamers can be classified as:

DNA and RNA: chains composed of traditional oligonucleotides.

XNA: structures similar to those of nucleic acids but showing changes in some of the three parts that make up them in the skeleton of phosphate phosphodiester, the pentose sugar, or the nucleobase. In an analog any of these three parts may be altered. Generally, phosphate or sugar is replaced. In the case of phosphate replacement, analogs usually have no charge, so they are more hydrophobic. When sugar is replaced or modified, products are more resistant to degradation. In both cases, affinity coefficients are usually higher than the natural ones since analogs are, normally, more rigid.

Peptides: one or more variable short domains of peptides linked, on both sides, to a protein structure. They have more chemical groups capable of giving sensitivity to different molecules (Fig. 2.8).

Aptamers are widely used today as bio-recognition elements and are subject to a depth research in the area of nanobiosensors. They are small in size, which allows them to be easily anchored to nanoparticles with low risk of altering their

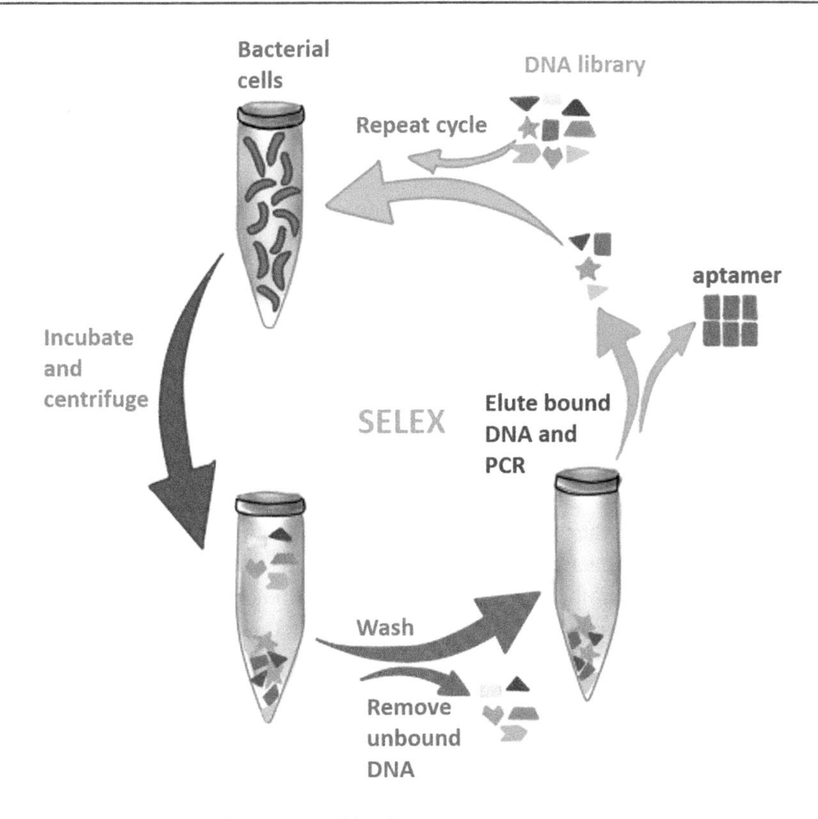

Fig. 2.7 SELEX method of aptamer purification

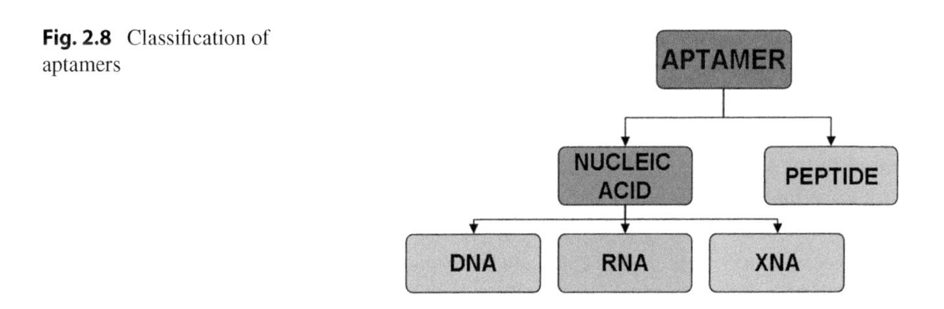

Fig. 2.8 Classification of aptamers

conformation. They also show good chemical stability. They can be synthesized as complementary or self-complementary structures to achieve interesting effects of self-hybridization (hairpin) and competition for the affinity for the analyte. They can present molecules in their end side, facilitating their anchoring to the support and avoiding the drawbacks observed in other bio-recognition molecules.

2.3 Role of Nanomaterials in Nanobiosensors

2.3.1 Introduction

Nanomaterials form a fundamental part in nanobiosensors. The excellent analytical response achieved after its addition is mainly attributed to the physicochemical properties of nanomaterials and the strategy of integrating this property to the final response of the sensor.

Nanomaterials have intermediate properties between individual molecules and massive materials, although they have unique characteristics. As known, individual molecules have energy restrictions to reach their different excited states, being able to absorb energy from the environment only in predetermined quantities (or quantum) according to their chemical nature. On the other hand, bulk materials evidence an overlap of the states, i.e., each atom or molecule comprising it affects the state of the neighbor, generating a continuum where the effects of quantization are practically diluted. On the other hand, the particles of nanoscopic size are formed by a reduced number of atoms. These atoms have quantized states that overlap with their neighbors, but their number is not large enough to reach a continuum like bulk materials. Therefore, quantum properties are manifested in nanoparticles, as well as in individual molecules, but in the former, these properties are a function of the size (and shape) of the particle, giving rise to new alternatives for taking advantage of these effects.

Among the main unique characteristics of nanomaterials regarding the area of nanobiosensors, we can include quantum confinement, local surface plasmon resonance, and superparamagnetism. A further fundamental characteristic is its high area/volume ratio in relation to the same bulk material.

2.3.2 Quantum Confinement

When materials are small, their optical and electronic properties differ substantially from bulk materials. If a particle is large, the confinement dimension is long compared to the wavelength of the particle. During this state, the bandgap corresponds to its original energy due to a state of continuous energy. However, when the confinement dimension decreases and reaches a certain limit, the energy of the spectrum becomes discrete. As a result, the bandgap becomes dependent on size. This brings about a blue shift of the emission spectrum as the size of the particles decreases. Quantum confinement can be observed when the diameter of the nanomaterial is of the same magnitude as the De Broglie wavelength of the electron wave function. Specifically, this effect describes the resulting phenomenon from electrons and electron holes (both constitute an exciton) that are compressed in a dimension called radius of the Bohr exciton. A good approximation of the behavior of exciton in a nanoparticle is a 3D model of a particle in a box. The solution to this problem provides a connection between the mathematical solution of state energy

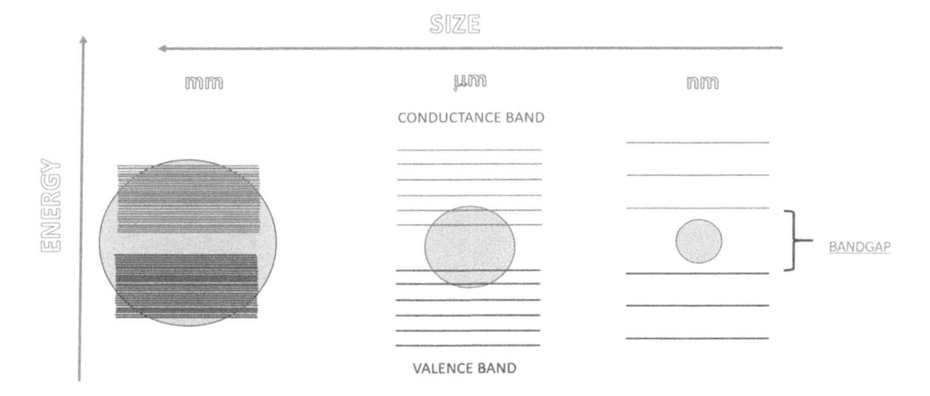

Fig. 2.9 The following diagram represents the effect of quantum confinement in different size particles

and the dimension of space. The decrease in volume or dimensions of available space increases the energy of the states (Cahay 2001) (Fig. 2.9).

2.3.3 Local Surface Plasmon Resonance (LSPR)

The local surface plasmon resonance is a collective oscillation of electronic charge in metallic nanoparticles by the effect of an electron or photon beam. The oscillations exhibit an increase in the field amplitude near the resonance wavelength. This field is highly localized in the nanoparticle and decays rapidly with distance.

As with conventional surface plasmon resonance, this phenomenon is carried out at the interface between negative permittivity (e.g., metals) and positive materials (dielectrics such as water or air). To excite surface plasmons in a resonant manner, the incident beam must have the same moment as the plasmon, and this can be achieved using polarized light parallel to the plane of incidence.

Experimentally this is possible by passing light through a glass prism to increase the wave number and the moment of the beam. When the beam of parallel polarized light hits a prism at an angle greater than the critical, the total internal reflection of the beam and an evanescent wave are produced. The latter can excite the surface plasmon of the metal in resonant form. The surface propagation wave has a propagation constant β, expressed by the following equation:

$$\beta = \omega / c\left[\varepsilon M \varepsilon D / \left(\varepsilon M + \varepsilon D\right)\right]^{1/2} \tag{2.5}$$

where ω is the angular frequency, c is the speed of light in vacuum, and εD and εM are the dielectric constants of the dielectric material and the metal, respectively.

Nanoparticles differ from normal surface plasmons because plasmonic nanoparticles can exhibit scattering, absorbance, and coupling properties based on their geometries and relative positions. Plasmons form a dipole in the material due to

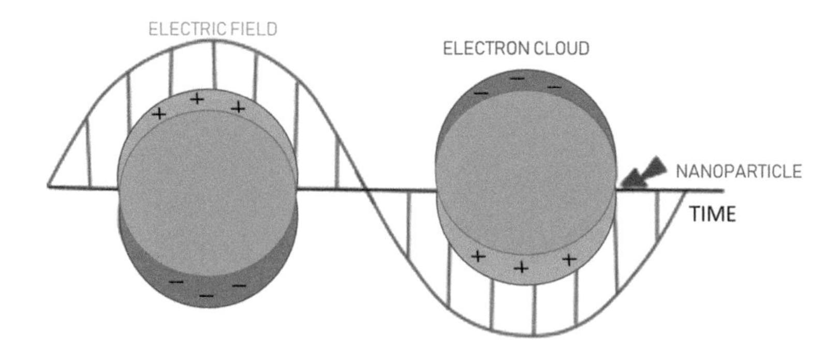

Fig. 2.10 Representation of LSPR coupling with an electromagnetic beam

electromagnetic waves. Electrons migrate in the material to restore their initial state. However, light waves oscillate, giving a constant shift in the dipole that forces electrons to oscillate at the same frequency of light. This coupling is produced only when the frequency of the light is equal to or less than that of the plasma and is called resonant frequency. In addition, although the wave is at the boundary of the conductor and the external environment, these oscillations are very sensitive to any change in the near environment, such as adsorption of molecules on the conductive face. Thus, this principle is used as a method to increase the sensitivity of interaction between the analyte and bio-recognition molecule. In other words, if a bio-recognition molecule is anchored to the surface of a nanoparticle, the analyte recognition event will produce an important change in the localized surface plasmon, this effect being fundamentally important for the development of sensors (Hutter and Flendler 2004) (Fig. 2.10).

2.3.4 Effect of Plasmons on surface-enhanced Raman scattering (SERS)

Raman is a spectroscopic technique used to study the vibrational, rotational, and other low-frequency modes of molecules. To excite these modes, a monochromatic laser is used in the visible, ultraviolet, or near-infrared range producing an inelastic dispersion of the photons (Raman scattering). As a result of the interaction of the photons with the different modes of the system, the energy of the laser photons is run up or down, allowing information on the structure and composition of the system.

However, when this technique is performed on molecules associated with a nanostructure, under certain conditions, a surface increase in the Raman dispersion can be produced by a factor of 10^{10} times. Although the origin of the process is not precisely known, the most plausible explanation indicates that it is produced by excitation of the localized surface plasmons of the nanoparticle. This effect is produced when the molecule is chemically associated with the nanoparticle, but it can also take place when the excited molecule is relatively far away from it.

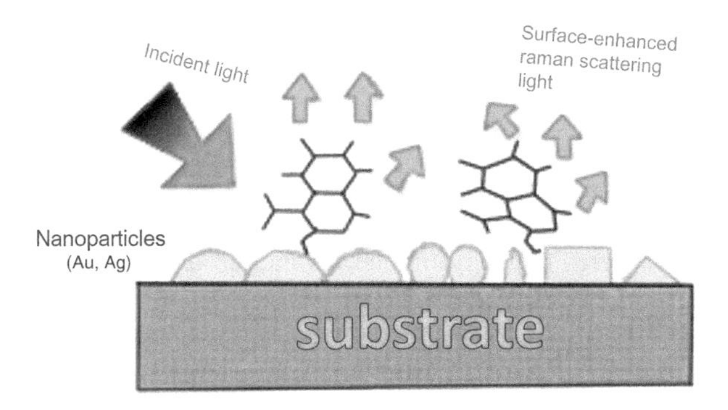

Fig. 2.11 Scheme of surface enhancement RAMAN produced by nanoparticles

The increase in the field is a maximum when the frequency of the plasmon is in resonance with the radiation, which requires that the oscillations of the plasmon are perpendicular to the surface. Therefore, rough surfaces or arrays of nanoparticles are required as these surfaces provide the area wherein these collective oscillations can occur. The field increase is not the same in all frequencies. For those frequencies in which the Raman signal is only slightly displaced from the incident light, both the incident and Raman light can be close to the resonance frequency of the plasmon, allowing a maximum increase. When the shift is marked, incident light and Raman cannot be in resonance, resulting in minor field increase. Generally, gold or silver nanoparticles are used for these tests since the plasmon frequency of these metals can be excited by the electromagnetic radiation usually used in Raman experiments. SERS has great relevance in the development of nanobiosensors since biomolecular coupling events can be determined in the vicinity of the nanoparticle in extremely low concentrations. Even weak couplings that hardly produce changes in the center symmetry of the molecules and alter the rules of spectroscopic selection changing their RAMAN absorptivity (Le Ru and Etchegoin 2009) can be detected (Fig. 2.11).

2.3.5 Effect of Quantum Confinement on Fluorescence

To understand these phenomena, we need to know certain basic concepts related to fluorescence. For fluorescence to occur, the presence of a component in the molecule called fluorochrome or fluorophore is required. This is a functional group of the molecule that absorbs energy of a specific wavelength and will re-emit it in another determined wavelength, i.e., of lower energy. The relationship between the energy emitted and the energy received is called quantum yield. The amount of re-emitted energy and its wavelength depend on both fluorochrome itself and its chemical environment. Among fluorophores we can find natural molecules, such as luciferin (present in fireflies) and quantum dots (<10 nm). Metal nanoparticles can strongly confine electromagnetic fields through their coupling to surface plasmons or propagated. This interaction allows increasing the speed of excitation and quantum yield and controlling the fluorescent light emitted by organic dyes and quantum dots. The fluorescence signal produced as a result of molecular binding events can be increased several orders of magnitude. Thus, detection times are decreased, and the sensitivity of the sensor is increased several orders.

On the other hand, the term fluorescent quenching or deactivation refers to any process that produces a decrease in the intensity of fluorescence emitted by a substance. A variety of processes can cause fluorescent deactivation, such as reactions in an excited state, energy transfer, complex formation, and quenching by molecular collisions.

In nanobiosensors, a strategy widely used to amplify the detection events of biorecognition molecules is the transfer of fluorescent energy between chromophores, in which one or both are nanoparticles. The Forster resonance energy transfer (FRET) is based on the fact that the excitation of a chromophore can be transferred to a nearby one by means of a dipole-dipole interaction, generally located in the same molecule.

This deactivation mechanism is dynamic since transfer occurs while the donor is in an excited state. In the FRET the energy transfer is not radiant and takes place between a donor fluorophore and an acceptor when the donor emission spectrum and the acceptor excitation overlap sufficiently so that the donor, once excited, transfers energy that allows response of the acceptor with the subsequent fluorescence emission. The efficiency of FRET decays rapidly with the distance between the fluorophores (1/r6), called the Forster radius, which is the distance at which the efficiency of the transfer decreases to 50%. Transfer is most effective when there is high superposition between the excitation and emission spectra of the donor and acceptor. It also depends on the relative spatial orientation of the transient dipole moments of the donor and acceptor fluorophores. In general, for the transfer to occur, fluorophores must be at a distance of less than 10 nm.

Nanobiosensors have been developed which, before their recognition event, their fluorescent components present FRET and, after interaction, fluorescent emission increases (or vice versa). In general, these phenomena occur when, as a consequence of the recognition event, an acceptor fluorochrome is fluorescent and therefore incapable of absorbing photons, increasing donor intensity. This process is called FRET inhibition. In this case, by exciting the donor fluorochrome, it increases its intensity as it does not lose energy by transfer to the acceptor.

Nanomaterials can be used either as quenchers or as fluorescent agents. As quenchers, it is produced when a fluorescent molecule is near the surface of a metallic nanomaterial due to transfer of energy, which decreases its emission. On the other hand, there are nanoparticles that exhibit fluorescence, such as graphene oxide, and this can be quenched by conductive polymers. They can also be used simultaneously as donors and acceptors, and the emission of quantum dots depends on their size; hence, they can be used in combination with bio-recognition agents and produce FRET after the detection event (Demir et al. 2016) (Fig. 2.12).

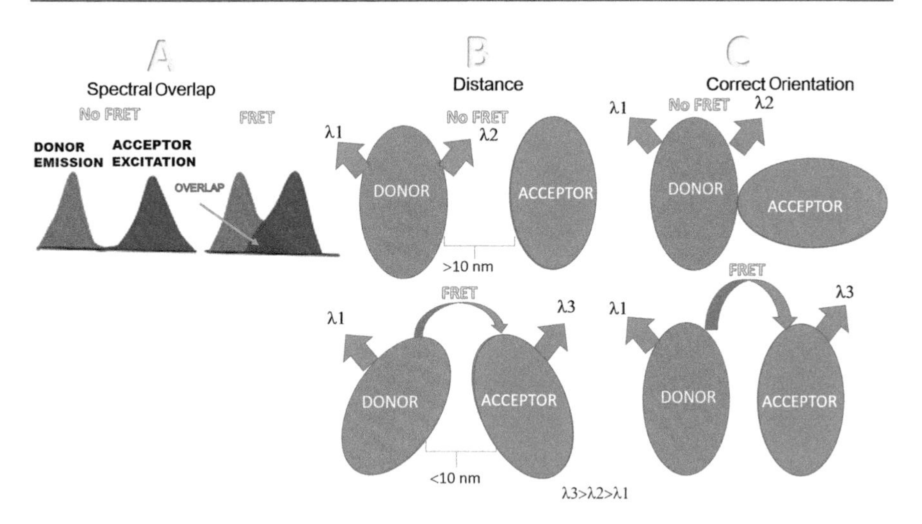

Fig. 2.12 Scheme of FRET properties based on (**a**) spectral overlap, (**b**) distance between donor and acceptor, and (**c**) orientation between molecules

2.3.6 Superparamagnetism

In ferromagnetic nanoparticles ranging between 2 and 50 nm, magnetization can randomly change the direction under the influence of temperature. Conventional ferromagnetic materials transition to paramagnetic above the Curie temperature (Tc). In contrast, in ferromagnetic nanoparticles, this transition can occur at a lower temperature. Due to their small size, particles can have only a single magnetic domain and two stable antiparallel orientations separated by a temperature-dependent energy barrier. The change of orientation may be caused at a time called Néel relaxation time. This relaxation time is an exponential function of particle size and inverse to temperature. This means that, below a certain size, ferromagnetic particles behave analogously to paramagnetic materials. As seen, particle size is particularly important in superparamagnetic properties. The effect of particle size on magnetic properties can be expressed in the following equation:

$$r = \left(6k_B T_b / K_u\right)^{1/3} \tag{2.6}$$

where r is the particle radius, k_B is the Boltzmann constant, T is the temperature, and K_u is the anisotropy constant.

In addition, T_b is the blocking temperature, which represents the temperature at which a superparamagnetic ordering generally exists.

The magnetic nanoparticles can be manipulated with magnetic fields, being very important as a bioaffinity platform. Such characteristic is key in the use of magnetic methods of preconcentration, separation, and purification of molecules such as

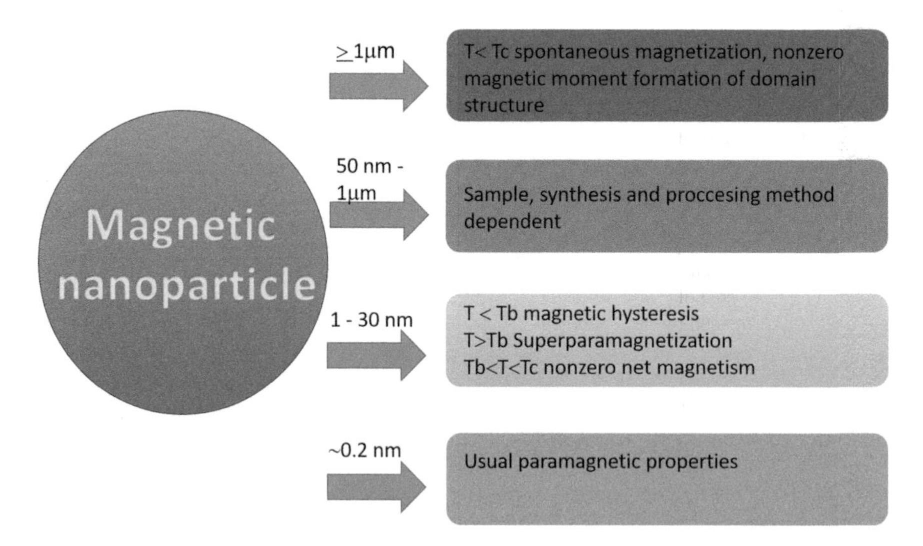

Fig. 2.13 Dependence of magnetic properties on particle size

DNA, proteins, and cells, allowing their transport, separation, positioning, organization, etc. (Mohammed et al. 2017) (Fig. 2.13).

2.3.7 Effects Given by Increase in the Surface

As already mentioned, much of the improvement made in the analytical response of the sensors is attributed not only to the unique quantum characteristics they show but also to the fact that they behave as extraordinarily large and rough surfaces in small-sized environments, allowing a large contact area in molecules immobilized on its surface. Figure 2.12 provides the typical example that allows us to understand this feature. The overall area is increased in the nanoparticle around 10^7 times! In a large number of heterogeneous chemical reactions dependent on interaction surface such as electrochemical reactions, the extraordinary increase in the area decreases the times of encounter between the analyte and the bio-recognition molecule, consequently decreasing response times and increasing the analytical signal. The immobilization of biomolecules such as enzymes to nanoparticles can be a process that affects the stability of the former. However, there are strategies that propose first functionalizing nanoparticles with a biomimetic environment, as the decoration with hydrophilic dendrimers, and then such arrangements would allow taking advantage from the increase of surface area, minimizing the conformational effects on the enzyme. Some nanomaterials have been incorporated into fluorescence-based immunoassays due to their ability to dramatically increase surface area in a small volume of liquid; when dispersed, they decrease incubation time by reduction of diffusional distances (Xiong et al. 2011) (Fig. 2.14).

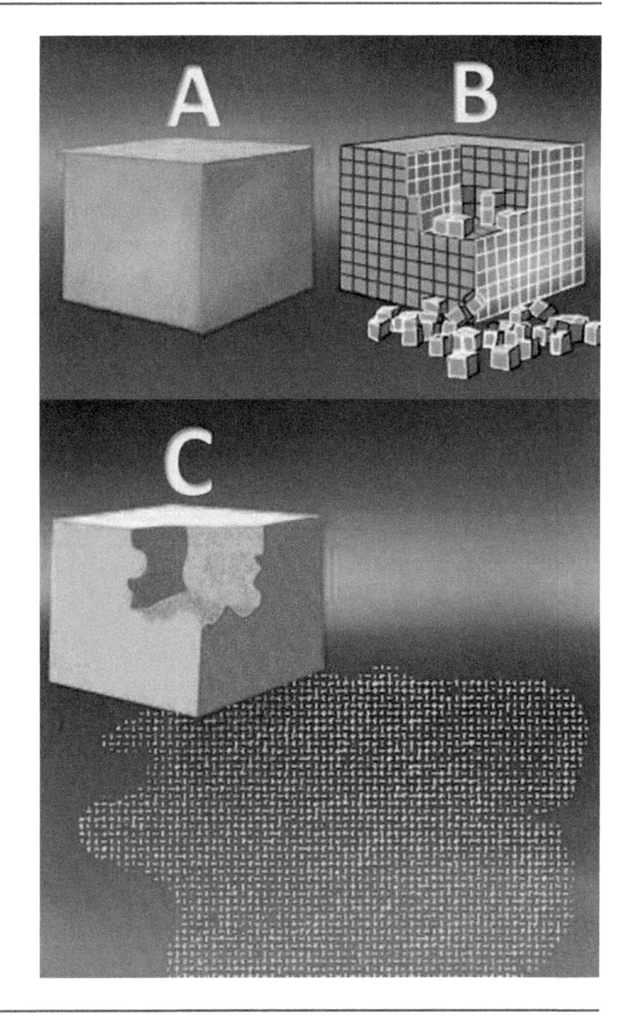

Fig. 2.14 Schematic representation of the increase in surface of nanoparticles in relation to macroscopic materials of the same overall volume. (**a**) Cube of 1 m on each side, total surface area of 6 m^2. (**b**) Cube formed by cubes of 10 cm on each side, total surface area of 60 m^2. (**c**) Cube formed by cubes of 1 nm on each side, total surface area of 60,000,000 m^2 (60 Km^2)

2.4 Classification of Nanobiosensors Based on Their Principle of Transduction

2.4.1 Electrochemical Nanobiosensors

Electrochemical nanobiosensors are those most commonly used due to the instrumental simplicity required for their operation, portability, and low cost as compared to other systems. In these biosensors, the species can be detected directly or through reactions, and, as a result, it produces electrical changes in the interface of one or several electrodes. Based on the different types of transducers, they can be divided into three main types: voltammetric, impedimetric, and photoelectrochemical.

2.4.1.1 Voltammetric

In voltammetry a potential difference is applied between a working electrode and a reference electrode. Then, the current flowing between the working electrode and a counter electrode (whose function is to close the circuit) is measured. Electrodes are placed in the electrochemical cell containing a solution in which the analyte to be determined is located. Electroactive species produce electron transfer reactions through the working electrode, and a current proportional to the number of electrons transferred is accordingly measured. There are different ways to apply the potential difference, and, therefore, different voltammetric techniques are derived. When the applied potential is a linear increasing and decreasing ramp (or vice versa), it is called cyclic voltammetry. If the applied potential follows a square wave, the technique is named square wave voltammetry. When the applied potential is a series of regular pulses superimposed on the linear sweep ramp, it is called differential pulse voltammetry. On the other hand, when the applied potential is constant throughout the test, it is called amperometry. This last is the most frequently used technique in biosensors due to its simplicity. Although it provides little mechanistic information, quantifications are easily performed on the basis of the measurement of the current in the working electrode. The Cottrell equation is a simple mathematical expression that allows relating the detected current (i) in the working electrode (of plane geometry) to different times (t) with the concentration of the analyte (C) in the solution:

$$i(t) = nFAC^{1/2} / \left(\pi^{1/2} t^{1/2}\right) \tag{2.7}$$

where n is the number of electrons exchanged in the redox process, F is the Faraday constant, and A is the area of the electrode.

In this type of biosensors, nanoparticles are used as a means to increase the area of the electrode. The electrical current, which is proportional to the area, is greatly increased by incorporating nanoparticles. They also improve response times by decreasing the mass transport distance required by the analyte to reach the electrode or nanoparticle. Normally, these sensors use carbon nanoparticles such as nanotubes or graphene, although quantum dots and metal nanoparticles have been used for the same purposes. In other cases, molecules are used as a template during the synthesis of a conductive polymer on the surface of the electrode (printed polymer). Enzymes are commonly used as bio-recognition molecules, although the use of aptamers has also been reported. In the case of enzymes (generally oxidases), they convert the analyte into an electroactive product. Its oxidation or reduction produces a current proportional to its concentration. On the other hand, the detection event of the aptamers is usually used as a means to block the access and subsequent electrochemical reaction of electroactive species to nanoparticles electrically connected to the electrode. Similar systems based on antibodies for the construction of immunosensors have also been developed (Zhang and Wei 2016) (Fig. 2.15).

2.4.1.2 Electrochemical Impedance Spectroscopy (EIS)

In this technique, the sample is placed between two electrodes, and a small electrical disturbance is applied to the system, which is usually sinusoidal of low potential and variable frequency. During this frequency sweep, the complex impedance of the system and its resistive, capacitive, and inductive components are determined. The

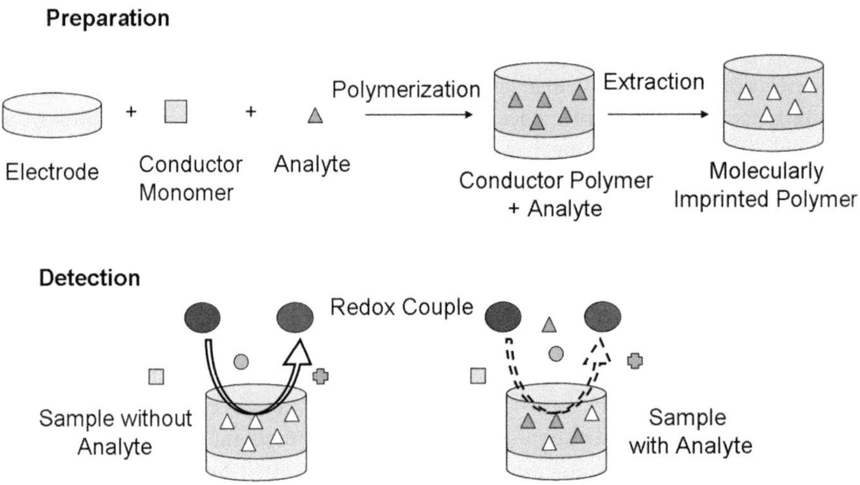

Fig. 2.15 Voltamperometric nanobiosensor based on molecularly imprinted polymer

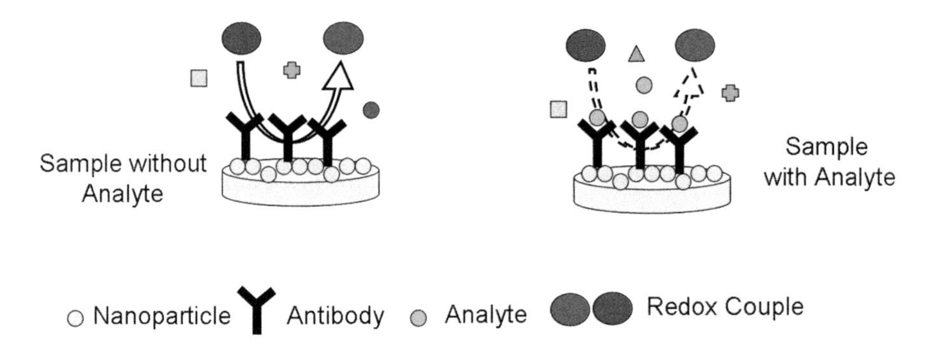

Fig. 2.16 Impedimetric nanobiosensor based on antibody-binding reaction and redox couple as an indicator

most important phenomena that can be studied by this technique include electrical charge transport, capacitance of the double layer, and diffusion of species. Although it is a technique developed for the investigation of electrochemical processes and the mechanism of electronic transfer, it has become significant due to its sensitivity to molecular interactions such as those occurring in an antibody or aptamer. However, it has been observed that the addition of nanoparticles improves the response of this type of sensors since they facilitate the kinetics of electronic transfer (Bahadır and Sezgintürk 2016) (Fig. 2.16).

2.4.2 Piezoelectric

Piezoelectricity is a property that has certain types of materials; thus, when deformed they produce a potential difference in their faces. Conversely, the application of a potential difference causes mechanical deformation in its structure. Silicon is a typical material with these features, and it is obtained commercially in disks with AT crystalline cut, covered by two metallic layers. This property has been used for years for the development of time bases in clocks and timer circuits since the mechanical deformation in the crystal shows a characteristic resonance frequency. In general, the quartz crystals used in sensors have a basal resonance frequency ranging between 5 and 10 MHz. A fundamental characteristic of these crystals that allows their use as sensors is that their resonance frequency is proportional to the mass of the crystal and the mass deposited on it (under certain conditions of elasticity, deposited section, and mass interval). The general mathematical expression that allows quantification of mass with these systems is called Sauerbrey and relates frequency displacement with mass deposited on the crystal:

$$\Delta f = -\left[2 f_0^{\,2} / A\left(\rho_q \mu_q\right)^{1/2}\right] \Delta m \tag{2.8}$$

where f_0 is the resonant frequency of the crystal (Hz), Δf is the frequency shift (Hz), Δm is the mass change (g), A is the active piezoelectric area of the crystal (cm^2), ρ_q is the density of the quartz (2.648 g/cm^3), and μ_q is the cutting module of the cutting glass AT (2.947×10^{11} $g/cm.s^2$).

Usually, these silicon disks are used with one of the faces exposed to a solution with the analyte, and the metal face is used as an electrochemical working electrode. The development of piezoelectric biosensors based on nanomaterials allows combining the high sensitivity of the frequency displacement measurement with the high contact area nanomaterials, which enables the immobilization of a large number of bio-recognition molecules, improving the analytical response of the system (Formisano et al. 2015) (Fig. 2.17).

2.4.3 Colorimetric

Colorimetric nanobiosensors are based on the color change occurring as a result of the analyte detection event. The measurement is based on conventional photometry that involves irradiating the material with a light beam of a single wavelength or a sweep of wavelengths, determining the absorbance of the sample. They can also be semiquantitative, in which color intensity can be determined visually on a predetermined scale, knowing the approximated analyte concentration. As seen previously, the absorption properties of nanoparticles, such as gold, depend on the surface plasmon, and the absorption spectrum is a function of the particle size. For this reason, if the particle is modified with a bio-recognition element in the presence of the analyte of interest, an aggregation of the nanoparticle will take place, consequently changing the absorption spectrum of the resulting aggregate (Ghadeer et al. 2015) (Fig. 2.18).

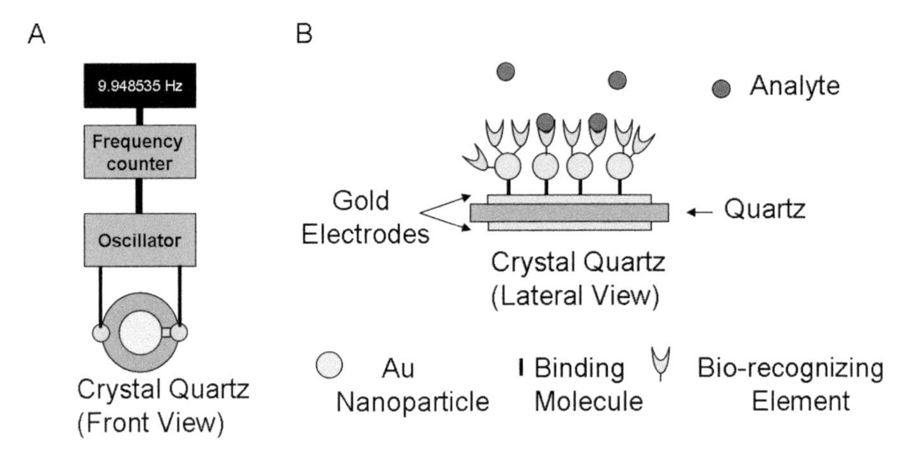

Fig. 2.17 (**a**) Experimental setup for piezoelectric determinations. (**b**) Typical nanobiosensor configuration

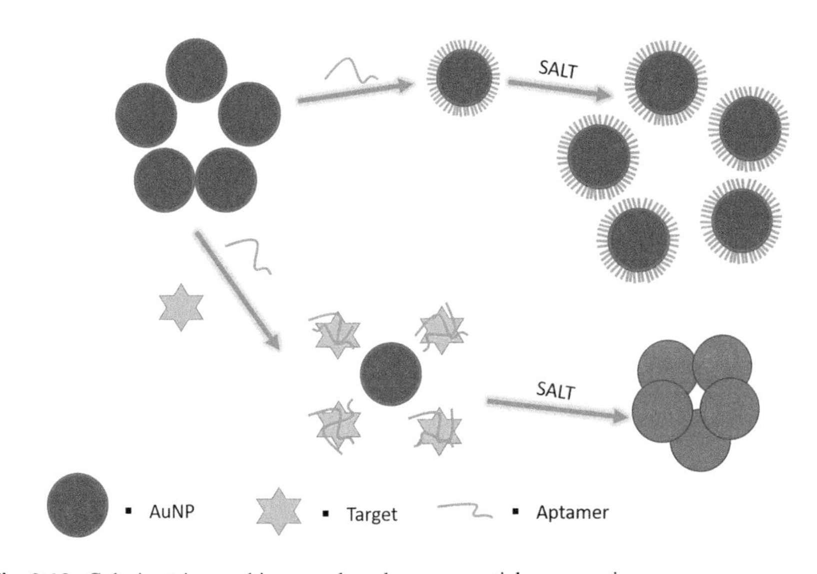

Fig. 2.18 Colorimetric nanobiosensor based on nanoparticle aggregation

2.4.4 Fluorescent

The emission of fluorescence is determined in a complex optical arrangement in which a light beam passing through a diffraction network allows selecting the wavelength of fluorescent excitation that strikes the sample. On the other hand, at a certain angle of the sample (generally at 90°), another diffraction network selects the wavelength of fluorescent emission from the sample, and an optical detector of either diode array or photomultiplier allows converting the optical signal in electrical signal to be quantified and processed in a computer.

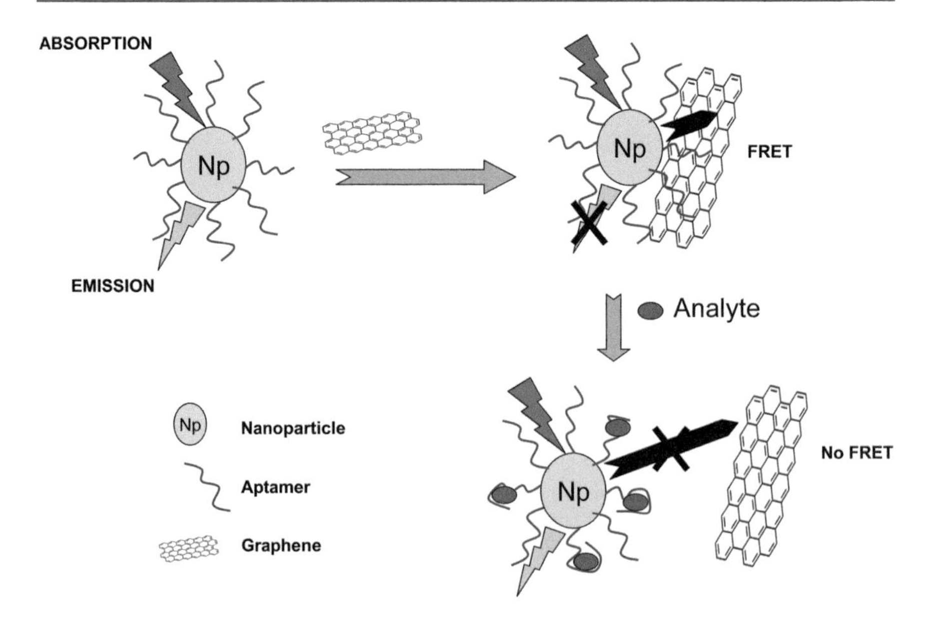

Fig. 2.19 Fluorescence nanobiosensor based on FRET blocking

As detailed before, fluorescent biosensors rely primarily on FRET to control the fluorescent emission between a fluorophore and a quencher which, under certain conditions and given the presence or absence of different analyte concentrations, changes the distance between the pair components and therefore the resulting fluorescent emission. Nanoparticles can be used both as fluorophore and quencher.

These systems usually use aptamers, which can be intercalated and bind nanoparticles or fluorescent dyes to the quencher. The quencher is usually a conductive polymer or graphene and, due to its small size, allows FRET to occur between both species. The presence of the analyte changes the conditions of the complex, releasing the fluorophore that, far from the quencher, produces fluorescence that can be detected and quantified. Although multiple strategies are reported in the literature, most of the methods used are based on these detailed principles (Aranda et al. 2018) (Fig. 2.19).

2.4.5 Chemiluminescent

In chemiluminescence a chemical reaction is produced after the detection event. In general this chemical reaction is redox type and occurs with rapid reaction kinetics and in several stages; thus, the emission is prolonged in time. Unlike fluorescence, which requires a source of excitation, chemiluminescent reactions produce emission by themselves, so they can be quantified by an optical device called a luminometer that consists of an optical detector able to convert the sample light signal into an electric current. They can have several channels to simultaneously quantify

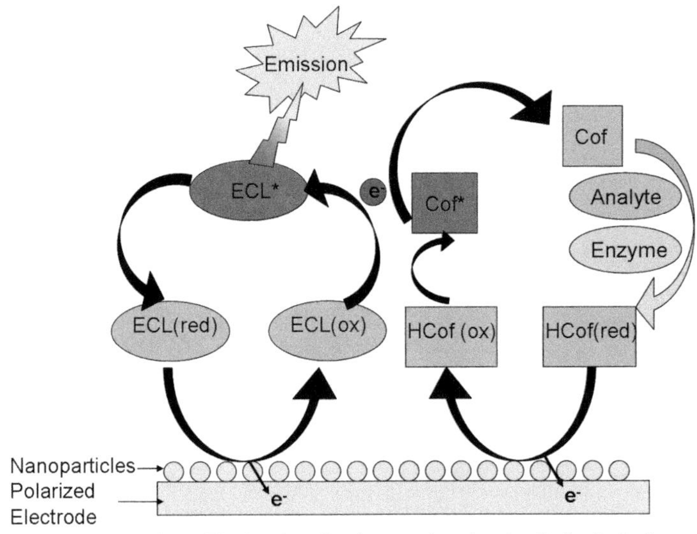

Fig. 2.20 Electrochemiluminescent nanobiosensor reactions

different emission wavelengths within the sample and are generally smaller than fluorescent and photometric equipment, since they require no source of light emission. Chemiluminescent biosensors are usually combined with nanomaterials such as carbon nanotubes, quantum dots, silica nanoparticles, and metallic nanoparticles achieving high sensitivity devices. A combination that has proved to be particularly suitable is the development of electrochemiluminescent sensors based on chemiluminescent reactions caused by the transfer of an electron with the polarized surface of an electrode. In these sensors, the presence of NPs greatly increases the interaction area between the molecules, therefore improving the system sensitivity (Jiao et al. 2008). As shown in the figure, the presence of the analyte (catalyzed by an enzyme) allows the regeneration of the cofactor (Cof) to its protonated reduced form (HCof (red)). This is oxidized at the electrode and then delivers an electron to an oxidized electrochemiluminescent molecule (ECL) which goes into excited state and emits a photon. Then, the oxidized ECL is regenerated again in the polarized electrode (Fig. 2.20).

2.4.6 Nanobiosensors Based on LSPR

The SPR, in biosensors based on this principle, is measured using an expensive optical device which includes a source of light emission that, after crossing a lens array, is polarized and focused on a prism, to finally make the beam strike on a sample in the proper configuration. The light beam that leaves the prism after internal reflection is detected in a CCD. As previously explained, the generated evanescent wave

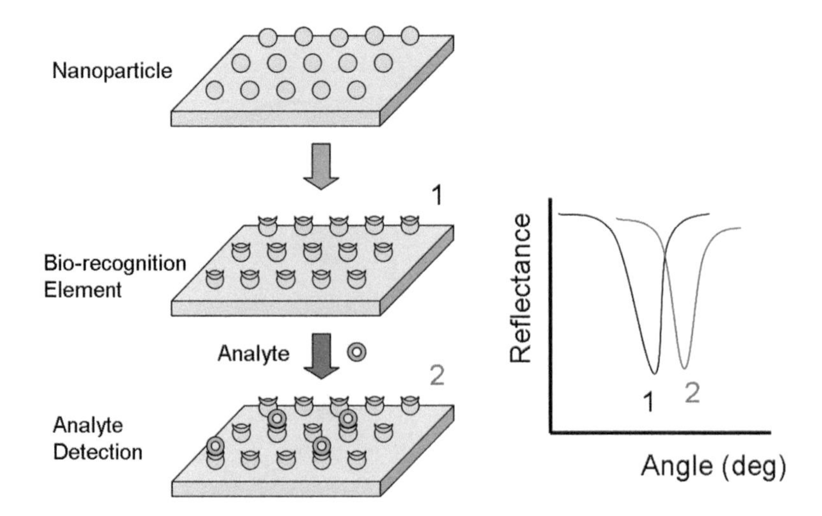

Fig. 2.21 Left: Configuration of a typical LSPR nanobiosensor. Right: Representation of signal shift after analyte detection

excites the surface plasmon of a metal surface in a resonant manner. As a consequence, this technique is very sensitive to the refractive index of the dielectric attached to the metal surface; thus, molecules bound to its surface can be directly detected, as well as the bio-recognition events of these molecules with others from the surrounding medium. SPR biosensors usually immobilize various biological components including enzymes, antibodies, nucleic acids, and aptamers bound to the metal surface as well as their binding reactions. In addition, if nanoparticles are used during the binding event, an increase in the signal is observed due to the LSPR effect (Sepúlveda et al. 2009) (Fig. 2.21).

2.4.7 Nanobiosensors Based on SERS

The development of biosensors based on this technique has been possible thanks to the increased Raman dispersion in the surface allowed by the presence of nanoparticles. On the other hand, with the traditional technique, the Raman signal is very weak due to the low probability that the light is dispersed inelastically by a molecule. Raman signals of reporter molecules are amplified using nanomaterials such as gold, silver, and other nanostructures. In this sense, the characteristic spectral signals of nanomaterials are absorbed on the surface of SERS substrates, and the increase comes from the excitation of localized SPR, which are collective oscillations of conducting electrons of metallic nanoparticles in the electromagnetic field. During the analyte detection event by the bio-recognition element, the competition between the reporter molecules occurs by the nanoparticle; therefore, the Raman signal is modified proportionally to the concentration of the molecule to be quantified as shown in the figure (Ilkhani et al. 2016) (Fig. 2.22).

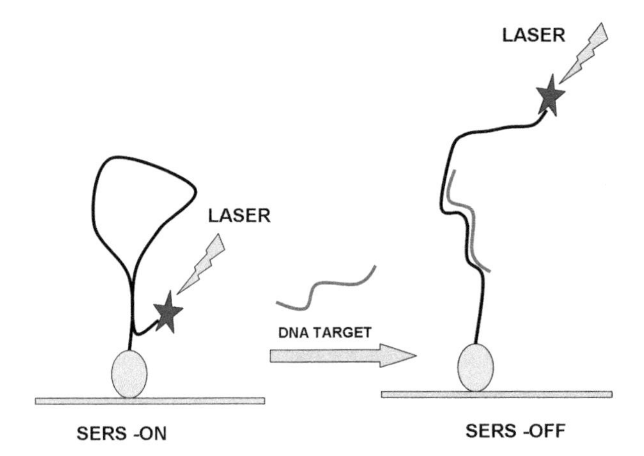

Fig. 2.22 Example of a SERS nanobiosensor for DNA detection using a reporter molecule coupled with a nanoparticle

2.5 Specific Applications of Nanobiosensors in Diagnostics

Nanotechnology has made it possible to improve the analytical characteristics of sensors based on the incorporation and integration of nanomaterials. Nanobiosensors have produced a great impact in medicine, expanding the detection horizons of species for the diagnosis of pathologies. With nanobiosensors it is possible to detect an innumerable amount of biological analytes such as DNA, RNA, enzymes, antibodies or proteins in general, carbohydrates, lipids, hormones, and metabolites. It can also determine drugs, ions, or large-sized complex structures such as viruses, bacteria, fungi, and tumor cells. In all cases, markers indicating the presence of cells with alterations in the cell cycle or antigens found in infectious organisms can be quantified.

Although many conventional biosensors allow determining the same species as nanobiosensors, the large difference lies in the high sensitivity observed in sensors combined with nanomaterials. In the latter, detection limits are observed for species in the order of picomolar to atomolar which are difficult to reach by conventional sensors. In addition, a direct consequence of high sensitivity is the decrease in the incubation times needed to reach a quantifiable response.

On the other hand, it has been observed that most recent works belong to studies of electrochemical nanobiosensors and correspond to biosensors based on fluorescence coupled to FRET processes, while the rest minority corresponds to the other methodologies studied in this chapter.

From the point of view of the biomarkers that allow the diagnosis of different pathologies, it should be noted that these molecules show differences of around ten orders of magnitude between the most concentrated and the most diluted in biological fluids. Certainly, nanobiosensors must be exploited for the quantification of markers that are in concentrations of the order of traces.

In this sense, it is reported in the literature that 1/3 of research work in which nanobiosensors are developed for the diagnosis of pathologies is conducted in order to quantify tumor markers. These chemical species are present in very low concentrations, particularly in the early stages of the disease, in which the expectation of the survival of patients is the most favorable.

Among them we find biosensors developed to determine small molecules of simple chain RNA (micro RNA or miRNA), working as negative key regulators for posttranscriptional modulation. These markers are overexpressed in breast cancer (Ebrahimi et al. 2018; Azimzadeh et al. 2016; Tian et al. 2018) or gastric cancer (Zhou et al. 2018; Zheng et al. 2019; Daneshpour et al. 2018). On the other hand, nanobiosensors have been developed for the determination of widely known biomarkers such as the prostatic antigen (Yazdani et al. 2019), p53 (Zheng et al. 2019), alpha-fetoprotein, and carcinoembryonic antigen (Khan et al. 2018). Other important markers include adenomatous polyposis coli (APC) (Darestani-Farahani et al. 2018), colon cancer secreted protein-2 (CCSP-2) (Lee et al. 2018), cell adhesion protein (Pallela et al. 2016), indicator of inflammation and cancer by arginase activity (Malalasekera et al. 2017), tumor cell nucleosin receptor MCF-7 (Borghei et al. 2016), and mucin-1 biomarker (MUC-1) (Yousefi et al. 2019) (Fig. 2.23).

Sensors have also been developed for the detection of infarcts such as myoglobin (Suprun et al. 2010) and troponin (Radha Shanmugam et al. 2017).

On the other hand, the third part of research into nanobiosensors aims at the detection of antigenic markers found in infectious microorganisms such as viruses, bacteria, fungi, and parasites. We can cite, among them, sensors designed for the detection of human papillomavirus (Shamsipur et al. 2017; Azizah et al. 2017), dengue (Cheng et al. 2012; Nguyen et al. 2012), H1N1 influenza (Takemura et al. 2017), and hepatitis virus (Soleymani et al. 2018). With the rest of the microorganisms, the use of sensors for antigens of *Aspergillus* (Bhatnagar et al. 2018), *Salmonella* (Dao et al. 2018), malaria (Brince Paul et al. 2016), anthrax (Yoon 2009), *Helicobacter* (Mojtaba et al. 2011), and *E. coli* (Singh et al. 2017; Suaifan et al. 2017) was reported.

A less important place is occupied by the sensors that determine hormones and conventional metabolites present in plasma. These include thyroid-stimulating hormone (TSH) (Salahvarzi et al. 2017), prolactin (Faridli et al. 2016), acetylcholine (Martín-Barreiro et al. 2018), and chemokines (Chung et al. 2018). In addition, conventional species that can also be determined by conventional biosensors include glucose (Wu et al. 2010), creatinine (Pundir et al. 2013), cholesterol (Rahim et al. 2018), and hemoglobin (Zhou et al. 2018).

It is important to note that nanobiosensors have also been developed for the detection of neurochemical indicators that mark the early stages of Alzheimer's disease. Such markers consist mainly of peptides released by the proteolytic action on the amyloid precursor protein and released in cerebrospinal fluid and plasma. These nanobiosensors allow detection of very low concentrations of these species and grant the possibility of carrying out early treatments for this serious disease (Nazem and Mansoori 2009; de la Escosura-Muñiz et al. 2015).

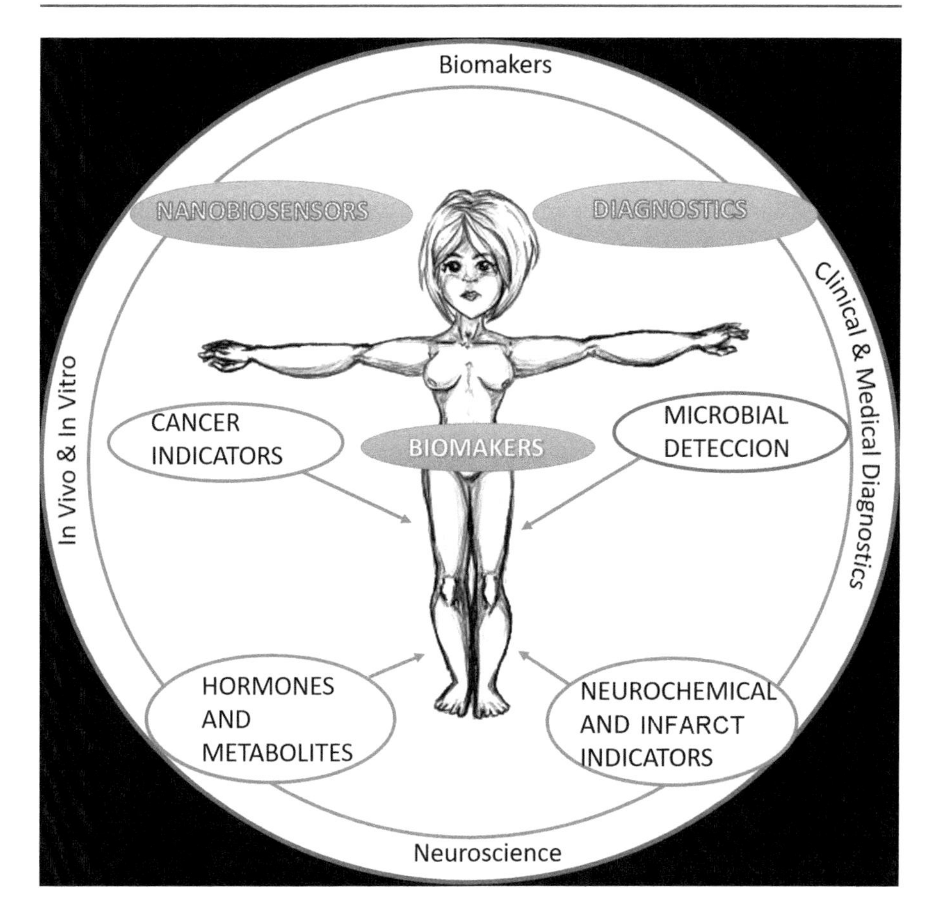

Fig. 2.23 Biomarkers present in traces depend on highly sensitive analytical techniques such as nanobiosensors to achieve sufficient predictive power in the diagnosis of pathologies

2.6 Conclusions and Future Perspectives

The use of nanoparticles has opened a new era in the development of nanobiosensors capable of achieving analytical responses that compete with the most powerful instrumental techniques. The application of the quantum properties of nanomaterials in nanobiosensors requires the joint work of multidisciplinary teams capable to adequately control the particle properties such as shape, size, and composition. Thus, the greater sensitivity achieved in nanobiosensors allows decreasing the size of the devices, enabling excellent responses in microscopic systems. The current development of nanobiosensors indicates that the new areas of research will be focused on the development of sensors that can be placed within the human body by microsurgeries or simply by means of a simple injection. Once inside the organism, they will allow monitoring different chemical species for a long period of time, communicating through wireless transmission with technological devices like smartphones.

One of the current challenges is to achieve sensors that are degradable and non-toxic and that can be energetically integrated into the body to avoid the need to use batteries. The complementation of sensors with actuators will allow replacing or compensating the functional deficiencies of organs such as the kidney, liver, or pancreas. On the other hand, they will allow diminishing the time and resources required for the diagnosis of pathologies. Regardless of the evolution of their development, in the future, it is envisioned that nanobiosensors will play a fundamental role in diagnosing diseases and improving the human quality of life.

Acknowledgments We would like to thank Agustina A. Romero for her assistance with the edition of the illustrations of this chapter. We would also like to thank Prof. Pranjal Chandra for his invitation to join us in his book project. Financial support from FCQ-UNC and IPQA - CONICET (Argentina) is gratefully acknowledged.

References

Alberts B, Johnson A, Lewis J, Raff M, Roberts K, Walter P (2014) Chapter 4: DNA, chromosomes and genomes. In: Molecular biology of the cell, 6th edn. Garland Science, New York/Abingdon, UK. ISBN 978-0-81534-432-2

Aranda PR, Messina GA, Bertolino FA, Pereira SV, Fernández Baldo MA, Raba J (2018) Nanomaterials in fluorescent laser-based immunosensors: review and applications. Microchem J 141:308–323

Azimzadeh M, Rahaie M, Nasirizadeh N, Ashtari K, Naderi-Manesh H (2016) An electrochemical nanobiosensor for plasma miRNA-155, based on graphene oxide and gold nanorod, for early detection of breast cancer. Biosens Bioelectron 77:99–106

Azizah N, Hashim U, Gopinath SCB, Nadzirah S (2017) A direct detection of human papillomavirus 16 genomic DNA using gold nanoprobes. Int J Biol Macromol 94(Pt A):571–575

Bahadır EB, Sezgintürk MK (2016) A review on impedimetric biosensors. Artif Nanomed Biotechnol 44(1):248–262

Bhakta AS, Evans E, Benavidez TE, Garcia CD (2015) Protein adsorption onto nanomaterials for the development ofbiosensors and analytical devices: A review. Anal Chim Acta 872:7–25

Bhatnagar I, Mahato K, Ealla KKR, Asthana A, Chandra P (2018) Chitosan stabilized gold nanoparticle mediated self-assembled gliP nanobiosensor for diagnosis of Invasive Aspergillosis. Int J Biol Macromol 110:449–456

Borghei YS, Hosseini M, Dadmehr M, Hosseinkhani S, Ganjali MR, Sheikhnejad R (2016) Visual detection of cancer cells by colorimetric aptasensor based on aggregation of gold nanoparticles induced by DNA hybridization. Anal Chim Acta 904:92–97

Brince Paul K, Kumar S, Tripathy S, Vanjari SRK, Singh V, Singh SG (2016) A highly sensitive self-assembled monolayer modified copper doped zinc oxide nanofiber interface for detection of Plasmodium falciparum histidine-rich protein-2: targeted towards rapid, early diagnosis of malaria. Biosens Bioelectron 80:39–46

Cahay M (2001) Quantum confinement VI: nanostructured materials and devices: proceedings of the international symposium. The Electrochemical Society, Pennington. ISBN:978-1-56677-352-2

Chandra P (2016) Nanobiosensors for personalized and onsite biomedical diagnosis. IET, London. ISBN: 978-1-84919-950-6

Chandra P, Tan YN, Singh SP (2017) Next generation point-of-care biomedical sensors technologies for cancer diagnosis. Springer Nature, Singapore, p 9. ISBN:78-981-10-4726-8

Cheng MS, Ho JS, Tan CH, Wong JPS, Ng LC, Toh CS (2012) Development of an electrochemical membrane-based nanobiosensor for ultrasensitive detection of dengue virus. Anal Chim Acta 725:74–80

Chung S, Chandra P, Koo JP, Shim YB (2018) Development of a bifunctional nanobiosensor for screening and detection of chemokine ligand in colorectal cancer cell line. Biosens Bioelectron 100:396–403

Daneshpour M, Karimi B, Omidfar K (2018) Simultaneous detection of gastric cancer-involved miR-106a and let-7a through a dual-signal-marked electrochemical nanobiosensor. Biosens Bioelectron 109:197–205

Dao TNT, Yoon J, Jin CE, Koo B, Han K, Shin Y, Lee TY (2018) Rapid and sensitive detection of Salmonella based on microfluidic enrichment with a label-free nanobiosensing platform. Sens Actuator B Chem 262:588–594

Darestani-Farahani M, Faridbod F, Ganjali MR (2018) A sensitive fluorometric DNA nanobiosensor based on a new fluorophore for tumor suppressor gene detection. Talanta 190:140–146

de la Escosura-Muñiz A, Plichta Z, Horák D, Merkoçi A (2015) Alzheimer's disease biomarkers detection in human samples by efficient capturing through porous magnetic microspheres and labelling with electrocatalytic gold nanoparticles. Biosens Bioelectron 67:162–169

Demir HV, Hernández Martínez PL, Govorov A (2016) Förster-type resonance energy transfer (FRET): applications. Springer, Singapore

Ebrahimi A, Nikokar I, Zokaei M, Bozorgzadeh E (2018) Design, development and evaluation of microRNA-199a-5p detecting electrochemical nanobiosensor with diagnostic application in Triple Negative Breast Cancer. Talanta 189:592–598

Faridli Z, Mahani M, Torkzadeh-Mahani M, Fasihi J (2016) Development of a localized surface plasmon resonance-based gold nanobiosensor for the determination of prolactin hormone in human serum. Anal Biochem 495:32–36

Formisano N, Jolly P, Bhalla N, Cromhout M, Estrela P (2015) Optimisation of an electrochemical impedance spectroscopy aptasensor by exploiting quartz crystal microbalance with dissipation signals. Sensors Actuators B Chem 220:369–375

Hermanson GT, Smith P, Mallia K (1992) Immobilized affinity ligand techniques. Publisher Academic Press, San Diego ISBN:9780123423306

Hutter E, Flendler JH (2004) Exploitation of surface plasmon resonance. Adv Mater 16:1685–1706

Ilkhani H, Hughes T, Li J, Zhong CJ, Hepel M (2016) Nanostructured SERS-electrochemical biosensors for testing of anticancer drug interactions with DNA. Biosens Bioelectron 80(15):257–264

Janeway C (2001) Immunobiology, 5th edn. Garland Publishing, New York. ISBN 978-0-8153-3642-6

Jiao T, Leca-Bouvier BD, Boullanger P, Blum LJ, Girard-Egrot AP (2008) Electrochemiluminescent detection of hydrogen peroxide using amphiphilic luminol derivatives in solution. Colloids Surf A Physicochem Eng Asp 321(15):143–146

Khan NU, Feng Z, He H, Wang Q, Liu X, Li S, Shi X, Wang X, Ge B, Huang F (2018) A facile plasmonic silver needle for fluorescence-enhanced detection of tumor markers. Anal Chim Acta 1040:120–127

Lee HJ, Do EJ, Jeun M, Park S, Sung YN, Hong SM, Kim SY, Kang JY, Kim DH, Son HN, Joo J, Hwang SW, Park SH, Yang DH, Ye BD, Byeon JS, Choe J, Yang SK, Lee KH, Myung SJ (2018) Novel blood-based detection of colorectal cancer and adenoma using a nanobiosensor targeting CCSP-2 (Colon cancer secreted protein-2). Gastroenterology 154:1062

Malalasekera AP, Wang H, Samarakoon TN, Udukala DN, Yapa AS, Ortega R, Shrestha TB, Alshetaiwi H, McLaurin EJ, Troyer DL, Bossmann SH (2017) A nanobiosensor for the detection of arginase activity. Nanomedicine 13(2):383–390

Martín-Barreiro A, de Marcos S, de la Fuente JM, Grazú V, Galbán J (2018) Gold nanocluster fluorescence as an indicator for optical enzymatic nanobiosensors: choline and acetylcholine determination. Sens Actuator B Chem 277:261–270

Mohammed L, Gomaa HG, Ragab D, Zhu J (2017) Magnetic nanoparticles for environmental and biomedical applications: a review. Particuology 30:1–14
Mojtaba S, Maryam S, Afshin M, Sadegh H, Nazanin P, Mahsa S (2011) A new and highly sensitive nanobiosensor for detection of Helicobacter based on fluorescence resonance energy transfer. Clin Biochem 44(13):S222
Nazem A, Mansoori GA (2009) Screening for Alzheimer's pathology through nanobiosensors. Pathology 41(1):78
Nguyen BTT, Peh AEK, Chee CYL, Fink K, Chow VTK, Ng MML, Toh CS (2012) Electrochemical impedance spectroscopy characterization of nanoporous alumina dengue virus biosensor. Bioelectrochemistry 88:15–21
Pallela R, Chandra P, Noh HB, Shim YB (2016) An amperometric nanobiosensor using a biocompatible conjugate for early detection of metastatic cancer cells in biological fluid. Biosens Bioelectron 85:883–890
Pundir CS, Yadav S, Kumar A (2013) Creatinine sensors. TrAC 50:42–52
Radha Shanmugam N, Muthukumar S, Chaudhry S, Anguiano J, Prasad S (2017) Ultrasensitive nanostructure sensor arrays on flexible substrates for multiplexed and simultaneous electrochemical detection of a panel of cardiac biomarkers. Biosens Bioelectron 89.(Pt 2:764–772
Rahim MZA, Govender-Hondros G, Adeloju SB (2018) A single step electrochemical integration of gold nanoparticles, cholesterol oxidase, cholesterol esterase and mediator with polypyrrole films for fabrication of free and total cholesterol nanobiosensors. Talanta 189:418–428
Romero MR, Baruzzi AM, Garay F (2012a) Mathematical modeling and experimental results of a sandwich-type amperometric biosensor. Sensors Actuators B 162:284–291
Romero MR, Baruzzi AM, Garay F (2012b) How low does the oxygen concentration go within a sándwich-type amperometric biosensor? Sensors Actuators B 174:279–284
Romero MR, Peralta D, Alvarez Igarzabal CI, Baruzzi AM, Strumia MC, Garay F (2017) Supramolecular complex based on MWNTs/Boltorn H40 provides fast response to a Sandwich-type amperometric lactate biosensor. Sensors Actuators B Chem 244:577–584
Ru EL, Etchegoin P (2009) Principles of surface-enhanced raman spectroscopy and related plasmonic effects, 1st edn. Elsevier Science, Amsterdam/Heidelberg. https://doi.org/10.1016/B978-0-444-52779-0.X0001-3. ISBN:9780444527790
Salahvarzi A, Mahani M, Torkzadeh-Mahani M, Alizadeh R (2017) Localized surface plasmon resonance based gold nanobiosensor: Determination of thyroid stimulating hormone. Anal Biochem 516:1–5
Sefah K, Shangguan D, Xiong X, O'Donoghue MB, Tan W (2010) Development of DNA aptamers using Cell-SELEX. Nat Protoc 5:1169. https://doi.org/10.1038/nprot.2010.66
Sepúlveda B, Angelomé PC, Lechuga LM, Liz-Marzán LM (2009) LSPR-based nanobiosensors. Nano Today 4(3):244–251
Shamsipur M, Nasirian V, Mansouri K, Barati A, Veisi-Raygani A, Kashanian S (2017) A highly sensitive quantum dots-DNA nanobiosensor based on fluorescence resonance energy transfer for rapid detection of nanomolar amounts of human papillomavirus 18. J Pharm Biomed Anal 136:140–147
Singh KP, Dhek NS, Nehra A, Ahlawat S, Puri A (2017) Applying graphene oxide nano-film over a polycarbonate nanoporous membrane to monitor E. coli by infrared spectroscopy. Spectrochim Acta A Mol Biomol Spectrosc 170:14–18
Soleymani J, Hasanzadeh M, Somi MH, Jouyban A (2018) Nanomaterials based optical biosensing of hepatitis: recent analytical advancements. TrAC 107:169–180
Suaifan GARY, Alhogail S, Zourob M (2017) Paper-based magnetic nanoparticle-peptide probe for rapid and quantitative colorimetric detection of Escherichia coli O157:H7. Biosens Bioelectron 92(15):702–708
Suprun E, Bulko T, Lisitsa A, Gnedenko O, Ivanov A, Shumyantseva V, Archakov A (2010) Electrochemical nanobiosensor for express diagnosis of acute myocardial infarction in undiluted plasma. Biosens Bioelectron 25(7):1694–1698
Takemura K, Adegoke O, Takahashi N, Kato T, Li TC, Kitamoto N, Tanaka T, Suzuki T, Park EY (2017) Versatility of a localized surface plasmon resonance-based gold nanoparticle-alloyed

quantum dot nanobiosensor for immunofluorescence detection of viruses. Biosens Bioelectron 89(2):998–1005

Tian L, Qian K, Qi J, Liu Q, Yao C, Song W, Wang Y (2018) Gold nanoparticles superlattices assembly for electrochemical biosensor detection of microRNA-21. Biosens Bioelectron 99:564–570

Wu W, Zhou T, Aiello M, Zhou S (2010) Construction of optical glucose nanobiosensor with high sensitivity and selectivity at physiological pH on the basis of organic-inorganic hybrid microgels. Biosens Bioelectron 25(12):2603–2610

Xiong S, Qi W, Cheng Y, Huang B, Wang M, Li Y (2011) Modeling size effects on the surface free energy of metallic nanoparticles and nanocavities. Phys Chem Chem Phys 13:10648–10651

Yazdani Z, Yadegari H, Heli H (2019) A molecularly imprinted electrochemical nanobiosensor for prostate specific antigen determination. Anal Biochem 566:116–125

Yoon MY (2009) Novel application to the diagnostic nanobiosensor for anthrax. New Biotechnol 25:S33

Yousefi M, Dehghani S, Nosrati R, Zare H, Evazalipour M, Mosafer J, Tehrani BS, Pasdar A, Mokhtarzadeh A, Ramezani M (2019) Aptasensors as a new sensing technology developed for the detection of MUC1 mucin: a review. Biosens Bioelectron 130:1. https://doi.org/10.1016/j.bios.2019.01.015

Zhang Y, Wei Q (2016) The role of nanomaterials in electroanalytical biosensors: a mini review. J Electroanal Chem 781:401–409

Zheng XT, Goh WL, Yeow P, Lane DP, Ghadessy FJ, Tan YN (2019) Ultrasensitive dynamic light scattering based nanobiosensor for rapid anticancer drug screening. Sens Actuator B Chem 279:79–86

Zhou T, Ashley J, Feng X, Sun Y (2018) Detection of hemoglobin using hybrid molecularly imprinted polymers/carbon quantum dots-based nanobiosensor prepared from surfactant-free Pickering emulsion. Talanta 190:443–449

3 Carbon Quantum Dots: A Potential Candidate for Diagnostic and Therapeutic Application

S. Sharath Shankar, Vishnu Ramachandran, Rabina P. Raj, T. V. Sruthi, and V. B. Sameer Kumar

Abstract

Among various quantum dots, carbon quantum dots (CQDs) are getting much more attention due to their nontoxicity and high water solubility. CQDs with or without functionalization/passivation possess different fluorescent properties. Moreover, simple and economical procedure involved in their preparation makes them a reliable substitute for nano-materials/semiconductor quantum dots. Due to their tunable fluorescent properties, these carbon quantum dots find application in anticancer studies, tissue engineering, nano-medicine and targeted drug delivery, electrochemical sensing, bio-sensing, and bio-imaging. Also, these CQDs are capable of showing high conductivity and therefore can be useful for improving the electrical/conductive nature of different materials. The CQDs can find significant role in LEDs. CQDS can be prepared by top-down approaches or bottom-down approaches. Top-down approaches involve arc discharge method, laser ablation, electrochemical oxidation, and chemical oxidation, and the bottom-down approaches include combustion, pyrolysis, microwave-assisted method, ultrasonic method, oxidative acid treatment, and hydrothermal methods. In this chapter we have tried to give an insight to the preparation and biological applications of CQDs.

Keywords

Carbon quantum dots · Fluorescent nano particles · Applications of Carbon quantum dots · Electrochemical/ bio-sensing · Bio-imaging

S. S. Shankar · V. Ramachandran · R. P. Raj · T. V. Sruthi · V. B. S. Kumar (✉)
Department of Biochemistry & Molecular Biology, Central University of Kerala, Kasaragod, Kerala, India

P. Chandra, R. Prakash (eds.), *Nanobiomaterial Engineering*,
https://doi.org/10.1007/978-981-32-9840-8_3

3.1 Introduction

Carbon quantum dots (CQDs) are a new class of fluorescent carbon nano-materials of less than 10 nm size. CQDs are also referred to as carbon nano-dots (CNDs) and carbon dots (CDs/C-dots), which can be defined as having a carbogenic nucleus with functional surface groups. Discovery of CQDs was serendipity, when Xu and co-workers in 2004 saw some residual fluorescent nanoparticles during the purification process of single-wall carbon nanotubes (SWNTs) (Xu et al. 2004). It was Sun and team who named these particles as carbon quantum dots (Sun et al. 2006). The chemical character of CQDs varies from amorphous to nanocrystalline and has predominant sp^2 carbon, the spacing of which consists of graphitic or turbostatic carbon (Baker and Baker 2010; Zheng et al. 2011). CQDs are typically composed of 53.93% carbon, 2.56% hydrogen, 1.30% nitrogen, and 40.33% oxygen. The intriguing features of carbon dots are their chemical inertness, low photo-bleaching and photodegradation, low cytotoxicity, and excellent biocompatibility. Due to the existence of oxygen moieties, the dots have proven to be an extremely soluble material in aqueous solution (Baker and Baker 2010; Zheng et al. 2011). Functional groups such as –OH, -COOH, and $-NH_2$ and other CQD surface groups make them more water soluble (Cao et al. 2007). For these reasons, CQDs had drawn broad attention for various ranges of applications in technology, engineering, and particularly biomedical fields.

One of the most interesting characteristics of CQDs is the size-dependent optical absorption. Carbon dots have strong absorption in both ultraviolet and visible region (Li et al. 2010a). CQD's luminescence characteristics include electroluminescence and photoluminescence (PL), the most prominent of which is the latter. Carbon quantum dots have outstanding optical properties such as elevated fluorescence stability, non-blinking, adjustable excitation, and wavelengths of emission (Li et al. 2010a; Jia et al. 2012; Wang et al. 2013a, 2014). The precise understanding behind the emitting mechanism of CQDs is uncertain, but scientists speculated it involved quantum containment impact, surface trap stabilization, or exciton recombination radiation (Li et al. 2012). The presence of surface energy traps is stated to result in surface passivation emissions ascribed to CQD's photoluminescence (PL) (Sun et al. 2006). Moreover the photo-excited CQDs act as efficient electron donors and acceptors; therefore, electron acceptor or electron donor molecules in alternatives can be readily quenched. The CQDs also act as an appropriate candidate for light energy conversion, photovoltaic devices, and associated applications owing to their capacity to display photo-induced electron transfer property (Shen et al. 2012). CQDs with their simplistic methods of synthesis displayed a plethora of immense implications in the field of chemistry, engineering, biology, as well as medicine. Various precursors such as chemical, green, and waste products can be used to synthesize CQDs. Carbon dot synthesis is widely divided into two classifications, top-down and bottom-up. Top-down technique usually includes splitting bigger carbonaceous materials (Xu et al. 2004), whereas bottom-up synthesis process includes supported path and carbonization where a smaller precursor molecule forms nano-carbon dots through a series of chemical reaction (Baker and Baker

2010). Supported route is referred as blocking of nanoparticle agglomeration during high-temperature treatment (Wang et al. 2015a). Both approaches of synthesis require particle size control, proper measures against agglomeration during carbonization, and surface passivation. Method of arc discharge, chemical oxidation, laser ablation, electrochemical oxidation, etc. are included under top-down approaches. Generally dots synthesized by means of these methods lack fluorescence; an additional passivation is necessary to produce light emission. Bottom-up approaches include combustion, pyrolysis, oxidative acid treatment, microwave, ultrasonic, and hydrothermal methods, which serve more attraction and attention in recent decades due to their simplest preparation routes and inexpensive carbon sources (Li et al. 2012). Nowadays green synthesis is gaining more for the CQD synthesis. In the present chapter, we mainly focus on various methods of synthesis of CQDs and discuss their potential applications in different fields of therapeutics.

3.2 Synthesis

During the synthesis of single-walled carbon nanotubes by arc discharge technique (Xu et al. 2004), Xu et al. found carbon nano-dots in 2004. Laser ablation and arc discharge are altered technique of physical vapor deposition (PVD) involving the condensation of the strong carbon atom formed by warm gaseous carbon atom. Laser ablation is a simplistic, environmentally friendly, and effective technique for producing CQDs in which the surface conditions of particles can be adjusted. In this process, the carbon material is irradiated under elevated pressure and temperature with a laser beam in the presence of an inert gas as a carrier. Usually, carbon dots are prepared by dispersing the carbon precursor in a solvent by ultra-sonication and dropping the resultant suspension onto a glass cell; later those glass cells are treated as carbon target for laser irradiation/ablation. The surface of CQDs may be changed by choosing appropriate organic solvents to achieve tunable light emission. After laser irradiation, for the development of fluorescent carbon dots, the reaction mixture is subjected to centrifugation, purification, and surface passivation. Li et al. declared an easy laser ablation method to produce CQDs employing a common solvent as the liquid medium (such as acetone, ethanol, or water) and nano-carbon materials (less than 50 nm) as the carbon precursor (Thongpool et al. 2012; Qu et al. 2013). By adding nitryl with arc discharging fuming nitric acid, Xu et al. enhanced the hydrophilicity of the synthesized dots. The electrophoretic separation eventually resulted in three groups of fluorescent nanoparticles emitting blue-green, yellow, and orange at distinct molecular weights. Hu et al. in 2011 merged the multistep synthesis process into one-step reaction for simplifying the reaction as well as obtaining CQDs with better fluorescence. They combined laser ablation and passivation using polyethylene solution under laser radiation for 2 h and finally obtained fluorescent CQDs of size 3.2 nm with a quantum yield of 12.2%. Furthermore they developed CQDs of varied fluorescence by choosing different organic solutions (Hu et al. 2011).

Electrochemical synthesis is one of the CQD synthesis approaches by applying direct current from comparatively big carbon products such as graphene, graphite, carbon fiber, etc. The significant factors for generating CQDs through the electrochemical oxidation method are hydroxyl groups and an alkaline atmosphere. In electrochemical synthesis, three electrode systems are imparted; two Pt sheets were used as operating and counter-electrodes, and one calomel electrode installed on the lugging capillary was used as reference electrodes. In such a manner that the distance between the two Pt sheets is about 3 cm, the electrodes are fixed with a rubber plug. The precursor solution is to mix alcohol and water in a fundamental medium, and the length of the response is until the transparent solution becomes black. Later, equal ethanol volume is added to salting out sodium hydroxide (added for alkalinity). The mixture is kept overnight for evaporation of solvent, and a solid product is formed which after separation and purification yields carbon quantum dots. Zhao and Xie effectively synthesized CQDs using electrochemical method employing a 3.0 V electro-oxidized graphite column electrode (GE) with a Pt counter electrode in 0.1 M potassium dihydrogen phosphate aqueous solution as the electrolyte supporting solution (Zhao and Xie 2017). Subsequently, the resulting oxidant solution needs ultra-sonication, ultra-filtration through a 22.0 mm filter membrane, washing, and drying. Kang and his colleagues revealed another electrochemical operation for producing CQDs of 1–4 nm using alkali-assisted electrochemical techniques (Li et al. 2010a). Another simple strategy of producing high-quality CQDs is the use of small graphite pieces by accurate cutting into ultrafine particles of a graphite honeycomb layer. In 2016, Canaveri and his colleagues discovered that the CQD characteristics depended considerably on the moment spent in the method of electrolysis. The benefit of this synthesis is that by tuning the intensity of the applied present, size-dependent photoluminescent characteristics can be accomplished. Zhou et al. developed 2.8 nm large fluorescent carbon dots from multi-walled carbon nanotubes as the working electrode and later cultivated in carbon diaphragm using chemical vapor deposition (CVD) method and variable cycle voltage (Zhou et al. 2007). In another study, Hou et al. acquired water-soluble functionalized fluorescent CQDs with two Pt sheets (1.5×2 cm^{-2}) as positive and negative electrodes through electrochemical carbonization of sodium citrate and urea (Hou et al. 2015). Despite of all the advantages, major disadvantages in electrochemical synthesis are the tedious purification procedures of the synthesized carbon nanoparticle and lesser production yield.

Other strategies in carbon quantity dots synthesis involve (Fig. 3.1) hydrothermal method (Wang et al. 2013b; Xu et al. 2014; Dong et al. 2014; Gao et al. 2013; Shen et al. 2018), microwave-assisted pyrolysis method (Tang et al. 2014; Zhu et al. 2009; Wang et al. 2012; Rodríguez-Padrón et al. 2018; Yang et al. 2012), ultrasonic method (Oza et al. 2015), and acid dehydration method (LeCroy et al. 2014), and among those, the former two are the most widely used ones. Hydrothermal method of CQD preparation was first introduced by Liu and co-workers in 2007. It is the most common and easiest technique for the synthesis of carbon dots using both chemical and green precursors. Typically, in the presence of either air or organic solvent, the precursor is carbonized in an autoclave at elevated temperature. The blend is permitted to cool at room temperature after full carbonization, and products

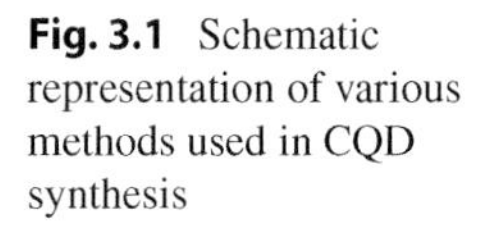

Fig. 3.1 Schematic representation of various methods used in CQD synthesis

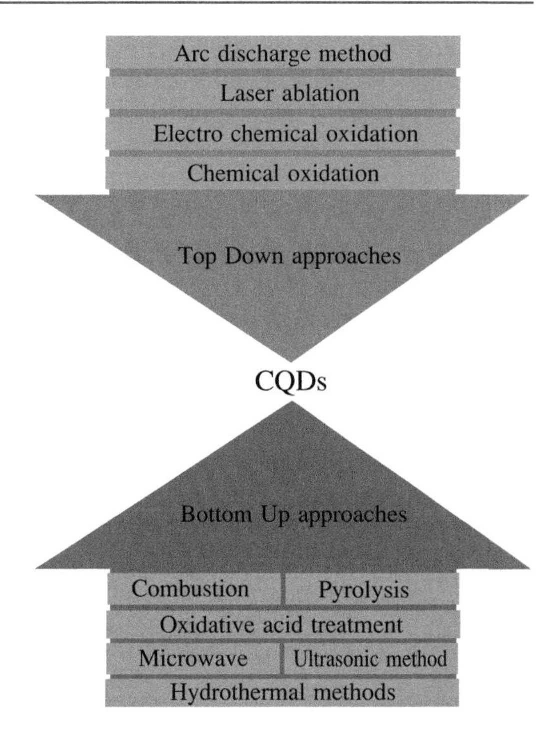

can be extracted with an organic solvent. They chose candle ash as a carbon source and blended it with nitric acid via heat reflex and produced a black homogeneous solution and purified it further through a sequence of centrifugation, dialysis, and gel electrophoresis to achieve CQDs of distinct particle sizes. They discovered it even under the same excitation wavelength (λex, 315 nm); particles of different sizes showed different emission wavelengths in its fluorescence which signifies the better optical character of florescent CQDs (called elementary excitation and multi-emission). Many CQDs were subsequently synthesized using hydrothermal methods and using various carbon sources that considerably improved the quantity output of CQDs. Depending upon the precursor, hydrophilicity and hydrophobicity of the synthesized florescent CQDs can be manipulated. Normally surface passivation is not required; however, there is limited control over the particle sizes. In 2013, Gao et al. used C^{60} as a carbon source and CTAB as a passivator to prepare elevated fluorescence CQDs with a quantum yield of up to 60%.

Microwave method of CQD synthesis was introduced in 2009 by Zhu and co-workers, to synthesize novel CQDs showing that fluorescence depends up on the reaction time. Sugar moieties are generally used as a source of carbon and polymeric oligomers as a response medium. In their experiment, Zhu et al. chose carbohydrates as a source of carbon and PEG-200 as a solvent and coating agent, and the response took 2–10 min under 500 W microwave power radiation (Zhu et al. 2009). The colorless solution transforms into pale yellow and then dark brown in a couple of minutes, depicting the formation of CQDs. Wang and co-workers synthesized CQDs from eggshells by initially burning eggshell into ashes and mixing it with sodium hydroxide solution and treating in microwave radiation for 5 min (Wang et al. 2012). Chandra et al. synthesized CQDs of shining bright green fluorescence in 3 min 40 s from sucrose using microwave radiation in phosphoric acid environment (Chandra et al. 2011). These low-cost economic syntheses of carbon dots are now widely used for various industrial and biological applications. The carbon dot synthesis of oxidative acid therapy utilizes waste soot as a carbon precursor. Waste soot is a mass of carbon particles from incomplete hydrocarbon combustion. The therapy includes reflux ionization in acidic medium of waste soot, followed by centrifugation, neutralization, and purification. Typically, burned waste soot was mixed with nitric acid and refluxed for 12 h. The carbon particles are collected by centrifugation after room temperature is reached. The collected mixture undergoes neutralization with sodium carbonate and extensive dialysis against water for obtaining pure carbon dots. The advantages of these acid treatments are efficient introduction of functional groups like carbonyl, carboxyl, amines, and epoxy to the synthesized CQDs which bestow them requisite properties for imaging application. However, the storage stability of thus synthesized carbon dots is minimal.

To remove amorphous carbon, catalyst, and other impurities introduced during the synthesis, it is essential to purify carbon nano-material for biological implementation. The typical purification processes used after synthesis are plasma oxidation, high-temperature annealing, and some chemical treatments. Surface passivation and functionalization are two critical steps for selective implementation in bio-imaging, drug delivery, etc. in post synthesis of carbon dots. Surface passivation decreases the damaging impact of surface contamination on their optical properties and gives them a high level of fluorescence. It is also a significant step in achieving PL property for CQDs with a size of 1.5–2 nm (Dimos 2016). It is typically performed by covering the carbon quantum dots surface with a thin layer of oligomers (PEG), thionyl chloride, thiols, and spiropyrans. Both the acid therapy and the passivation of the surface enhance the quantum yield of synthesized carbon dots as well as enhanced florescent emissions. Similarly, surface functionalization is very important for the synthetic carbon dots' florescent behavior. Wet chemical treatment involves acid reflux, and dry treatments such as RF-plasma treatment and treatment with strong acids such as H_2SO_4, HNO_3, $KMnO_4$, OsO_4, $K_2Cr_2O_7$, CCl_4, CF_4, $O_2(g)$, and SF_6 are employed. It introduces functional groups like carbonyl, ketone, hydroxyl, ester, alcohol, fluorine, thiol, amines, and carboxyl which can function as surface energy traps and lead to fluorescence emission behavior variations in CQDs. Numerous studies to functionalize and modify the surface of CQDs such as covalent bonding have been carried out to date (Dong et al. 2012; Yin et al. 2013) along with p–p interactions (Li et al. 2011), sol–gel (Mao et al. 2012; Wang et al. 2011), coordination (Zhao et al. 2011), which renders them solubility, biocompatibility and low toxicity. CQDs have big quantities of oxygen-containing groups that enable them to bind covalently to other functional groups. The latest strategy for surface alteration of CQDs is covalent bonding with chemical agents comprising amine groups to recover the photo luminescence of CQDs, which has been shown to be highly efficient in changing the characteristics of CQDs (Liao et al. 2013). Li et al. studied that the PL of prepared CQDs originated from carboxylate groups produced on the particle surface (Li et al. 2010b). They stated that on the surface of the preliminary carbon precursors generated by laser irradiation, several oxygen-containing radicals could potentially be the cause of PL. The primary features of these functional CQDs are high stability and outstanding photo-reversibility. Another attractive advantage for creating functional group is that it can load smaller therapeutic molecule by covalent conjugation or non-covalent adsorption onto the CQDs. Doping with heteroatoms, sulfur, nitrogen, and magnesium ion, other than surface passivation and functionalization, can increase the quantity yield up to 83%. The most preferential use of nitrogen as a dopant is that it can donate its electron to carbon dots so that the change in electronic configuration contributes in increased quantum yield. It can either be accomplished by selecting precursor-containing nitrogen or by post-functionalization.

3.3 Applications

Since the discovery of CQDs and their fluorescent properties and tenable surface-related properties, research has been done with the material for its implementation in different fields of applications (Fig. 3.2). All the current fields of applications of CQDs are discussed below.

3.3.1 Bio-imaging

Live cell bio-imaging is an imperative tool for elucidating the dynamics of biological mechanism. Maintaining the cellular ambience is very critical in live cell imaging, as the environmental factors can duly impose abiotic stresses. When light radiation

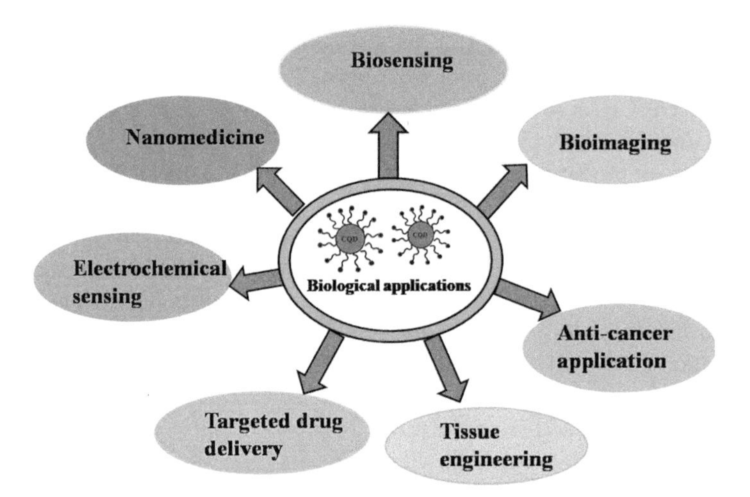

Fig. 3.2 Biological implementation of carbon quantum dots; CQDs display adjustable physical and chemical properties and could therefore be used in different biological areas such as bio-sensing, bio-imaging, etc.

interacts with the cell, temperature rises owing to the excitation of light active molecules that can boost free radical formation. It is therefore essential to require minimum energy or a source of low intensity light radiation to decrease oxidative stress. Traditionally, fluorescent probes were used in bio-imaging studies, to understand cellular processes (Luo et al. 2013). Rhodamine 6G is one of the best commercially available organic fluorophores with quantum yield up to 80%; however, the major drawbacks of such probes are photo-bleaching, less photostability, narrow excitation, and emission wavelength and cytotoxicity. One of the remarkable properties of doped CQDs that distinguish them from conventional organic dyes is their resistance from photo-bleaching and photodegradation. The unique fluorescent properties, excitation-dependent multicolor emission, high photostability, high aqueous stability, superior cell permeability, surface modification capability, and good biocompatibility have made carbon dots a versatile material for bio-imaging and sensing. Unlike semiconductor nanocrystal-like quantum dots, CQDs are nontoxic in nature. Semiconductor quantum dots have been advantageous over florescent samples but are less biocompatible because most QDs have toxic heavy metals such as Cd, Pb, and Hg in their structure. Even if they have ground coverage, the likelihood that toxic metals will be released into the medium is not negligible (Jaishankar et al. 2014). CQDs could address effectively all of the aforementioned complications. Internalization of carbon dots by the cell is a temperature-dependent process, and it takes place at ambient temperature of 37 °C. They translocate into the cell via endocytosis; hence, coupling with membrane-translocating peptides can enhance the process. CQDs are used as fluorescent markers to test multiple cellular organs such as the endoplasmic reticulum, Golgi body, nucleolus, mitochondria, lysosome/endosome, etc. (Fig. 3.3), and CQDs can be efficiently removed from the body via excretion, within an hour of their intravenous injection (Longmire et al. 2008).

Carbon dots are very specific in their visible lengths of excitation and emission waves, and they also have elevated brightness at individual dot concentrations. Yang et al. first discovered the applicability of using CQDs in bio-imaging through the deployment of CQDs as a FL contrast agent in 2009. Several studies revealed the use of different CQDs in cellular fluorescence imaging using different cell lines, including Caco-2 (Dias et al. 2019), Ehrlich ascites carcinoma (Ray et al. 2009), HEK293 (Luo et al. 2013), pig kidney (Ray et al. 2009), B16F11 (Luo et al. 2013), murine P19 (Liu et al. 2009), BGC823 (Wang et al. 2011), and HeLa (Xu et al. 2015). Yang et al. used pegylated CQDs with adequate contrast for in vivo optical imaging (440 μg in 200 μL) by subcutaneously injecting CQDs into mice (Yang et al. 2009). Tao et al. used the same protocol and then collected images from 455 nm to 704 nm at distinct wavelengths after subcutaneous injection of aqueous CQD solution (Tao et al. 2012). In 2012, Hahn et al. developed diamine-capped pegylated CQDs to label B16F1 and HEK293 (Tao et al. 2012). In addition, real-time bio-imaging was produced possible by Hang's and colleagues' job, where they reported that CQDs could be used for real-time bio-imaging by delivering hyaluronic acid (HA) derivatives specifically for the purpose (Goh et al. 2012). HA/CQD hybrids are synthesized through amidation reaction between amino groups on the surface of CQDs and carboxylic groups of HA. The in vivo imaging

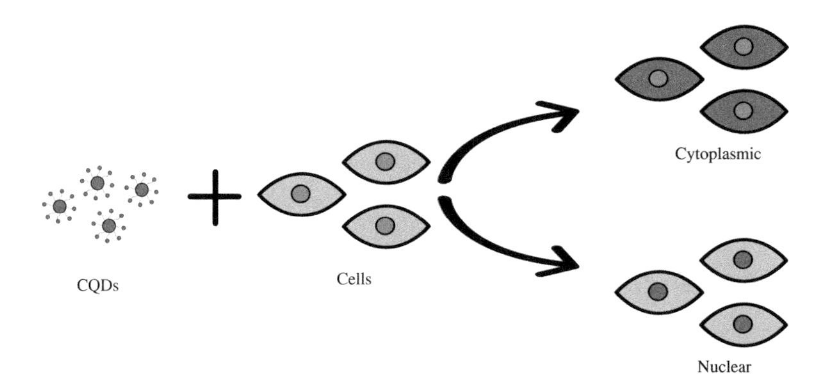

Fig. 3.3 CQDs' ability as a fluorophore in bio-imaging: The CQDs exhibit fluorescence under UV lights. Due to their smaller size, they can easily enter into the cells. Based on its functionality, it may locate into the nucleus or in the cytoplasm. Hence, the nucleus or cytoplasm can be easily visualized through the fluorescent microscope

results revealed that these hybrids could employ target-specific delivery in the liver (Goh et al. 2012).

A combo of optical imaging (OI) and magnetic resonance imaging (MRI) is the latest appealing technology in bio-imaging using CQDs, where optical imaging provides quick screening, while MRI provides physiological and anatomical data and elevated spatial resolution (Lee et al. 2012). Because of its small cytotoxicity compared to commercial Gadovist VR, Gd(III)-doped CQDs could be used in biomedical research for multimodal imaging. In 2012, Bourlinos AB et al. prepared ultrafine water-dispersible Gd(III)-doped CQDs with dual MRI/FL personality through heat decomposition (Bourlinos et al. 2012) and thus acquired particles showing bright FL in the visible region and showing powerful T1-weighted MRI (Bourlinos et al. 2012). In the presence of small Fe_3O_4 nanoparticles for MR/FL multi-imaging, a 6 nm iron oxide double CQD was recently synthesized by pyrolysis of organic molecules (Srivastava et al. 2012). The CQDs produced by hydrothermal chitosan carbonization with amino acids were used to evaluate the cyto-compatibility of human lung adenocarcinoma cells (A549) (Yang et al. 2012). Zhang and Huang et al. revealed a tiny molecular fluorescent carbogenic complex that could selectively stain the nucleolus rich in RNA (He et al. 2016). Recently it has been revealed that CQDs generated by the hydrothermal one-pot response of m-phenylenediamine and L-cysteine can efficiently target the nucleolus. It offers high-quality nucleolus imagery in living cells, enabling nucleolus-related biological behavior to be tracked in real time, as opposed to SYTO RNASelect, a widely used commercial nucleolus imagery in fixed cells (Hua et al. 2018). These multifunctional CQDs prevent lysosomal/endosomal trapping and target nucleolus selectively and bring protoporphyrin IX (PpIX, a commonly used photosensitizer) into the nucleus effectively (Hua et al. 2018).

By its excitation-dependent fluorescence emission, numerous studies revealed the cell imaging ability of doped CQDs. CQDs are shown as multicolor nano-samples due to excitation-dependent fluorescence emission, which can be excited with distinct excitation wavelengths. For example, on 543, 488 and 405 nm excitation, Zhai et al. (2012) reported that C-dots incubated with L929 cells could emit red, green, and blue fluorescence (Zhai et al. 2012). The in vivo fluorescence image of CQDs injected into the nude mouse with different agitation wavelengths shows a better contrast between the signal and background signal over 595 nm long, which is also efficient for in vivo imaging (Yang et al. 2009; Tao et al. 2012) (Timur 2018).

3.3.2 Bio-sensing

Biosensors are the type of analytical device which converts biological signals/ response to electrical signals (Shankar et al. 2012, 2013). There are various types of biosensors present at the moment, like DNA sensor, enzyme sensors, protein sensors, etc. Nano-biosensors are considered to be the most important of different biosensors because they are capable of detecting biochemical signals and/or biophysical signals related to a specific disease at the level of one or a few molecules (Chandra et al. 2017). Nano-biosensors were effectively used to detect several biomolecules in vitro as well as in vivo. This technology is anticipated to revolutionize point-of-care and personalized diagnostics and be highly useful for early identification of disease (Chandra and Segal 2016). CQDs are capable of serving as either great electron donors or electron acceptors, making them suitable for monitoring/ sensing different materials and parameters such as cell iron, copper, pH, proteins, enzymes, vitamins, and nucleic acid (da Silva and Gonçalves 2011). The mechanism behind the detection of ions is through the surface functional groups on CQDs, which show distinctive affinities to different target ions, resulting in an electron or energy transfer process quenching of PL intensity (Fig. 3.4). When the interaction is broken by external force, the PL is restored. Qu et al. established CQDs in 2013 using dopamine as the raw material that could be used as a Fe^{3+} detector with a stronger 0.32 μM detection threshold (Qu et al. 2013). For example, Xu et al. created a CQD-based ultrasensitive sensor in 2015 through a one-step hydrothermal therapy of potatoes for phosphate detection (Xu et al. 2015). Recently, Yang et al. built a fluorescent turn-on scheme in which a novel oligodeoxyribonucleotide (ODN)-CDs and graphene manufactured the optical sensor to identify Hg^{2+}. The method behind the sensing was FRET, graphene oxide could quench the fluorescence of ODN-CQDs via FRET, and luminescence was later retrieved with the addition of Hg^{2+} by removing ODN-Ds from graphene oxide owing to the creation of T-Hg^{2}-T duplex (Cui et al. 2015). Tian and colleagues produced aminomethylphenylterpyridine (AE-TPY) CQDs to determine changes in the pH value of physiological circumstances such as tumors. Thus, manufactured PL sensors with great selectivity and sensitivity could be used to monitor pH value gradients in a range of 6.0–8.5 (Kong et al. 2012). They showed the applicability of these sensors in mice's living cells and tumor tissues, which in the near future will require in vitro and in vivo applications. The CQDs have recently been used to identify the importance of intracellular pH in living pathogenic fungal cells (Cui et al. 2015).

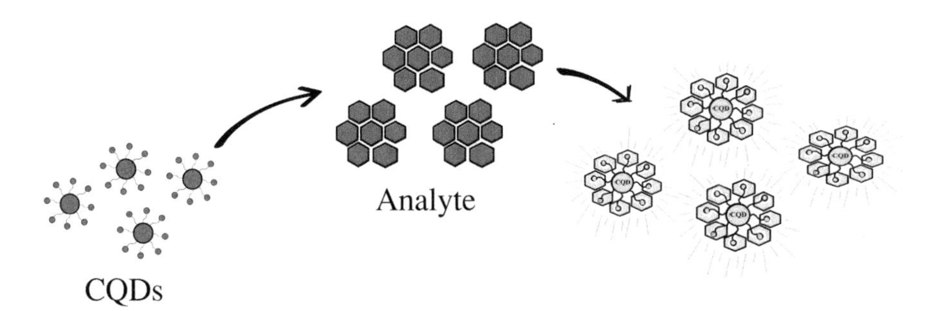

Fig. 3.4 Functional groups present in the functionalized CQDs have the ability to target certain biological molecules; binding to these molecules could result in the quenching of PL of CQDs which forms the basis of bio-sensing application

Bio-sensing of nucleic acid is based on adsorption by CQDs (as a fluorescent quencher) of the fluorescent single-stranded DNA (ssDNA) probe through p–p interactions. When ssDNA is hybridized with its target and double-stranded DNA (dsDNA) is formed, the desorption of the hybridized dsDNA from the surface of the CQDs causes its florescence to recover, resulting in the detection of the target DNA (Li et al. 2011). DNA-labeled CQDs have been created to detect proteins and enzymes; Maiti and colleagues have created CD-dsDNA sensors for histone detection (Maiti et al. 2013). The PL of this sensor would be quenched in the presence of dsDNA, while PL could be retrieved with the addition of histone owing to the powerful connection between histone and dsDNA through unwinding of dsDNA from CQDs. Quantitative protein detection can therefore be calculated from restored fluorescence. C-dot-based sensors widely detect biomolecules such as amino acids and vitamins. Xu et al. and Wang et al. in 2015 detected riboflavin and vitamin B12 using surface functionalized CQDs. Ratio-metric sensing protocol was adopted for detection, and CQD-based ratio metric sensors could detect riboflavin with high sensitivity even in 1.9 nM (Xu et al. 2015; Wang et al. 2015b). Novel nano-sensors composed of CQDs and gold particles were developed to detect cysteine, by Liu and Zang et al. in 2015 (Deng et al. 2015). In addition, Jana et al. in 2015 fabricated a turn-on CQD-based sensors that could detect cysteine with better sensitivity and selectivity (Jana et al. 2015).

3.3.3 Electrochemical Sensors

The biomolecules which are electrochemically active can be easily determined by means of electrochemical techniques such as cyclic voltammetry (Shankar and Swamy 2014), differential pulse voltammetry, etc. (Shankar et al. 2009). In the field of electrochemical sensing, the electrochemical characteristics of CQDs have recently been studied. In the presence of other molecules such as ascorbic acid and serotonin, Shereema and crew launched CQD-based carbon paste electrode for the electrochemical determination of dopamine (Shankar et al. 2015). The electrochemical and capacitive properties of CQDs were reported by another team. They observed that rGO's capacity was enhanced by adding a suitable quantity of CQDs to the material. They claimed that the CD/rGO electrode in 1 M H_2SO_4 had good reversibility, excellent rate capability, fast charging, and high specific capacity (Dang et al. 2016). He et al. claimed that a composite of CQDs and poly-ortho-aminopyridine was capable of improving the electrochemical sensing properties of glassy carbon electrode. Further, they used this fabricated electrode for simultaneous detection of guanidine and adenine in a mixture (He et al. 2018). Kakaei's team studied the potential CQDs toward the oxidation-reduction reaction of methanol (Kakaei et al. 2016). Identification of metanil yellow and curcumin from a mixture of real sample was made possible with the help of CQD-fabricated electrodes, thus opening an application in food industry, for the detection of commonly employed adulterant (Fig. 3.6) (Shereema et al. 2018). Rene Kizek et al. manufactured CQDs for the electrochemical determination of anticancer drug, i.e., etoposide (ETO), and modified glass carbon electrode (GCE). They also suggested three distinct methods to modify GCE with CQDs. The electrode reported was capable of detecting etoposide with the smallest LOD relative to current detectors (Nguyen et al. 2016). Shunxing Li et al. in 2014 synthesized CQDs/octahedral Cu_2O nanocomposites and coated it on the GCE surface. Surface and electrochemical properties of the developed electrode were carried out and reported that the CQDs/octahedral Cu_2O-based electrode exhibits low electron transfer resistance. They also launched the electrode to detect glucose and hydrogen peroxide electrochemically (Li et al. 2015). Houcem Maaoui and his colleagues built and reported similar types of CQDs/Cu_2O/GCE in 2016 and used the same to determine glucose in patient serum (Maaoui et al. 2016). The Liu Junkang group reported a detailed investigation into the modification strategy of glassy carbon electrode with a hybrid material consisting of graphene aerogel and octadecylamine-functionalized carbon dots and their application in the electrochemical determination of acetaminophen (Ruiyi et al. 2018). Sundramoorthy and his team recently revealed electrochemically exfoliated carbon quantum electrodes modified in 2018 (Devi et al. 2018). In their method, electrochemical exfoliation method synthesized CQD using graphite rods in alcoholic NaOH was cast on a GCE or screen-printed carbon electrode (SPCE) for electrode manufacture. Electrochemical studies with these electrodes disclosed that the reported electrode could act as a dopamine electrochemical sensor with a 0.099 μM detection threshold (Fig. 3.5).

3.3.4 Nano-medicine and Targeted Drug Delivery

CQDs have been proved to be nontoxic and been reported for in vivo studies; this made it valid for nano-medicine application. Bechet and colleagues indicated that CQDs can be used in photodynamic therapy, a clinical procedure used for surface tumor treatment. This technique includes the localization and accumulation of photosensitizers in the tumor tissue, followed by irradiation using an accurate wavelength, producing single oxygen species (Bechet et al. 2008). Andrius et al. in Du145 cells indicated implementation of CQDs in radiotherapy, where PEG-CQDs

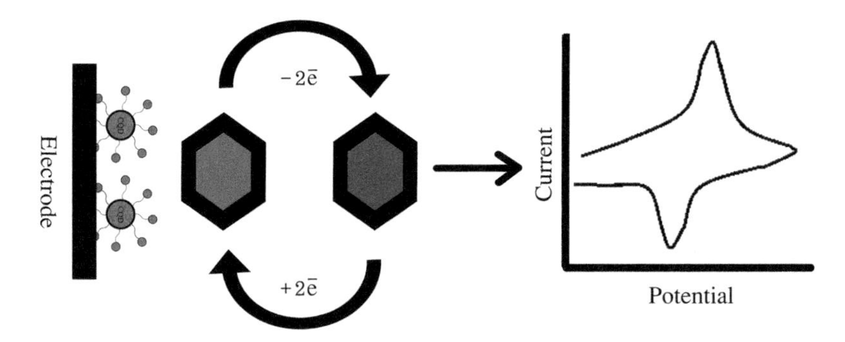

Fig. 3.5 Electrochemical sensing mechanism of electroactive molecules with CQDs: The CQDs catalyze the electrochemical redox process of biomolecules, and the resulting current is plotted against the potential

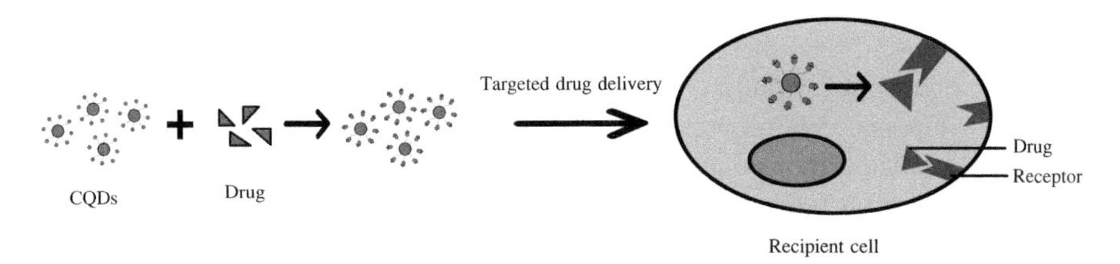

Fig. 3.6 Schematic representation of CQD-guided drug delivery system: Drugs conjugate with CQDs through its functional groups and directly deliver to the recipient cell. The released drug binds to the receptor site and acts specifically

covered with a silver shell were used as radio sensitizers (Kleinauskas et al. 2013). When exposed to low-energy X-rays, electrons were ejected from the coated silver shell, forming free radicals in the cancer cells adjacent to the CQDs-bPEI-coated CQDs (bPEI-CQDs), which have a large number of amino acids on their surface to condense DNA, making them suitable for gene carrier function (Hu et al. 2014). The nanotechnology-based drug delivery system has been gradually developed in recent years. The most explored drug delivery scheme is based primarily on AuNPs, but because of its toxicity, it limits its clinical treatment applications (Alkilany and Murphy 2010). Additionally, the requirement of thiol group for drug loading and the difficulty in tracking the AuNPs in in vivo system due to the quenching of fluorophores impose more limitation of drug choice (Kumar et al. 2013; Dulkeith et al. 2002). Therefore, CQDs serve as a good substitute for AuNPs, and different functionalizations could lead to better options for drug conjugation in conjunction with target agents and thus to a rise in drug delivery decisions (Fig. 3.6) (Kumar et al. 2013).

Zheng and team synthesized CQDs conjugated with platinum-based anticancer prodrug oxidized oxaliplatin through chemical coupling and were able to demonstrate the distribution of the conjugated CQDs by tracking the fluorescence signal (Zheng et al. 2014). The introduction of quinolone (photosensitive molecule) into the CQD-based drug delivery scheme revealed a controlled release mechanism (Zheng et al. 2014). In this system, fluorescent characteristics of CQDs helped to activate photo-regulated drug release as a monitoring mechanism for drug delivery and quinoline molecules. Similarly, the release of pH-induced drugs was also recorded (Zheng et al. 2014; Karthik et al. 2013). A study on controlled release mechanism of DOX conjugated CQDs in HeLa cells showed that it drastically increased the drug efficacy. The CQDs-DOX conjugate did not exhibit any visible negative effect on normal cells, unlike on cancer cells (Gogoi and Chowdhury 2014). In addition, PEG-functionalized CQDs enabled longer circulation time in physiological systems and localized treatment there (Lai et al. 2012).

3.3.5 Tissue Engineering

CQDs hold excellent platform for tissue engineering, particularly mesenchymal stem cell (MSC)-based therapy (Shao et al. 2017a). CQDs were also apparently researched for implementation in bone tissue engineering, where the synthetic viable nanocomposite exhibited elevated load-bearing capacity and bioactivity. It is collectively called carbon dot decorated hydroxyapatite nanohybrid, a manufactured novel biomaterial. *Colocasia esculenta*'s aqueous extract of corms was taken as a precursor to CQDs and CaO from the shell of the egg served as a precursor to hydroxyapatite. It showed great biocompatibility, proliferation of cells, and activity of alkaline phosphatase against an osteoblast cell line, MG-63. The final nanocomposite verified its effectiveness as a bone-regenerating material by further enhancing its mechanical characteristics (Gogoi et al. 2016). In 2017, Dan Shao and colleagues created rat bone marrow mesenchymal stem cells (rBMSCs) based on citric acid carbon dots and their derivatives. The CQDs support long-term monitoring and differentiation of rBMSCs into osteoblasts by encouraging osteogenic transcription and mineralization of matrixes without influencing cell viability (Shao et al. 2017b). Alginate-derived CQDs have excellent gene delivery applicability as well as a fluorescent sample for visualizing the process of gene uptake. For the one-step green synthesis method, sodium alginate, a polysaccharide separated from seaweed, was used as a carbon source (Zhou et al. 2016). These CQDs which are positively charged have low toxicity, the ability to condense plasmid DNA, and the ability of CQDs produced by microwave-assisted pyrolysis to deliver SOX9 plasmid DNA in a non-viral manner. The successful delivery of SOX9 gene to the mouse embryonic fibroblasts (MEFs) induces chondrogenesis after in vitro transfection (Cao et al. 2018). Recently, collagen-derived carbon quantum dots with outstanding photo stability were also reported. Synthesized CQDs are reported to be deficient in photo-bleaching and are successfully introduced for the cell imaging in the 3D printed scaffold. It has an excellent application in the field of tissue engineering to monitor the efficiency and the success of regenerative medicine (Dehghani et al. 2018).

3.3.6 Anticancer Study

CQDs with hydroxyl group synthesized from styrene soot have been reported for its anticancer property. In this study, it was found that the CQDs have deleterious effect on the cancer cells (A549) and it showed negligible influence in non-cancerous cell (HEK 293T) (Fig. 3.7). The angiogenic study using these dots showed that it possesses anti-angiogenic properties too, as evidenced by the downregulation of angiogenic markers and fewer vessel density, when analyzed by CAM assay (Shereema et al. 2015). Walnut oil-synthesized CQDs through green synthesis also demonstrated anticancer activity against cell lines MCF-7 and PC-3. By activating caspase-3, the CQDs caused apoptosis in these cells (Arkan et al. 2018). CQDs produced from *L. plantarum* LLC-605 are capable of inhibiting the development of *E. coli* biofilm through a one-step hydrothermal carbonization technique and act as highly compatible and less toxic anti-biofilm material (Lin et al. 2018). Chang et al. produced CQDs from fresh tender ginger juice having on the surface groups of hydroxyl and carboxylate. They explain the efficient function of prepared CQDs in selective growth inhibition of hepatocarcinoma cell lines without influencing the growth of normal mammalian epithelial cells (MCF 10A) and liver cells (FL83B) in the suggested research. The flow cytometric assessment disclosed that a significant uptake of CQDs and the production of reactive oxygen species have been observed and induce apoptosis in HepG2 cells and are pointless in A549, MDA-MB-231, and HeLa cells (Li et al. 2014). In vivo studies were also carried out to prove the efficacy of CQDs to reduce tumor size that proposed the new strategy for therapeutic applications in liver cancer. Arkan et al.'s research showed that CQDs produced from walnut oil using hydrothermal techniques have cytotoxic and apoptotic potential on cells of prostate and breast cancer. Studies were conducted with PC-3 and MCF-7 cell lines, and the impact of prepared CQDs on caspase-3 and caspase-9 activity was verified. The findings acquired indicate that the CQDs specifically

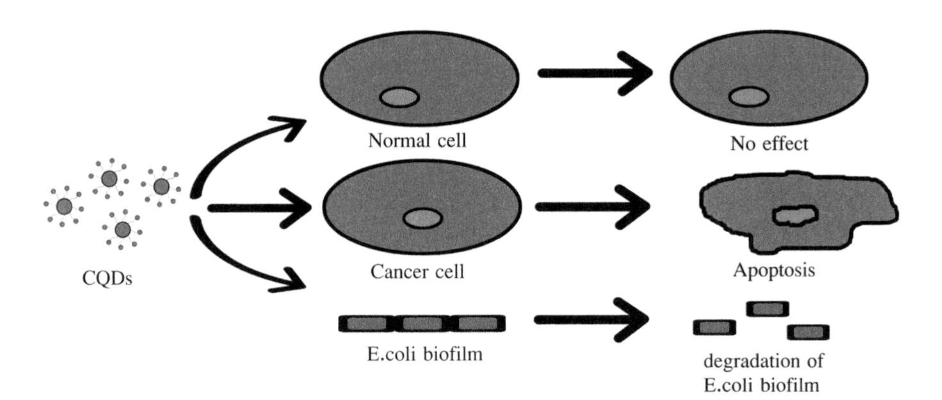

Fig. 3.7 The interaction of CQDS with anticancer property and ability to distinguish between a normal cell and cancer cell make it a perfect candidate for anticancer therapy, CQDs also possess potential antibiotic properties

boost the MCF-7 and PC3 cell lines caspase-3 activity. In addition, CQD levels also play a vital role in the operation of caspase-3. On the other hand, CQDs in the MCF-7 and PC-3 cell lines could not increase the activation of caspase-9 (Arkan et al. 2018). All these studies suggest that the starting material decides the functionality of the synthesized CQDs to a great extent and hence their capability to inhibiting the growth of different cancer cells. Other than the applications mentioned in this chapter, CQDs have been exploited for a plethora of application in different fields of science but mostly for its tunable fluorescence property. The application of CQDs holds promising prospects, and a plenty of applications are still to be explored in both biological and non-biological field.

3.4 Conclusion

- This carbon material possesses size less than 10 nm, which on surface passivation shows strong and tunable fluorescence emission properties and physical properties.
- High stability, excellent conductivity, and low toxicity among others are among their features.
- Different chemical, electrochemical, and physical methods can achieve CQD synthesis, which can be widely categorized into two kinds: top-down and bottom-up. Other methods, such as laser ablation, electrochemical synthesis methods, hydrothermal method, microwave-assisted pyrolysis method, ultrasonic method, acid dehydration method, etc., are frequently used.
- CQDs get great attraction due to their optical stability, nontoxicity, and solubility and can be easily functionalized with various functional group moieties or can be doped with heteroatoms on the surface, making them suitable candidate for biological applications like bio-imaging, nano-medicine and targeted drug delivery, bio-sensing, and electrochemical sensing.
- CQDs can be used as a significant material for in vitro and in vivo live cell imaging studies due to fluorescence and cell viability.
- It directly tags to the cellular organelles and reduces the drawbacks of organic fluorophore with high toxicity, photo-bleaching, etc.
- It is also known as multicolor nano-probe, due to excitation at different excitation wavelengths. Through the surface functional group they intract with different molecules in different way, this intraction helps to sense the presence of various materials, making it another promising application of CQDs known as bio-sensing.
- The ability to work as a drug delivery system makes CQDs an important factor in clinical and anticancer applications. Electrochemical sensor application of CQDs recently gained more attraction; tissue engineering and anticancer study is an important application.

- Due to the diversity in synthesis, surface functionalization and biological application make CQDs an important target for future applications.
- In a nutshell, CQDs are proving to be a good replacement for a plethora of nanomaterials present at this time period. CQDs have overcome the limitation with the currently used materials in all the mentioned applications. Thus, in a period of time, these dots may bring more refinement and more applications.

References

Alkilany AM, Murphy CJ (2010) Toxicity and cellular uptake of gold nanoparticles: what we have learned so far? J Nanopart Res 12(7):2313–2333

Arkan E, Barati A, Rahmanpanah M, Hosseinzadeh L, Moradi S, Hajialyani M (2018) Green synthesis of carbon dots derived from walnut oil and an investigation of their cytotoxic and Apoptogenic activities toward cancer cells. Adv Pharma Bull 8(1):149

Baker SN, Baker GA (2010) Luminescent carbon nanodots: emergent nanolights. Angew Chem Int Ed 49(38):6726–6744

Bechet D, Couleaud P, Frochot C, Viriot ML, Guillemin F, Barberi-Heyob M (2008) Nanoparticles as vehicles for delivery of photodynamic therapy agents. Trends Biotechnol 26(11):612–621

Bourlinos AB, Bakandritsos A, Kouloumpis A, Gournis D, Krysmann M, Giannelis EP, Polakova K, Safarova K, Hola K, Zboril R (2012) Gd (III)-doped carbon dots as a dual fluorescent-MRI probe. J Mater Chem 22(44):23327–23330

Cao L, Wang X, Meziani MJ, Lu F, Wang H, Luo PG, Lin Y, Harruff BA, Veca LM, Murray D, Xie SY (2007) Carbon dots for multiphoton bioimaging. J Am Chem Soc 129(37):11318–11319

Cao X, Wang J, Deng W, Chen J, Wang Y, Zhou J, Du P, Xu W, Wang Q, Wang Q, Yu Q (2018) Photoluminescent cationic carbon dots as efficient non-viral delivery of plasmid SOX9 and chondrogenesis of fibroblasts. Sci Rep 8(1):7057

Chandra P, Segal E (2016) Nanobiosensors for personalized and onsite biomedical diagnosis. The Institution of Engineering and Technology, London

Chandra S, Das P, Bag S, Laha D, Pramanik P (2011) Synthesis, functionalization and bioimaging applications of highly fluorescent carbon nanoparticles. Nanoscale 3(4):1533–1540

Chandra P, Tan YN, Singh SP (eds) (2017) Next generation point-of-care biomedical sensors technologies for cancer diagnosis. Springer, Singapore

Cui X, Zhu L, Wu J, Hou Y, Wang P, Wang Z, Yang M (2015) A fluorescent biosensor based on carbon dots-labeled oligodeoxyribonucleotide and graphene oxide for mercury (II) detection. Biosens Bioelectron 63:506–512

da Silva JCE, Gonçalves HM (2011) Analytical and bioanalytical applications of carbon dots. TrAC Trends Anal Chem 30(8):1327–1336

Dang YQ, Ren SZ, Liu G, Cai J, Zhang Y, Qiu J (2016) Electrochemical and capacitive properties of carbon dots/reduced graphene oxide supercapacitors. Nanomaterials 6(11):212

Dehghani A, Ardekani SM, Hassan M, Gomes VG (2018) Collagen derived carbon quantum dots for cell imaging in 3D scaffolds via two-photon spectroscopy. Carbon 131:238–245

Deng J, Lu Q, Hou Y, Liu M, Li H, Zhang Y, Yao S (2015) Nanosensor composed of nitrogen-doped carbon dots and gold nanoparticles for highly selective detection of cysteine with multiple signals. Anal Chem 87(4):2195–2203

Devi NR, Kumar TV, Sundramoorthy AK (2018) Electrochemically exfoliated carbon quantum dots modified electrodes for detection of dopamine neurotransmitter. J Electrochem Soc 165(12):G3112–G3119

Dias C, Vasimalai N, Sárria MP, Pinheiro I, Vilas-Boas V, Peixoto J, Espiña B (2019) Biocompatibility and bioimaging potential of fruit-based carbon dots. Nanomaterials 9(2):199
Dimos K (2016) Carbon quantum dots: surface passivation and functionalization. Curr Org Chem 20(6):682–695
Dong Y, Wang R, Li H, Shao J, Chi Y, Lin X, Chen G (2012) Polyamine-functionalized carbon quantum dots for chemical sensing. Carbon 50(8):2810–2815
Dong H, Kuzmanoski A, Gößl DM, Popescu R, Gerthsen D, Feldmann C (2014) Polyol-mediated C-dot formation showing efficient Tb 3+/Eu 3+ emission. Chem Commun 50(56):7503–7506
Dulkeith E, Morteani AC, Niedereichholz T, Klar TA, Feldmann J, Levi SA, Van Veggel FCJM, Reinhoudt DN, Möller M, Gittins DI (2002) Fluorescence quenching of dye molecules near gold nanoparticles: radiative and nonradiative effects. Phys Rev Lett 89(20):203002
Gao MX, Liu CF, Wu ZL, Zeng QL, Yang XX, Wu WB, Li YF, Huang CZ (2013) A surfactant-assisted redox hydrothermal route to prepare highly photoluminescent carbon quantum dots with aggregation-induced emission enhancement properties. Chem Commun 49(73):8015–8017
Gogoi N, Chowdhury D (2014) Novel carbon dot coated alginate beads with superior stability, swelling and pH responsive drug delivery. J Mater Chem B 2(26):4089–4099
Gogoi S, Kumar M, Mandal BB, Karak N (2016) A renewable resource based carbon dot decorated hydroxyapatite nanohybrid and its fabrication with waterborne hyperbranched polyurethane for bone tissue engineering. RSC Adv 6(31):26066–26076
Goh EJ, Kim KS, Kim YR, Jung HS, Beack S, Kong WH, Scarcelli G, Yun SH, Hahn SK (2012) Bioimaging of hyaluronic acid derivatives using nanosized carbon dots. Biomacromolecules 13(8):2554–2561
He H, Wang Z, Cheng T, Liu X, Wang X, Wang J, Ren H, Sun Y, Song Y, Yang J, Xia Y (2016) Visible and near-infrared dual-emission carbogenic small molecular complex with high RNA selectivity and renal clearance for nucleolus and tumor imaging. ACS Appl Mater Interfaces 8(42):28529–28537
He S, He P, Zhang X, Zhang X, Dong F, Jia L, Du L, Lei H (2018) Simultaneous voltammetric determination of guanine and adenine by using a glassy carbon electrode modified with a composite consisting of carbon quantum dots and overoxidized poly (2-aminopyridine). Microchim Acta 185(2):107
Hou Y, Lu Q, Deng J, Li H, Zhang Y (2015) One-pot electrochemical synthesis of functionalized fluorescent carbon dots and their selective sensing for mercury ion. Anal Chim Acta 866:69–74
Hu S, Liu J, Yang J, Wang Y, Cao S (2011) Laser synthesis and size tailor of carbon quantum dots. J Nanopart Res 13(12):7247–7252
Hu L, Sun Y, Li S, Wang X, Hu K, Wang L, Liang XJ, Wu Y (2014) Multifunctional carbon dots with high quantum yield for imaging and gene delivery. Carbon 67:508–513
Hua XW, Bao YW, Zeng J, Wu FG (2018) Ultrasmall all-in-one Nanodots formed via carbon dot-mediated and albumin-based synthesis: multimodal imaging-guided and mild laser-enhanced Cancer therapy. ACS Appl Mater Interfaces 10(49):42077–42087
Jaishankar M, Tseten T, Anbalagan N, Mathew BB, Beeregowda KN (2014) Toxicity, mechanism and health effects of some heavy metals. Interdiscip Toxicol 7(2):60–72
Jana J, Ganguly M, Pal T (2015) Intriguing cysteine induced improvement of the emissive property of carbon dots with sensing applications. Phys Chem Chem Phys 17(4):2394–2403
Jia X, Li J, Wang E (2012) One-pot green synthesis of optically pH-sensitive carbon dots with upconversion luminescence. Nanoscale 4(18):5572–5575
Kakaei K, Javan H, Mohammadi HB (2016) Synthesis of carbon quantum dots nanoparticles by cyclic voltammetry and its application as methanol tolerant oxygen reduction reaction electrocatalyst. J Chin Chem Soc 63(5):432–437
Karthik S, Saha B, Ghosh SK, Singh NP (2013) Photoresponsive quinoline tethered fluorescent carbon dots for regulated anticancer drug delivery. Chem Commun 49(89):10471–10473
Kleinauskas A, Rocha S, Sahu S, Sun YP, Juzenas P (2013) Carbon-core silver-shell nanodots as sensitizers for phototherapy and radiotherapy. Nanotechnology 24(32):325103

Kong B, Zhu A, Ding C, Zhao X, Li B, Tian Y (2012) Carbon dot-based inorganic–organic nanosystem for two-photon imaging and biosensing of pH variation in living cells and tissues. Adv Mater 24(43):5844–5848

Kumar V, Toffoli G, Rizzolio F (2013) Fluorescent carbon nanoparticles in medicine for cancer therapy. ACS Med Chem Lett 4(11):1012–1013

Lai CW, Hsiao YH, Peng YK, Chou PT (2012) Facile synthesis of highly emissive carbon dots from pyrolysis of glycerol; gram scale production of carbon dots/mSiO_2 for cell imaging and drug release. J Mater Chem 22(29):14403–14409

LeCroy GE, Sonkar SK, Yang F, Veca LM, Wang P, Tackett KN, Yu JJ, Vasile E, Qian H, Liu Y, Luo P (2014) Toward structurally defined carbon dots as ultracompact fluorescent probes. ACS Nano 8(5):4522–4529

Lee DE, Koo H, Sun IC, Ryu JH, Kim K, Kwon IC (2012) Multifunctional nanoparticles for multimodal imaging and theragnosis. Chem Soc Rev 41(7):2656–2672

Li H, He X, Kang Z, Huang H, Liu Y, Liu J, Lian S, Tsang CHA, Yang X, Lee ST (2010a) Water-soluble fluorescent carbon quantum dots and photocatalyst design. Angew Chem Int Ed 49(26):4430–4434

Li X, Wang H, Shimizu Y, Pyatenko A, Kawaguchi K, Koshizaki N (2010b) Preparation of carbon quantum dots with tunable photoluminescence by rapid *laser passivation in ordinary org*anic solvents. Chem Commun 47(3):932–934

Li H, Zhang Y, Wang L, Tian J, Sun X (2011) Nucleic acid detection using carbon nanoparticles as a fluorescent sensing platform. Chem Commun 47(3):961–963

Li H, Kang Z, Liu Y, Lee ST (2012) Carbon nanodots: synthesis, properties and applications. J Mater Chem 22(46):24230–24253

Li CL, Ou CM, Huang CC, Wu WC, Chen YP, Lin TE, Ho LC, Wang CW, Shih CC, Zhou HC, Lee YC (2014) Carbon dots prepared from ginger exhibiting efficient inhibition of human hepatocellular carcinoma cells. J Mater Chem B 2(28):4564–4571

Li Y, Zhong Y, Zhang Y, Weng W, Li S (2015) Carbon quantum dots/octahedral Cu2O nanocomposites for non-enzymatic glucose and hydrogen peroxide amperometric sensor. Sensors Actuators B Chem 206:735–743

Liao B, Long P, He B, Yi S, Ou B, Shen S, Chen J (2013) Reversible fluorescence modulation of spiropyran-functionalized carbon nanoparticles. J Mater Chem C 1(23):3716–3721

Lin F, Li C, Chen Z (2018) Bacteria-derived carbon dots inhibit biofilm formation of Escherichia coli without affecting cell growth. Front Microbiol 9:259

Liu R, Wu D, Liu S, Koynov K, Knoll W, Li Q (2009) An aqueous route to multicolor photoluminescent carbon dots using silica spheres as carriers. Angew Chem Int Ed 48(25):4598–4601

Longmire M, Choyke PL, Kobayashi H (2008) Clearance properties of nano-sized particles and molecules as imaging agents: considerations and caveats. Nanomedicine (London) 3(5):703–717

Luo PG, Sahu S, Yang ST, Sonkar SK, Wang J, Wang H, LeCroy GE, Cao L, Sun YP (2013) Carbon "quantum" dots for optical bioimaging. J Mater Chem B 1(16):2116–2127

Maaoui H, Teodoresu F, Wang Q, Pan GH, Addad A, Chtourou R, Szunerits S, Boukherroub R (2016) Non-enzymatic glucose sensing using carbon quantum dots decorated with copper oxide nanoparticles. Sensors 16(10):1720

Maiti S, Das K, Das PK (2013) Label-free fluorimetric detection of histone using quaternized carbon dot–DNA nanobiohybrid. Chem Commun 49(78):8851–8853

Mao Y, Bao Y, Han D, Li F, Niu L (2012) Efficient one-pot synthesis of molecularly imprinted silica nanospheres embedded carbon dots for fluorescent dopamine optosensing. Biosens Bioelectron 38(1):55–60

Nguyen HV, Richtera L, Moulick A, Xhaxhiu K, Kudr J, Cernei N, Polanska H, Heger Z, Masarik M, Kopel P, Stiborova M (2016) Electrochemical sensing of etoposide using carbon quantum dot modified glassy carbon electrode. Analyst 141(9):2665–2675

Oza G, Oza K, Pandey S, Shinde S, Mewada A, Thakur M, Sharon M, Sharon M (2015) A green route towards highly photoluminescent and cytocompatible carbon dot synthesis and its separation using sucrose density gradient centrifugation. J Fluoresc 25(1):9–14

Qu K, Wang J, Ren J, Qu X (2013) Carbon dots prepared by hydrothermal treatment of dopamine as an effective fluorescent sensing platform for the label-free detection of iron (III) ions and dopamine. Chem Eur J 19(22):7243–7249

Ray SC, Saha A, Jana NR, Sarkar R (2009) Fluorescent carbon nanoparticles: synthesis, characterization, and bioimaging application. J Phys Chem C 113(43):18546–18551

Rodríguez-Padrón D, Algarra M, Tarelho LA, Frade J, Franco A, de Miguel G, Jiménez J, Rodríguez-Castellón E, Luque R (2018) Catalyzed microwave-assisted preparation of carbon quantum dots from lignocellulosic residues. ACS Sustain Chem Eng 6(6):7200–7205

Ruiyi L, Haiyan Z, Zaijun L, Junkang L (2018) Electrochemical determination of acetaminophen using a glassy carbon electrode modified with a hybrid material consisting of graphene aerogel and octadecylamine-functionalized carbon quantum dots. Microchim Acta 185(2):145

Shankar SS, Swamy BK (2014) Detection of epinephrine in presence of serotonin and ascorbic acid by TTAB modified carbon paste electrode: a voltammetric study. Int J Electrochem Sci 9(3):1321–1339

Shankar SS, Swamy BK, Ch U, Manjunatha JG, Sherigara BS (2009) Simultaneous determination of dopamine, uric acid and ascorbic acid with CTAB modified carbon paste electrode. Int J Electrochem Sci 4:592–601

Shankar SS, Swamy BK, Chandrashekar BN (2012) Electrochemical selective determination of dopamine at TX-100 modified carbon paste electrode: a voltammetric study. J Mol Liq 168:80–86

Shankar SS, Swamy BK, Chandrashekar BN, Gururaj KJ (2013) Sodium do-decyl benzene sulfate modified carbon paste electrode as an electrochemical sensor for the simultaneous analysis of dopamine, ascorbic acid and uric acid: a voltammetric study. J Mol Liq 177:32–39

Shankar SS, Shereema RM, Prabhu GRD, Rao TP, Swamy KB (2015) Electrochemical detection of dopamine in presence of serotonin and ascorbic acid at tetraoctyl ammonium bromide modified carbon paste electrode: a voltammetric study. J Biosens Bioelectron 6(2):1

Shao D, Lu M, Xu D, Zheng X, Pan Y, Song Y, Xu J, Li M, Zhang M, Li J, Chi G (2017a) Carbon dots for tracking and promoting the osteogenic differentiation of mesenchymal stem cells. Biomater Sci 5(9):1820–1827

Shao D, Lu MM, Zhao YW, Zhang F, Tan YF, Zheng X, Pan Y, Xiao XA, Wang Z, Dong WF, Li J (2017b) The shape effect of magnetic mesoporous silica nanoparticles on endocytosis, biocompatibility and biodistribution. Actabiomaterialia 49:531–540

Shen J, Zhu Y, Yang X, Li C (2012) Graphene quantum dots: emergent nanolights for bioimaging, sensors, catalysis and photovoltaic devices. Chem Commun 48(31):3686–3699

Shen T, Wang Q, Guo Z, Kuang J, Cao W (2018) Hydrothermal synthesis of carbon quantum dots using different precursors and their combination with TiO2 for enhanced photocatalytic activity. Ceram Int 44(10):11828–11834

Shereema RM, Sruthi TV, Kumar VS, Rao TP, Shankar SS (2015) Angiogenic profiling of synthesized carbon quantum dots. Biochemistry 54(41):6352–6356

Shereema RM, Rao TP, Kumar VS, Sruthi TV, Vishnu R, Prabhu GRD, Shankar SS (2018) Individual and simultaneous electrochemical determination of metanil yellow and curcumin on carbon quantum dots based glassy carbon electrode. Mater Sci Eng C 93:21–27

Srivastava S, Awasthi R, Tripathi D, Rai MK, Agarwal V, Agrawal V, Gajbhiye NS, Gupta RK (2012) Magnetic-nanoparticle-doped carbogenic nanocomposite: an effective magnetic resonance/fluorescence multimodal imaging probe. Small 8(7):1099–1109

Sun YP, Zhou B, Lin Y, Wang W, Fernando KS, Pathak P, Meziani MJ, Harruff BA, Wang X, Wang H, Luo PG (2006) Quantum-sized carbon dots for bright and colorful photoluminescence. J Am Chem Soc 128(24):7756–7757

Tang Y, Su Y, Yang N, Zhang L, Lv Y (2014) Carbon nitride quantum dots: a novel chemiluminescence system for selective detection of free chlorine in water. Anal Chem 86(9):4528–4535

Tao H, Yang K, Ma Z, Wan J, Zhang Y, Kang Z, Liu Z (2012) In vivo NIR fluorescence imaging, biodistribution, and toxicology of photoluminescent carbon dots produced from carbon nanotubes and graphite. Small 8(2):281–290
Thongpool V, Asanithi P, Limsuwan P (2012) Synthesis of carbon particles using laser ablation in ethanol. Procedia Eng 32:1054–1060
Timur SA (2018) Doped carbon dots for sensing and bioimaging applications: a minireview. Nano 8:342–352
Wang F, Xie Z, Zhang H, Liu CY, Zhang YG (2011) Highly luminescent organosilane-functionalized carbon dots. Adv Funct Mater 21(6):1027–1031
Wang Q, Liu X, Zhang L, Lv Y (2012) Microwave-assisted synthesis of carbon nanodots through an eggshell membrane and their fluorescent application. Analyst 137(22):5392–5397
Wang K, Gao Z, Gao G, Wo Y, Wang Y, Shen G, Cui D (2013a) Systematic safety evaluation on photoluminescent carbon dots. Nanoscale Res Lett 8(1):122
Wang CI, Periasamy AP, Chang HT (2013b) Photoluminescent C-dots@ RGO probe for sensitive and selective detection of acetylcholine. Anal Chem 85(6):3263–3270
Wang L, Zhu SJ, Wang HY, Qu SN, Zhang YL, Zhang JH, Chen QD, Xu HL, Han W, Yang B, Sun HB (2014) Common origin of green luminescence in carbon nanodots and graphene quantum dots. ACS Nano 8(3):2541–2547
Wang J, Wei J, Su S, Qiu J (2015a) Novel fluorescence resonance energy transfer optical sensors for vitamin B 12 detection using thermally reduced carbon dots. New J Chem 39(1):501–507
Wang J, Su S, Wei J, Bahgi R, Hope-Weeks L, Qiu J, Wang S (2015b) Ratio-metric sensor to detect riboflavin via fluorescence resonance energy transfer with ultrahigh sensitivity. Phys E 72:17–24
Xu X, Ray R, Gu Y, Ploehn HJ, Gearheart L, Raker K, Scrivens WA (2004) Electrophoretic analysis and purification of fluorescent single-walled carbon nanotube fragments. J Am Chem Soc 126(40):12736–12737
Xu Y, Jia XH, Yin XB, He XW, Zhang YK (2014) Carbon quantum dot stabilized gadolinium nanoprobe prepared via a one-pot hydrothermal approach for magnetic resonance and fluorescence dual-modality bioimaging. Anal Chem 86(24):12122–12129
Xu J, Zhou Y, Cheng G, Dong M, Liu S, Huang C (2015) Carbon dots as a luminescence sensor for ultrasensitive detection of phosphate and their bioimaging properties. Luminescence 30(4):411–415
Yang ST, Wang X, Wang H, Lu F, Luo PG, Cao L, Meziani MJ, Liu JH, Liu Y, Chen M, Huang Y (2009) Carbon dots as nontoxic and high-performance fluorescence imaging agents. J Phys Chem C 113(42):18110–18114
Yang Y, Cui J, Zheng M, Hu C, Tan S, Xiao Y, Yang Q, Liu Y (2012) One-step synthesis of amino-functionalized fluorescent carbon nanoparticles by hydrothermal carbonization of chitosan. Chem Commun 48(3):380–382
Yin JY, Liu HJ, Jiang S, Chen Y, Yao Y (2013) Hyperbranched polymer functionalized carbon dots with multistimuli-responsive property. ACS Macro Lett 2(11):1033–1037
Zhai X, Zhang P, Liu C, Bai T, Li W, Dai L, Liu W (2012) Highly luminescent carbon nanodots by microwave-assisted pyrolysis. Chem Commun 48(64):7955–7957
Zhao Z, Xie Y (2017) Enhanced electrochemical performance of carbon quantum dots-polyaniline hybrid. J Power Sources 337:54–64
Zhao HX, Liu LQ, De Liu Z, Wang Y, Zhao XJ, Huang CZ (2011) Highly selective detection of phosphate in very complicated matrixes with an off–on fluorescent probe of europium-adjusted carbon dots. Chem Commun 47(9):2604–2606
Zheng H, Wang Q, Long Y, Zhang H, Huang X, Zhu R (2011) Enhancing the luminescence of carbon dots with a reduction pathway. Chem Commun 47(38):10650–10652
Zheng M, Liu S, Li J, Qu D, Zhao H, Guan X, Hu X, Xie Z, Jing X, Sun Z (2014) Integrating oxaliplatin with highly luminescent carbon dots: an unprecedented theranostic agent for personalized medicine. Adv Mater 26(21):3554–3560

Zhou J, Booker C, Li R, Zhou X, Sham TK, Sun X, Ding Z (2007) An electrochemical avenue to blue luminescent nanocrystals from multiwalled carbon nanotubes (MWCNTs). J Am Chem Soc 129(4):744–745

Zhou J, Deng W, Wang Y, Cao X, Chen J, Wang Q, Xu W, Du P, Yu Q, Chen J, Spector M (2016) Cationic carbon quantum dots derived from alginate for gene delivery: one-step synthesis and cellular uptake. Actabiomaterialia 42:209–219

Zhu H, Wang X, Li Y, Wang Z, Yang F, Yang X (2009) Microwave synthesis of fluorescent carbon nanoparticles with electrochemiluminescence properties. Chem Commun 34:5118–5120

Carbon Nanomaterials for Electrochemiluminescence-Based Immunosensors: Recent Advances and Applications

4

Nura Fazira Noor Azam, Syazana Abdullah Lim, and Minhaz Uddin Ahmed

Abstract
Electrochemiluminescence (ECL) technique is defined as luminescence emitted as a result from the occurrence of chemical reaction on the electrode's surface between luminophore (the light-emitting chemical species) and other species present in the same system when a small potential is applied to the electrode. ECL technique is oftentimes adopted for devising biosensors for the detection of various kinds of proteins by manipulating the interactions between antibody (Ab) and antigen (Ag) – these biosensors are also known as immunosensors. This technique is advantageous for immunosensors as it offers numerous benefits including straightforward operation and low background signal. Furthermore, the performance of these immunosensors can be elevated by integrating carbon nanomaterials (CNMs) into the biosensors and exploiting their excellent electrocatalytic properties for improving the sensitivity and specificity of the biosensors. This chapter comprises of an overview of ECL-based immunosensors integrated with CNMs, accentuating their recent developments and applications.

Keywords
Carbon nanomaterials · Electrochemiluminescence · Immunosensor · Biosensor

---

N. F. N. Azam · M. U. Ahmed (✉)
Biosensors and Biotechnology Laboratory, Chemical Science Programme, Faculty of Science, Universiti Brunei Darussalam, Gadong, Brunei Darussalam
e-mail: minhaz.ahmed@ubd.edu.bn

S. A. Lim
School of Applied Sciences and Mathematics, Universiti Teknologi Brunei, Gadong, Brunei Darussalam

P. Chandra, R. Prakash (eds.), *Nanobiomaterial Engineering*,
https://doi.org/10.1007/978-981-32-9840-8_4

4.1 Introduction

Early diagnosis of diseases, particularly of life-threatening nature, is crucial because treatment given at a curable stage will provide a significant impact on the quality of patients' life. A strategy that has a high potential to be employed as a preventive measure is by designing and developing point-of-care (POC) devices since these devices allow fast detection and continuous monitoring of diseases through instant data acquisition. Other advantages for utilisation of the instruments are that these can be used directly without any prior training (Lee et al. 2018) and they are portable and low cost, thus extremely useful in outlying areas and developing countries (Ge et al. 2014; Ahmed et al. 2014). In addition, POC devices allow users to operate the devices using a small amount of samples with reliable and precise results (Ahmed et al. 2016). Thereafter, these merits of POC devices could improve the efficiency of healthcare services where time and effort should be invested for further development as to make these devices to be wearable and also embeddable in vivo, allowing real-time observation of the specified parameters such as change in biomarker concentrations (Siontorou et al. 2017).

Incorporating biosensor in POC diagnostic devices is an attractive prospect in order to meet the demand of providing modern and sophisticated analytical devices for routine detection of biomarker(s). Biomarkers are commonly used in biomedical field to provide essential information that reflects the health status of patients. Biomarkers or biological markers as described by WHO (2001) are "a substance, structure or process that can be measured in the body or its products and influence or predict the incidence of outcome or disease". Monitoring the concentration of these biomarkers in the patients' serum or saliva can assist in averting the aggravation of diseases as any decrease or upsurge in the biomarkers' level might signify the related disease (Bertoncello et al. 2014).

Immunosensors are widely recognised as the analytical compact type of biosensors that utilised antibodies as bioreceptors, which are immobilised on or within the transducing element (Ju et al. 2017). Meanwhile, electrochemiluminescence (ECL) technique is progressively gaining more attention attributable to its attractive merits which include offering the opportunities for the ECL signal enhancement by modifying the electrode's surface with nanomaterials (Fang et al. 2017). Herewith, this chapter highlights the recent developments and applications of ECL immunosensors that incorporated carbon nanomaterials (CNMs).

4.2 Biosensor

The development of biosensors is instigated by the introduction of glucose detector by Clark and Lyons (1962), and ever since, the interest in inventing biosensors has been flourishing. According to the International Union Pure and Applied Chemistry (IUPAC), a biosensor is a device which comprises of two major modules: a bioreceptor (biological recognition element) and a transducer as means of detecting the target analytes (Farzin and Shamsipur 2017) (Fig. 4.1). This device detects any

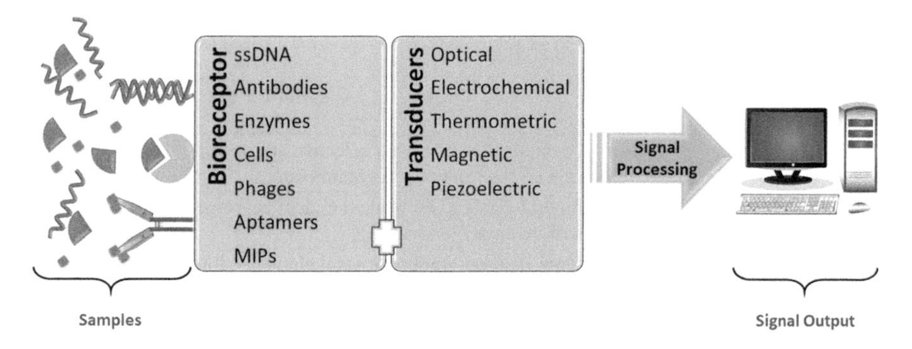

Fig. 4.1 Schematic representation of the basic components of biosensors

biological and/or chemical reactions produced by analytes, and subsequently signals obtained from these biochemical changes are computed and displayed as quantitative and semi-quantitative signals (Bhalla et al. 2016).

Biosensors are able to deliver results accurately and precisely (Ugo and Moretto 2017). Biosensors can be classified as (a) biocatalytic and (b) bioaffinity-based biosensors (Luppa et al. 2001). Biocatalytic-based biosensors mainly rely on enzymes as the biological intermediate that catalyses the generation of the signals from the reactions involved. On the other hand, bioaffinity biosensors involve the direct observation of the binding between the specified bioreceptor and the analyte as a means to generate the signal of the detection. Biosensors that are based on antibodies–antigen interactions are also known as immunosensors. Table 4.1 outlines important aspects of desirable biosensors for commercialisation.

In addition to these features, other preferable characteristics for a biosensor also include its portability and straightforward operation and also should be inexpensive for it to be competently commercialised (da Silva et al. 2017).

4.2.1 Basic Fundamental of ECL Immunosensors

Immunosensors are one of the well-known types of biosensors that employ antibodies (Abs) as the bioreceptors as a means to detect the target analytes. Abs and the corresponding antigens (Ag) interact with each other and form immuno-complexes (Liu and Saltman 2015). The formation of immuno-complexes induces changes in signal responses (e.g. in the form of potential or colour) or in the complexes' attributes (such as change in mass or density). These changes are detected by a specific transducer depending on the nature of the signal generated by the immuno-complexes. Currently, transducers that are widely used are based on optical and electrochemical changes (Moina and Ybarra 2012). Nevertheless, the focus in this chapter is electrochemiluminescence immunosensor that will be concisely defined in the next subsections.

Table 4.1 Characteristics which are preferred to be possessed by biosensor devices (Chakraborty and Hashmi 2017; Thakur and Ragavan 2013)

Aspects/ characteristics	Definition
Specificity	Biosensors are required to be highly selective and specific towards the target analytes with the least or no contaminants
Reproducibility	The device should be able to replicate the identical results of the same concentration of the target analyte
Stability	As biosensors are incorporating biological elements, it is crucial that these elements are able to withstand long-term storage without adversely affecting the device's overall performance
Sensitivity	Biosensors are expected to be able to detect trace amount of analytes without any prerequisite/pretreatment steps for the samples
Linearity	A linear range of concentrations of the target analyte should be detected by the biosensor, providing a quantitative detection
Response and recovery time	Prompt response is always desired especially for POC-based biosensors as to deliver real-time detection, and it is highly beneficial for the biosensor to be effectively reused with quick recovery time

4.2.1.1 Antibodies as the Bioreceptor

An immunosensor is an affinity-based biosensor since antibodies are known to provide high affinity, specificity and sensitivity towards their target proteins (Mathieu 2010). Other advantages for the employment of antibodies used in sensing include their flexibility for modification (e.g. for label-based biosensors) and their commercial obtainability (Rogers 2000). Immunoglobulin (Ig) or Ab is intricately composed of hundreds of separate amino acids, arrayed in the highest-ordered sequences (Vo-Dinh and Cullum 2000). They are formed by B lymphocytes, expressed as a protein and responsible as the antigen (target) receptor in the cell (Donahue and Albitar 2010).

The structure of antibody has a "Y"-like configuration, consisting of two analogous heavy polypeptide chains with the molecular weight of ~50 kDa for each chain and two other analogous light polypeptide chains (~25 kDa each). The respective pair of heavy and light chain is linked to the other pair via a disulphide bridge (Felix and Angnes 2017). Figure 4.2 signifies the structure of an Ab molecule, and Fab is the unit where the antigen binds to the antibody.

Abs can be further classified into monoclonal antibody (MAb) and polyclonal antibody (PAb). MAbs are commonly formed with the hybridoma technology in mice, and they are more responsive towards a single epitome (the binding site whereby Ab interacts with the corresponding antigen or Ag). MAbs possess better affinity and contribute more towards the specificity compared to the PAbs (Omidfar et al. 2013). Meanwhile, PAbs are habitually produced in goats, rabbits or sheep, and they are instinctively, heterogeneously reactive towards several epitomes. This feature jeopardises the overall specificity of the immunosensor and consequently provides less specificity for the detection of the target antigen (Byrne et al. 2009).

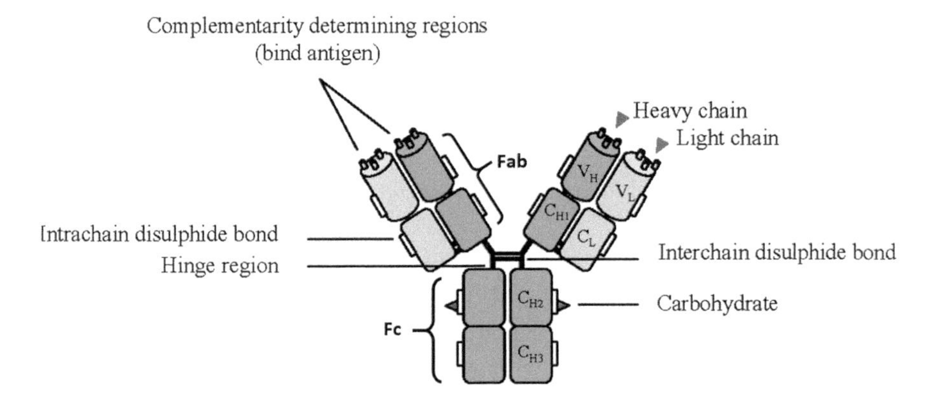

Fig. 4.2 Structure of immunoglobulin. Blue-coloured fragments represent the light chain, whereas the green-coloured fragments denote the heavy chain of the Ab. V_H variable heavy, V_L variable light, C_H constant heavy and C_L constant light. (Reproduced and modified from Byrne et al. 2009)

4.2.1.2 Electrochemiluminescence-Based Transducer

The studies of electrogenerated chemiluminescence, also commonly known as electrochemiluminescence (ECL), were firstly initiated in the 1960s (Hercules 1964; Santhanam and Bard 1965). This technique employs luminophores (molecular species) and fuses electrochemistry with chemiluminescence (CL). When a small amount of potential is applied onto an electrode surface, this will subsequently generate chemical species capable of emitting light without producing heat (Miao 2008).

The basic principle behind electrochemiluminescence-based method for detecting target analytes relies on the occurrence of homogeneous chemical reaction between a minimum of two chemical species in the same system. Both electron acceptor species and an electron donor species are imperatively required for ECL generation. These two species are the products of the chemical reaction that takes place at an electrode's surface and subsequently involved in an electron transfer activity, yielding a species with an excited state (Valenti et al. 2018). Luminol and tris(2,2-bipyridyl)-ruthenium(II) ($[Ru(bpy)_3]^{2+}$) are two of the most prevalent ECL luminophore in formulating POC-based biosensors (Azam et al. 2018; Roy et al. 2016).

There are two principal approaches in ECL technique: ion annihilation and co-reactant-based approach (Kerr et al. 2016). Ion annihilation involves electrochemical production of excited states from the two-step potentials at the electrode's surface, whereas co-reactant approach involves the luminophore and the corresponding co-reactant, which formed a radical species due to the oxidation or reduction of both species on the electrode's surface (Benoit and Choi 2017). With the purpose of elucidating the mechanism behind the two ECL routes, $[Ru(bpy)_3]^{2+}$ is used to exemplify the reactions (Hazelton et al. 2008).

In ion annihilation pathway, reactions are as follows:

$$\left[Ru\left(bpy\right)_3\right]^{2+} \rightarrow \left[Ru\left(bpy\right)_3\right]^{3+} \quad (4.1)$$

$$\left[Ru\left(bpy\right)_3\right]^{2+} + e^- \rightarrow \left[Ru\left(bpy\right)_3\right]^{+} \quad (4.2)$$

$$\left[Ru\left(bpy\right)_3\right]^{3+} + \left[Ru\left(bpy\right)_3\right]^{+} \rightarrow \left[Ru\left(bpy\right)_3\right]^{2+} + \left[Ru\left(bpy\right)_3\right]^{2+*} \quad (4.3)$$

$$\left[Ru\left(bpy\right)_3\right]^{2+*} \rightarrow \left[Ru\left(bpy\right)_3\right]^{2+} + h\nu \quad (4.4)$$

As potential is applied onto the electrode's surface, $[Ru(bpy)_3]^{2+}$ firstly undergoes consecutive reduction and oxidation reactions, generating both $[Ru(bpy)_3]^{3+}$ and $[Ru(bpy)_3]^{+}$ as depicted in Eqs. 4.1 and 4.2. These two resulting species then react with each other, forming an excited state of $[Ru(bpy)_3]^{2+}$ (Eq. 4.3), encouraging the subsequent annihilation process. The excited state will finally decay and return to its ground state and, hence, emit light (Eq. 4.4).

Contrastingly, co-reaction route involves the reactions as stated below:

$$\text{TPrA} \rightarrow \text{TPrA}^{\bullet+} + e^- \quad (4.5)$$

$$\text{TPrA}^{\bullet+} \rightarrow \text{TPrA}^{\bullet} + H^+ \quad (4.6)$$

$$\text{TPrA}^{\bullet} + \left[Ru\left(bpy\right)_3\right]^{2+} \rightarrow \left[Ru\left(bpy\right)_3\right]^{+} + \text{TPrA}^{+} \quad (4.7)$$

$$\left[Ru\left(bpy\right)_3\right]^{+} + \text{TPrA}^{\bullet+} \rightarrow \left[Ru\left(bpy\right)_3\right]^{2+*} + \text{products} \quad (4.8)$$

Oxalate-containing complexes and compounds with a tertiary amine group (e.g. TPrA) are known as the co-reactant for $[Ru(bpy)_3]^{2+}$. As demonstrated through Eqs. 4.5, 4.6, 4.7 and 4.8, the light emanation process was initiated by the formation of the TPrA radicals that react with $[Ru(bpy)_3]^{2+}$. Ultimately, light is emitted when the excited form of $[Ru(bpy)_3]^{2+}$ returns to its ground state (Fig. 4.3).

ECL is highly suitable to be integrated in a biosensor and will complement each other perfectly owing to numerous merits when combined. These merits include low background signal, rapid analysis can be easily achieved and requires less reagents as the reactive intermediates can be electro-regenerated, and better dynamic range can be determined (Sojic et al. 2017; Muzyka 2014). The electrode's surface is usually modified with biomolecules and/or nanomaterial in order to minimise the sample volume, reinforce better specificity, offer better sensitivity and ultimately better limit of detection (LOD) (Rizwan et al. 2018b; Wei et al. 2010). Employment of ECL is beneficial in developing biosensors as parameters can be easily calibrated according to the modification on the electrode's surface (Rizwan et al. 2018b).

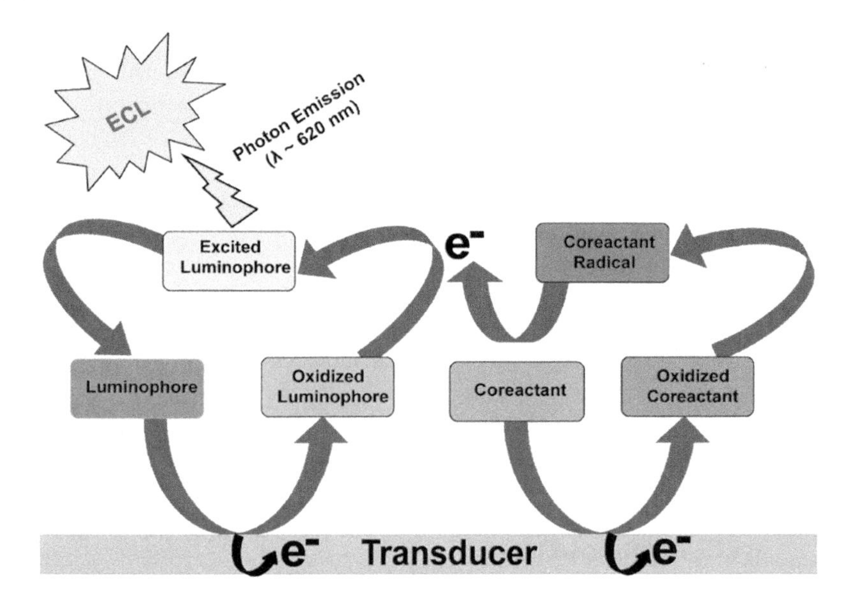

Fig. 4.3 Schematic illustration of the co-reactant-based ECL pathway's mechanism on the electrode's surface. (Adapted from Rizwan et al. 2018b)

4.3 Nanomaterials

Nanotechnology refers to the application of science and technology using materials in nanometre (10^{-9} m) scale in order to exploit their unique properties. Nanotechnology-based biosensor has been increasingly gaining recognition in many different industries such as agriculture (Nair et al. 2011), biomedical applications (Xu et al. 2017), cosmetics industry (Raj et al. 2012) and food industry (Neethirajan et al. 2018; Lim and Ahmed 2016b). It has been widely recognised that there are numerous benefits of using nanomaterials (NMs) in developing biosensors due to their optical, mechanical and electrical properties and also their sizes and shapes. The advantages include providing high biocompatibility in the biosensors and enhancing the electrical signal and rapid detection time.

There are four categories of NMs that are determined by their respective dimensions, which are 0D (zero-dimensional, spherical) NMs; 1D (one-dimensional) NMs such as nanotubes, nanowires and nanofibers; 2D (two-dimensional, e.g. multilayers and film) NMs; and 3D (three-dimensional, for instance, nanoflower and graphite) NMs (Machado et al. 2015; Quesada-González and Merkoçi 2018).

0D nanomaterials are simply delineated as spherical nanostructures, and fullerene (C_{60}), quantum dots (QDs) and metal nanoparticles (e.g. AuNPs and PdNPs) are some of the known 0D NMs. Additional classifications of the 0D NMs are magnetic, metallic and semi-conductor NMs. As for one-dimensional nanomaterials, they have structures with elongated shape with measurement exceeding nanometre range. Gold nanorod (AuNR), carbon nanotube (CNT) and gold nanowire (AuNW)

NMs are a few of the examples of 1D nanostructures. Contrastingly, 2D nanomaterials have layer-like frameworks, and some of these structures include graphene oxide (GO) and graphene nanoplatelets (GNPs). Lastly, 3D nanostructures are the nanomaterials with all of their dimensions transcend nanometre scale such as gold nanoflowers (AuNFs) (Lim and Ahmed 2016a).

Through covalent and non-covalent interactions, various nanocomposite/nanohybrid materials can be synthesised. Modification via covalent interactions engages with different organic functional groups (e.g. –CHO, -NH_2), free radicals, oxygen and dienophiles that cause disorder in the structure of the overall compound. In opposition, non-covalent modification does not perturb the structure of the core nanomaterials involved as they utilise biological moieties such as enzymes and proteins, biomimetic molecules, polymers (e.g. polyethylene glycol and polyvinylpyrrolidone) and other types of NMs (Kuila et al. 2012; Georgakilas et al. 2012).

4.3.1 Advances and Applications of Carbon Nanomaterials in ECL Immunosensors

Carbon nanotubes (CNTs) and graphene-based nanomaterials are two of the most widely used carbon nanomaterials in biosensing research efforts and explicitly for modifying the electrode's surface. Amongst the attractive traits of carbon nanomaterials (CNMs) are their excellent electroconductivity, remarkable compatibility with biological moieties, effortless methods offered for modifications and admirable mechanical strength (da Silva et al. 2017; Lim and Ahmed 2016c; Rizwan et al. 2018a; Zhang and Lieber 2015). On top of these, CNMs aid to elevate the adsorption of bioreceptors/analytes onto the surface of the electrode and, successively, enrich the sensitivity of the biosensor (Baig et al. 2019). Table 4.2 depicts some applications of reported ECL immunosensors that incorporated CNMs in their biosensors.

4.3.1.1 Carbon Nanotubes

Carbon nanotubes are hollow cylindrical tubes that are made up of rolled graphite sheets (sp^2-hybridised carbon units) with thickness in nanoscale range and length that can be within micrometre range. Carbon nanotubes can be rolled either into a single layer, known as single-walled carbon nanotubes (SWCNTs), or multilayers, known as multiwalled carbon nanotubes (MWCNTs). SWCNTs and MWCNTs are the two types of CNTs that are frequently utilised.

The vectors (n, m) of a SWCNT's cylindrical structure portray its electronic characteristic. The integers of the n and m determine the configuration of SWCNTs of being armchair, zigzag or chiral (Gupta et al. 2018). By relying on the chirality and the diameter of the particular SWCNTs, they can be either metallic or semiconductor. They facilitate high heterogeneous electron transfer (HET) with reduced surface obstruction. Functionalised CNTs with –COOH, -COH or –OH groups are more felicitous compared to bare CNTs as they have better affinity towards bio- or chemical receptors (Wohlstadter et al. 2003; Yang et al. 2015). SWCNTs are known

Table 4.2 Some of the reported (2016–2019) ECL immunosensors that employs CNMs in the fabrication process

Developed ECL immunosensor	Target analyte	Detection range	Sensitivity	Specificity	References
GCE/Au-rGO@CB-Ab_1	AFP	0.0001 to 30 ng/mL	33 fg/mL	High	Zhu et al. (2016)
GCE/cit-rGO-$BaYF_5$/PDDA/AuNPs/Ab/BSA/CEA	CEA	0.001 to 80 ng/mL	0.87 pg/mL	High	Zhao et al. (2016)
Magnet/GCE/Magnetic Beads-Ab_1/VV/GO-Ab_2	*Vibrio vulnificus*	$4 \sim 4 \times 10^8$ CFU/mL	1 CFU/mL	High	Guo et al. (2016)
GCE/BGNs/Fe3O4-Au/Ab_1/TTX/luminol-AuNPs/Ab_2	Tetrodotoxin	0.01 to 100 ng/mL	0.01 ng/mL	High	Shang et al. (2017)
GCE/Au-FONDs/Ab_1/BSA/PSA/Ru@FGAPd-Ab_2	PSA	0.0001 to 50 ng/mL	0.056 pg/mL	High	Yang et al. (2017a)
GCE/MWCNTs–Pt– Luminol/Chi/Ab_1/BSA	CA19–9	0.0001 U/mL to 10.0 U/mL	46 μU/mL	High	Zhang et al. (2017)
GCE/anti-CEA/Au-FrGO-CeO_2@TiO_2	CEA	0.01 pg/mL to 10 ng/mL	3.28 fg/mL	High	Yang et al. (2017b)
GCE/erGO/P5Fin/AuNP/Ab_1/CEA/Ab_2-GQDs@AuNP	CEA	0.1 pg/mL to 10 ng/mL	3.78 fg/mL	High	Nie et al. (2018)
GCE/Au/Ab_1/CEA/Ab_2/AuNPs/CQDs-Cu^{2+}-PTCA-graphene	CEA	0.001 fg/mL to 1 ng/mL	0.00026 fg/mL	High	Xu et al. (2018)
GCE/DMSA-CdTe QDs/chitosan/Ab_1/BSA/Ag/Ab_2-SWCNHs-HRP	NSE	1×10^{-9} g/L to 1×10^{-3} g/L	4.4×10^{-10} g/L	High	Ai et al. (2018)
GCE/GO/Ag-AuNRs/Sudan I/BSA/Au CSNs-Ab/CdSe-CdS QDs-PAMAM-Pd	Sudan I	0.001 to 500 ng/mL	0.3 pg/mL	High	Wang et al. (2018)
GCE/Au NPs-Lu/rGO/Ab_1/BSA/insulin/Ab_2-SiO_2@PDA	Insulin	0.0001 to 50 ng/mL	26 fg/mL	High	Xing et al. (2018)
GCE/Luminol-Au@Fe_3O_4-$Cu_3(PO_4)_2$/Ab_1/BSA/NT-proBNP/Ab_2-Au@CuS-rGO	NT-proBNP	0.5 pg/mL to 20 ng/mL	0.12 pg/mL	High	Li et al. (2018b)

(continued)

Table 4.2 (continued)

Developed ECL immunosensor	Target analyte	Detection range	Sensitivity	Specificity	References
GCE/Ce:ZnO/Ag-OVA/Ab/PdNPs/PEI-GO/QDs	DCF	0.001 to 1000 ng/mL	0.3 pg/mL	High	Chen et al. (2018)
GCE-AuNRs-DCF Ag	DCF	0.001 to 800 ng/mL	0.33 pg/mL	High	Wang et al. (2019)
GCE/anti-PSA-Lu-GS@Pt/BSA/PSA	PSA	1 pg/mL to 10 ng/mL	0.3 pg/mL	High	Khan et al. (2019)
GCE/Au NPs-luminol-LDH/Ab_1/PSA/Ab_2-GOx-PGA-Pt NPs	PSA	1.0 pg/mL to 150 ng/mL	0.6 pg/mL	High	Huo et al. (2019)
GCE/CdS/p53-Ab_1/p53/p53-Ab_2-tGO-AuNPs	p53 protein	20 to 1000 fg/ml	4 fg/mL	High	Heidari et al. (2019)

AFP α-fetoprotein, *CA19–9* tumour marker for pancreatic cancer, *CEA* carcinoembryonic antigen, *DCF* diclofenac, *GA* glutaraldehyde, *Gal-3* galectin-3, *NSE* neuron-specific enolase, *NT-proBNP* N-terminal pro-brain natriuretic peptide, *p53* tumour suppressor gene, *PSA* prostate serum antigen, *Sudan I* food colourant to have better electronic properties in contrast to MWCNTs making them an appealing option in devising biosensors despite its poor dispersal ability especially in polar solvents (Li et al. 2018a). SWCNTs' superb electroconductivity has been substantiated to be facilitating the electron transfer that subsequently improves the sensitivity of the detection. The impediment relating to the dispersal ability of SWCNTs can be resolved by functionalising the SWCNTs chemically with –OH, -COOH, -NH_2 or –SH (Zaman et al. 2012). SWCNTs and COOH have been evinced to possess better biocompatibility with the bioreceptors, and also more bioreceptors can be deposited onto the functionalised SWCNTs compared to the bare SWCNTs (Rezaei et al. 2016).

Single-walled carbon nanohorns (SWCNHs) are firstly identified by Iijima et al. (1999) by excising carbon with CO_2 laser at room temperature without requiring any metal catalysts. SWCNHs are also known as nanocones, attributable to their conical structure of sp^2-hybridised carbon atoms with a diameter of 2–5 nm and length of 40–50 nm (Karousis et al. 2016). To date, there are three kinds of SWCNHs discovered: "dahlia-like", "budlike" and "seedlike". Their properties are similar to SWCNTs with the advantage of being less toxic and undemanding to be mass produced compared to SWCNTs (Zhu and Xu 2010). More amount of biomolecules can be immobilised onto SWCNHs on the account of their large surface area when the pores of SWCNHs are opened, in contrast to SWCNTs (Farka et al. 2017).

Electrocatalytic characteristic of SWCNHs was manoeuvred for the construction of an ECL immunosensor for the detection of N-terminal brain-type natriuretic peptide (NT-proBNP), a biomarker for heart failure (Liu et al. 2017). This group immobilised the secondary Abs onto the SWCNHs decorated with PdCu (bimetal) nanocomposites (PdCu@SWCNHs), which have been conjugated with PTC-Lu (3,4,9,10-perylenetetracarboxylic acid-luminol) – the selected luminophore. It was ascertained that by employing PdCu@SWCNHs, the ECL intensity is further enhanced as the nanocomposite facilitates the production of ROS (reactive oxygen species) for luminol-H_2O_2 system. Thereafter, their biosensor can selectively detect NT-proBNP linearly from 0.0001 ng/mL to 25 ng/mL, and the sensitivity is determined to be 0.05 pg/mL. They have also effectually performed the real sample analysis with four different human serums with the reasonable %RSD (% relative standard deviation) of −5.0–6.0%.

MWCNTs are known as lightweight nanomaterials which possess superior tensile strength, excellent electroconductivity and large surface area. Owing to their unique properties, MWCNTs have been progressively amalgamated into biosensors. Zhang et al. (2017) designed an immunosensor for the detection of CA19-9, a tumour marker (pancreatic cancer), by combining MWCNTs, Pt, and luminol (luminophore) as a nanocomposite. They immobilised the nanocomposite on the glassy carbon electrode (GCE) and reinforced the nanocomposite by a layer of chitosan. In their study, they claimed that MWCNT-Pt functions as the catalyst which facilitates the electron transfer as well as a catalyst in the production of reactive oxygen species (ROS) and subsequently enhances the ECL intensity of luminol-H_2O_2 ECL system. H_2O_2 is known to be the oxidising agent for luminol to emit luminescence through the annihilation pathway. The resultant immunosensor is capable to detect CA19–9 linearly from 0.0001 U mL^{-1} to 10.0 U mL^{-1} with a high sensitivity of 0.000046 U mL^{-1}. They have also successfully applied it for the detection of target in real human serum with good percentage recovery (96.7–105.0%).

Aside from the increasing interest in the development of ECL immunosensors for applications in clinical diagnosis with CNMs, there are also biosensors being devised for applications in other areas. In clinical diagnosis, the growth in the biosensor fabrication is due to the requirement for analyses to be prompt in producing results with accuracy and precision. Therefore, it is crucial for the developed biosensors to explicitly detect the target analytes within the specified range (should include both healthy and unhealthy concentrations). A plethora of researches are being carried out to allow early screening for any anomalies in specified biomarker's concentration that signify the progress of related illness for the prevention of it to be further developed as the consequence might be terminal.

Other recent applications of ECL immunosensors that employed CNTs or CNHs include monitoring environmental situations as it is vital in order to ensure the safety of foods for consumption and as a step for preventing illness in humans, animals and plants. One example of environmental applications is supervising the concentration of pharmaceutical compounds diclofenac in the environment that can cause mortality in aquatic animals (Hu et al. 2018).

4.3.1.2 Graphene-Based Nanomaterials

Per contra, graphene maintains single-layer frameworks with honeycomb-like lattice that are made up of sp^2-hybridised carbon atoms with a plethora of delocalised π electrons within the structure. Graphene nanomaterials appeal researchers for the construction of POC biosensors as they have undeniably marvellous properties that allow rapid, on-site detections. Particularly on the defects of the edges of graphene, the electron transfer is brisker in comparison with that on the plane and, therefore, confirmed to have extraordinary electrocatalytic property. Moreover, graphene has a large surface area of 2630 m^2/g, supplying colossal area for immobilisation of bioreceptors (Adeel et al. 2018).

The oxidised derivatives of graphene – GO and rGO – are two of the most prominent nanomaterials utilised in the fabrication of biosensors. These are due to their exquisite properties as mentioned in the previous paragraph, with additional benefits of possessing exquisite optical transmittance, and excellent mechanical property (Farka et al. 2017). Additionally, the presence of oxygen groups contributes to the hydrophilicity of GO and rGO, and thereupon, they are easier to be dispersed (Erol et al. 2018). Nonetheless, as GO comprises of oxygen-containing functional groups, these groups unfavourably affected the electroconductivity of GO. One of the measures taken to tackle this issue is by loading metal nanoparticles on GO frameworks, as shown by Wang et al. (2018). Their group has reported a competitive type of ECL immunosensor by combining Au nanorods (AuNRs) with GO (AuNRs/GO) as the carrier for the Ab. Employment of AuNRs-GO helps to amplify the ECL signal of the immunosensor as integration of GO allows more AuNRs to be loaded onto the modified detection platform in comparison to the absence of GO.

Moreover, GO can also be converted to its other semi-conductor counterpart via chemical or thermal reactions, known as reduced graphene oxide (rGO). rGO is claimed to possess significantly higher electroconductivity as against to GO but two degree lesser as opposed to graphene (Mao et al. 2012). Therefore, by taking account of this fact, Xing et al. (2018) chose rGO as a scaffold for immobilisation of Abs as rGO has a large surface area. SnO_2 molecules were also decorated onto rGO structure prior to Abs immobilisation as to further improve the overall biosensor's performance. The resulting ECL signal was remarkably intensified as they modified SnO_2/rGO/Au NPs-Lu onto the GCE's surface which subsequently refined the sensitivity of the biosensor to 26 fg/mL with superlative specificity and stability.

GO/rGO-integrated ECL immunosensors have been implemented for various applications such as examining the environmental conditions, for example, detecting the presence of pathogens in water such as *Vibrio vulnificus* (Guo et al. 2016) is important as it is one of the widely utilised raw materials by the living things. On the other hand, the constituents of commercially available food products must be properly analysed, stated and labelled on the packaging for aiding the consumers in selecting the products according to their preferences. Toxins such as tetrodotoxins (one of the types of strong marine neurotoxins) and hazardous food colourant (e.g. Sudan I) might be present in the food products and can be fatal when ingested, and thus, these products should be screened before they are released for commercialisation (Shang et al. 2017; Wang et al. 2018).

Other fascinating nanomaterials that falls under graphene-based CNMs are graphene nanoribbons (GNRs) and graphene oxide nanoribbons (GONRs). The two aforementioned NMs are cognate CNMs that can be procured by unzipping the CNTs, forming stretched monolayer graphene sheets (Georgakilas et al. 2015). Electronic attributes of GNRs/GONRs depend on the width of their respective structures. The difference in the syntheses of GNRs and GONRs is the addition of a strong oxidising agent, causing the presence of oxygen-containing groups in GONRs (Kosynkin et al. 2009). Although GONRs have a large surface area for the immobilisation of bioreceptors, their electrochemical property has attenuated due to the high amount of oxygen functional groups. One of the solutions for this matter is by incorporating metal nanoparticles with GONRs as performed by Ismail et al. (2015). AuNPs were decorated onto the GONRs and facilitated the overall electrocatalytic performance of the ABEI (*N*-(aminobutyl)-*N*-(ethylisoluminol))-functionalised AuNP nanohybrids on the H_2O_2-based ECL system. Other types of carbon nanomaterials that recently gain the attention of the researchers are discussed in the next section.

4.3.1.3 Quantum Dots

Semi-conductor nanocrystals, familiarly known as quantum dots (QDs), are comprised of clusters made of 100–1000 atoms with magnitude ranging from 1 to 10 nm (Alivisatos et al. 1996). These nanocrystals are mostly exploited as both fluorophore and luminophore and, thus, desirable as modules in ECL-based biosensors (Krishna et al. 2018). They were first unearthed in 1983, and ever since then, they are extensively utilised in various fields including in developing electronic and optical devices (Brus 1984). Furthermore, they acquire attractive optical properties of adjustable emission bands, which vary depending on their size resulting in bandgap energy being inversely related to their size and composition (Chen and Park 2016). As a consequence, multiplex studies are possible as QDs possess broad emission wavelength and absorption spectra. They are also highly resistant towards photobleaching and chemical deterioration and, henceforth, are outstandingly photostable (Chen et al. 2017).

Carbon quantum dots (CQDs) and graphene quantum dots (GQDs) are the two organic nanoparticles composed of carbon atoms that are contemporary alternatives for the traditional QDs (e.g. CdSe and CdS). Contrastingly, these two carbon-based QDs have upper hands in several aspects compared to the conventional QDs that include their better biocompatibility for bioreceptor conjugations, low cytotoxicity and more superiority in photoluminescence property as it is tuneable (Xie et al. 2016; Zheng et al. 2015). CQDs are nanocrystal comprising of sp^2-/sp^3-hybridised carbon atoms with dimension of quasi-spherical, whereas GQDs consist of a single or several layers of sp^2-hybridised carbon atoms which are compressed into a planar form (Sun and Lei 2017; Nie et al. 2018).

Environment-friendly CQDs was opted by Li et al. (2017) as they possess superior photostability and fascinate electroconductivity, as the ECL luminophore in their study. The Ab_2 molecules were immobilised onto the CQD-GO-PEI nanohybrids (Li et al. 2017; Shi et al. 2018). The sandwiched interaction of CEA between Ab_1 and Ab_2 triggered the ECL reaction of CQDs, thereafter producing light as the signal. CQDs' ECL intensity in their work was elevated by the symbiosis performance of PEI-GO, AuNPs, AgNPs and polydopamine. This effect reflected on the immunosensor's accomplishment in sensing CEA dynamically from 5 pg/mL to 500 ng/mL and as low as 1.67 pg/mL with superlative specificity and stability.

Tian et al. (2019) have constructed an immunosensor with the aim of analysing the content of PSA (prostate-specific antigen) in human serum. GQDs fixated on the TiO_2 nanotubes (TiO_2 NTs) were utilised in this research study as an ECL probe as they withhold exceptional ECL characteristic attributable to their ability of catalysing photoluminescence reaction (Gupta et al. 2015). Persulfate ($K_2S_2O_8$) was selected as the co-reactant of the ECL probe. Their immunosensor has triumphantly detected PSA linearly from 1.0 fg/mL to 10 pg/mL with a sensitivity of 1 fg/mL and great specificity. This biosensor has also successfully detected the target analyte in human serum efficiently.

4.4 Conclusion and Future Prospects

ECL technique has gained significant interest from researchers for biosensing applications due to its advantages including simple operation and offer low background signal. Furthermore, an ECL signal can be further refined by incorporating nanomaterials. Unique properties such as large surface area and superb electroconductivity of CNMs contribute to the ability of CNMs in facilitating the operation of a particular immunosensor. Certain types of CNMs exhibit excellent optical property with notable photostability and thus are employed as the luminophore. Through the application of CNMs, the reported ECL immunosensors have been evinced to be able to detect their respective target with outstanding sensitivity, exceptional specificity and competent stability.

Nevertheless, there are still some areas in the development of these ECL immunosensors that require further studies before they can be applied for practical on-site detection. Limitations that need to be overcome include:

- Lengthy sample pretreatment process, particularly for human serums and food samples. This is due to the presence of complex substances that can impede and intervene with the bioreceptors of the immunosensor and the luminophore (Aydın et al. 2018; Terry et al. 2005).
- Inefficient adsorption of bioreceptors onto the surface of the electrode, compromising the efficiency of transducer's response (Chandra 2016).
- Unsatisfactory stability, reproducibility, response time and robustness of the immunosensors.
- Inadequacy in researches for the applications of ECL immunosensors that are modified with CNMs.

Therefore, to address these issues, better extraction protocols, preferably with only by necessitating one simple step or reagent without jeopardising the purity of the extracted proteins, could be developed as to ameliorate the application for on-site detections. Further studies on the properties of different CNMs whether as a stand-alone nanomaterial or by combining them with other nanomaterials as a nanocomposite (such as gold nanoparticles) can be done to effectively improve the overall performance of a particular immunosensor. This will eventually act as a stepping stone for the development of accurate and reliable POC devices, not only for detecting a single target but also enabling duplex or even multiplex detections.

References

Adeel M, Bilal M, Rasheed T, Sharma A, Iqbal HMN (2018) Graphene and graphene oxide: functionalization and nano-bio-catalytic system for enzyme immobilization and biotechnological perspective. Int J Biol Macromol 120:1430–1440

Ahmed MU, Saaem I, Wu PC, Brown AS (2014) Personalized diagnostics and biosensors: a review of the biology and technology needed for personalized medicine. Crit Rev Biotechnol 34(2):180–196

Ahmed MU, Hossain MM, Safavieh M, Wong YL, Rahman IA, Zourob M, Tamiya E (2016) Toward the development of smart and low cost point-of-care biosensors based on screen printed electrodes. Crit Rev Biotechnol 36(3):495–505. https://doi.org/10.3109/07388551.2014.992387

Ai Y, Li X, Zhang L, Zhong W, Wang J (2018) Highly sensitive electrochemiluminescent immunoassay for neuron-specific enolase amplified by single-walled carbon nanohorns and enzymatic biocatalytic precipitation. J Electroanal Chem 818:257–264

Alivisatos AP, Johnsson KP, Peng X, Wilson TE, Loweth CJ, Bruchez MP Jr, Schultz PG (1996) Organization of 'nanocrystal molecules' using DNA. Nature 382(6592):609

Aydın M, Aydın EB, Sezgintürk MK (2018) A disposable immunosensor using ITO based electrode modified by a star-shaped polymer for analysis of tumor suppressor protein p53 in human serum. Biosens Bioelectron 107:1–9. https://doi.org/10.1016/j.bios.2018.02.017

Azam NFN, Roy S, Lim SA, Uddin Ahmed M (2018) Meat species identification using DNA-luminol interaction and their slow diffusion onto the biochip surface. Food Chem 248:29–36

Baig N, Sajid M, Saleh TA (2019) Recent trends in nanomaterial-modified electrodes for electroanalytical applications. TrAC Trends Anal Chem 111:47–61

Benoit L, Choi JP (2017) Electrogenerated chemiluminescence of semiconductor nanoparticles and their applications in biosensors. ChemElectroChem 4(7):1573–1586

Bertoncello P, Stewart AJ, Dennany L (2014) Analytical applications of nanomaterials in electrogenerated chemiluminescence. Anal Bioanal Chem 406(23):5573–5587

Bhalla N, Jolly P, Formisano N, Estrela P (2016) Introduction to biosensors. Essays Biochem 60(1):1–8

Brus LE (1984) Electron–electron and electron-hole interactions in small semiconductor crystallites: the size dependence of the lowest excited electronic state. J Chem Phys 80(9):4403–4409

Byrne B, Stack E, Gilmartin N, O'Kennedy R (2009) Antibody-based sensors: principles, problems and potential for detection of pathogens and associated toxins. Sensors 9(6):4407–4445

Chakraborty M, Hashmi MSJ (2017) An overview of biosensors and devices. In: Reference module in materials science and materials engineering. Elsevier, Amsterdam

Chandra P (2016) Nanobiosensors for personalized and onsite biomedical diagnosis. The Institution of Engineering and Technology, London

Chen J, Park B (2016) Recent advancements in Nanobioassays and Nanobiosensors for foodborne pathogenic bacteria detection. J Food Prot 79(6):1055–1069

Chen Y, Zhou S, Li L, Zhu J-j (2017) Nanomaterials-based sensitive electrochemiluminescence biosensing. Nano Today 12:98–115

Chen W, Zhu Q, Tang Q, Zhao K, Deng A, Li J (2018) Ultrasensitive detection of diclofenac based on electrochemiluminescent immunosensor with multiple signal amplification strategy of palladium attached graphene oxide as bioprobes and ceria doped zinc oxide as substrates. Sensors Actuators B Chem 268:411–420

Clark LC, Lyons C (1962) Electrode systems for continuous monitoring in cardiovascular surgery. Ann N Y Acad Sci 102(1):29–45

da Silva ET, Souto DE, Barragan JT, de Giarola JF, de Moraes AC, Kubota LT (2017) Electrochemical biosensors in point-of-care devices: recent advances and future trends. ChemElectroChem 4(4):778–794

Donahue AC, Albitar M (2010) Antibodies in biosensing. In: Recognition receptors in biosensors. New York, Springer, pp 221–248

Erol O, Uyan I, Hatip M, Yilmaz C, Tekinay AB, Guler MO (2018) Recent advances in bioactive 1D and 2D carbon nanomaterials for biomedical applications. Nanomedicine 14(7):2433–2454

Fang C, Li H, Yan J, Guo H, Yifeng T (2017) Progress of the Electrochemiluminescence biosensing strategy for clinical diagnosis with Luminol as the sensing probe. ChemElectroChem 4(7):1587–1593

Farka Z, Juřík T, Kovář D, Trnková L, Skládal P (2017) Nanoparticle-based immunochemical biosensors and assays: recent advances and challenges. Chem Rev 117(15):9973–10042

Farzin L, Shamsipur M (2017) Recent advances in design of electrochemical affinity biosensors for low level detection of cancer protein biomarkers using nanomaterial-assisted signal enhancement strategies. J Pharm Biomed Anal 147:185–210

Felix FS, Angnes L (2017) Electrochemical Immunosensors–a powerful tool for analytical applications. Biosens Bioelectron 102:470–478

Ge L, Yu J, Ge S, Yan M (2014) Lab-on-paper-based devices using chemiluminescence and electrogenerated chemiluminescence detection. Anal Bioanal Chem 406(23):5613–5630

Georgakilas V, Otyepka M, Bourlinos AB, Chandra V, Kim N, Kemp KC, Hobza P, Zboril R, Kim KS (2012) Functionalization of graphene: covalent and non-covalent approaches, derivatives and applications. Chem Rev 112(11):6156–6214

Georgakilas V, Perman JA, Tucek J, Zboril R (2015) Broad family of carbon Nanoallotropes: classification, chemistry, and applications of fullerenes, carbon dots, nanotubes, graphene, Nanodiamonds, and combined superstructures. Chem Rev 115(11):4744–4822

Guo Z, Sha Y, Hu Y, Yu Z, Tao Y, Wu Y, Zeng M, Wang S, Li X, Zhou J (2016) Faraday cage-type electrochemiluminescence immunosensor for ultrasensitive detection of Vibrio vulnificus based on multi-functionalized graphene oxide. Anal Bioanal Chem 408(25):7203–7211

Gupta BK, Kedawat G, Agrawal Y, Kumar P, Dwivedi J, Dhawan SK (2015) A novel strategy to enhance ultraviolet light driven photocatalysis from graphene quantum dots infilled TiO2 nanotube arrays. RSC Adv 5(14):10623–10631. https://doi.org/10.1039/C4RA14039G

Gupta S, Murthy CN, Prabha CR (2018) Recent advances in carbon nanotube based electrochemical biosensors. Int J Biol Macromol 108:687–703

Hazelton S, Zheng X, Zhao J, Pierce D (2008) Developments and applications of electrogenerated chemiluminescence sensors based on micro-and nanomaterials. Sensors 8(9):5942–5960

Heidari R, Rashidiani J, Abkar M, Taheri RA, Moghaddam MM, Mirhosseini SA, Seidmoradi R, Nourani MR, Mahboobi M, Keihan AH, Kooshki H (2019) CdS nanocrystals/graphene oxide-AuNPs based electrochemiluminescence immunosensor in sensitive quantification of a cancer biomarker: p53. Biosens Bioelectron 126:7–14

Hercules DM (1964) Chemiluminescence resulting from electrochemically generated species. Science 145(3634):808–809

Hu L, Zheng J, Zhao K, Deng A, Li J (2018) An ultrasensitive electrochemiluminescent immunosensor based on graphene oxide coupled graphite-like carbon nitride and multiwalled carbon nanotubes-gold for the detection of diclofenac. Biosens Bioelectron 101:260–267

Huo X-L, Zhang N, Xu J-J, Chen H-Y (2019) Ultrasensitive electrochemiluminescence immunosensor with wide linear range based on a multiple amplification approach. Electrochem Commun 98:33–37

Iijima S, Yudasaka M, Yamada R, Bandow S, Suenaga K, Kokai F, Takahashi K (1999) Nano-aggregates of single-walled graphitic carbon nano-horns. Chem Phys Lett 309(3):165–170

Ismail NS, Le QH, Hasan Q, Yoshikawa H, Saito M, Tamiya E (2015) Enhanced electrochemiluminescence of N-(aminobutyl)-N-(ethylisoluminol) functionalized gold nanoparticles by graphene oxide nanoribbons. Electrochim Acta 180:409–418

Ju H, Lai G, Yan F (2017) Immunosensing for detection of protein biomarkers. Elsevier, Amsterdam

Karousis N, Suarez-Martinez I, Ewels CP, Tagmatarchis N (2016) Structure, properties, functionalization, and applications of carbon nanohorns. Chem Rev 116(8):4850–4883

Kerr E, Doeven EH, Barbante GJ, Hogan CF, Hayne DJ, Donnelly PS, Francis PS (2016) New perspectives on the annihilation electrogenerated chemiluminescence of mixed metal complexes in solution. Chem Sci 7(8):5271–5279

Khan MS, Zhu W, Ali A, Ahmad SM, Li X, Yang L, Wang Y, Wang H, Wei Q (2019) Electrochemiluminescent immunosensor for prostate specific antigen based upon luminol functionalized platinum nanoparticles loaded on graphene. Anal Biochem 566:50–57

Kosynkin DV, Higginbotham AL, Sinitskii A, Lomeda JR, Dimiev A, Price BK, Tour JM (2009) Longitudinal unzipping of carbon nanotubes to form graphene nanoribbons. Nature 458:872

Krishna VD, Wu K, Su D, Cheeran MC, Wang J-P, Perez A (2018) Nanotechnology: review of concepts and potential application of sensing platforms in food safety. Food Microbiol 75:47–54

Kuila T, Bose S, Mishra AK, Khanra P, Kim NH, Lee JH (2012) Chemical functionalization of graphene and its applications. Prog Mater Sci 57(7):1061–1105

Lee VBC, MOHD-NAIM NF, Tamiya E, Ahmed MU (2018) Trends in paper-based electrochemical biosensors: from design to application. Anal Sci 34(1):7–18

Li N-L, Jia L-P, Ma R-N, Jia W-L, Lu Y-Y, Shi S-S, Wang H-S (2017) A novel sandwiched electrochemiluminescence immunosensor for the detection of carcinoembryonic antigen based on carbon quantum dots and signal amplification. Biosens Bioelectron 89:453–460

Li Q, Jin J, Lou F, Xiao Y, Zhu J, Zhang S (2018a) Carbon nanomaterials-based electrochemical immunoassay with β-Galactosidase as labels for Carcinoembryonic antigen. Electroanalysis 30(5):852–858

Li X, Lu P, Wu B, Wang Y, Wang H, Du B, Pang X, Wei Q (2018b) Electrochemiluminescence quenching of luminol by CuS in situ grown on reduced graphene oxide for detection of N-terminal pro-brain natriuretic peptide. Biosens Bioelectron 112:40–47
Lim SA, Ahmed MU (2016a) Electrochemical immunosensors and their recent nanomaterial-based signal amplification strategies: a review. RSC Adv 6(30):24995–25014
Lim SA, Ahmed MU (2016b) A label free electrochemical immunosensor for sensitive detection of porcine serum albumin as a marker for pork adulteration in raw meat. Food Chem 206:197–203
Lim SA, Ahmed MU (2016c) A simple DNA-based electrochemical biosensor for highly sensitive detection of ciprofloxacin using disposable graphene. Anal Sci 32(6):687–693. https://doi.org/10.2116/analsci.32.687
Liu BL, Saltman MA (2015) Immunosensor technology: historical perspective and future outlook. Lab Med 27(2):109–115
Liu Y, Wang H, Xiong C, Chai Y, Yuan R (2017) An ultrasensitive electrochemiluminescence immunosensor for NT-proBNP based on self-catalyzed luminescence emitter coupled with PdCu@ carbon nanohorn hybrid. Biosens Bioelectron 87:779–785
Luppa PB, Sokoll LJ, Chan DW (2001) Immunosensors—principles and applications to clinical chemistry. Clin Chim Acta 314(1):1–26
Machado FM, Fagan SB, da Silva IZ, de Andrade MJ (2015) Carbon nanoadsorbents. In: Carbon nanomaterials as adsorbents for environmental and biological applications. Springer, New York, pp 11–32
Mao S, Pu H, Chen J (2012) Graphene oxide and its reduction: modeling and experimental progress. RSC Adv 2(7):2643–2662
Mathieu HJ (2010) Analytical tools for biosensor surface chemical characterization. In: Recognition receptors in biosensors. Springer, New York, pp 135–173
Miao W (2008) Electrogenerated chemiluminescence and its biorelated applications. Chem Rev 108(7):2506–2553
Moina C, Ybarra G (2012) Fundamentals and applications of immunosensors. In: Advances in immunoassay technology. InTech, Rijeka
Muzyka K (2014) Current trends in the development of the electrochemiluminescent immunosensors. Biosens Bioelectron 54:393–407
Nair R, Poulose AC, Nagaoka Y, Yoshida Y, Maekawa T, Kumar DS (2011) Uptake of FITC labeled silica nanoparticles and quantum dots by rice seedlings: effects on seed germination and their potential as biolabels for plants. J Fluoresc 21(6):2057
Neethirajan S, Ragavan V, Weng X, Chand R (2018) Biosensors for sustainable food engineering: challenges and perspectives. Biosensors 8(1):23
Nie G, Wang Y, Tang Y, Zhao D, Guo Q (2018) A graphene quantum dots based electrochemiluminescence immunosensor for carcinoembryonic antigen detection using poly (5-formylindole)/reduced graphene oxide nanocomposite. Biosens Bioelectron 101:123–128
Omidfar K, Khorsand F, Darziani Azizi M (2013) New analytical applications of gold nanoparticles as label in antibody based sensors. Biosens Bioelectron 43:336–347
Quesada-González D, Merkoçi A (2018) Nanomaterial-based devices for point-of-care diagnostic applications. Chem Soc Rev 47(13):4697–4709
Raj S, Jose S, Sumod U, Sabitha M (2012) Nanotechnology in cosmetics: opportunities and challenges. J Pharm Bioallied Sci 4(3):186–193. https://doi.org/10.4103/0975-7406.99016
Rezaei B, Ghani M, Shoushtari AM, Rabiee M (2016) Electrochemical biosensors based on nanofibres for cardiac biomarker detection: a comprehensive review. Biosens Bioelectron 78:513–523
Rizwan M, Elma S, Lim SA, Ahmed MU (2018a) AuNPs/CNOs/SWCNTs/chitosan-nanocomposite modified electrochemical sensor for the label-free detection of carcinoembryonic antigen. Biosens Bioelectron 107:211–217
Rizwan M, Mohd-Naim N, Ahmed M (2018b) Trends and advances in Electrochemiluminescence Nanobiosensors. Sensors 18(1):166
Rogers KR (2000) Principles of affinity-based biosensors. Mol Biotechnol 14(2):109–129

Roy S, Wei SX, Ying JLZ, Safavieh M, Ahmed MU (2016) A novel, sensitive and label-free loop-mediated isothermal amplification detection method for nucleic acids using luminophore dyes. Biosens Bioelectron 86:346–352
Santhanam KSV, Bard AJ (1965) Chemiluminescence of Electrogenerated 9,10-Diphenylanthracene anion Radical1. J Am Chem Soc 87(1):139–140
Shang F, Liu Y, Wang S, Hu Y, Guo Z (2017) Electrochemiluminescence Immunosensor based on functionalized graphene/Fe3O4-au magnetic capture probes for ultrasensitive detection of Tetrodotoxin. Electroanalysis 29(9):2098–2105
Shi X, Wei W, Fu Z, Gao W, Zhang C, Zhao Q, Deng F, Lu X (2018) Review on carbon dots in food safety applications. Talanta 194:809–821
Siontorou CG, Nikoleli G-PD, Nikolelis DP, Karapetis S, Tzamtzis N, Bratakou S (2017) Point-of-care and implantable biosensors in cancer research and diagnosis. In: Next generation point-of-care biomedical sensors technologies for cancer diagnosis. Springer, Singapore, pp 115–132
Sojic N, Arbault S, Bouffier L, Kuhn A (2017) Applications of electrogenerated chemiluminescence in analytical chemistry. In: Luminescence in electrochemistry. Springer, Cham, pp 257–291
Sun X, Lei Y (2017) Fluorescent carbon dots and their sensing applications. TrAC Trends Anal Chem 89:163–180
Terry LA, White SF, Tigwell LJ (2005) The application of biosensors to fresh produce and the wider food industry. J Agric Food Chem 53(5):1309–1316. https://doi.org/10.1021/jf040319t
Thakur M, Ragavan K (2013) Biosensors in food processing. J Food Sci Technol 50(4):625–641
Tian C, Wang L, Luan F, Zhuang X (2019) An electrochemiluminescence sensor for the detection of prostate protein antigen based on the graphene quantum dots infilled TiO2 nanotube arrays. Talanta 191:103–108
Ugo P, Moretto LM (2017) Electrochemical immunosensors and aptasensors. Multidisciplinary Digital Publishing Institute, Basel
Valenti G, Rampazzo E, Kesarkar S, Genovese D, Fiorani A, Zanut A, Palomba F, Marcaccio M, Paolucci F, Prodi L (2018) Electrogenerated chemiluminescence from metal complexes-based nanoparticles for highly sensitive sensors applications. Coord Chem Rev 367:65–81
Vo-Dinh T, Cullum B (2000) Biosensors and biochips: advances in biological and medical diagnostics. Fresenius J Anal Chem 366(6–7):540–551
Wang C, Hu L, Zhao K, Deng A, Li J (2018) Multiple signal amplification electrochemiluminescent immunoassay for Sudan I using gold nanorods functionalized graphene oxide and palladium/aurum core-shell nanocrystallines as labels. Electrochim Acta 278:352–362
Wang C, Jiang T, Zhao K, Deng A, Li J (2019) A novel electrochemiluminescent immunoassay for diclofenac using conductive polymer functionalized graphene oxide as labels and gold nanorods as signal enhancers. Talanta 193:184–191
Wei F, Lillehoj PB, Ho C-M (2010) DNA diagnostics: nanotechnology-enhanced electrochemical detection of nucleic acids. Pediatr Res 67:458
WHO (2001) International programme on chemical safety biomarkers in risk assessment: validity and validation. http://www.inchem.org/documents/ehc/ehc/ehc222.htm
Wohlstadter JN, Wilbur JL, Sigal GB, Biebuyck HA, Billadeau MA, Dong L, Fischer AB, Gudibande SR, Jameison SH, Kenten JH (2003) Carbon nanotube-based biosensor. Adv Mater 15(14):1184–1187
Xie R, Wang Z, Zhou W, Liu Y, Fan L, Li Y, Li X (2016) Graphene quantum dots as smart probes for biosensing. Anal Methods 8(20):4001–4016
Xing B, Zhu W, Zheng X, Zhu Y, Wei Q, Wu D (2018) Electrochemiluminescence immunosensor based on quenching effect of SiO2@PDA on SnO2/rGO/Au NPs-luminol for insulin detection. Sensors Actuators B Chem 265:403–411
Xu Z, Liao L, Chai Y, Wang H, Yuan R (2017) Ultrasensitive Electrochemiluminescence biosensor for MicroRNA detection by 3D DNA walking machine based target conversion and distance-controllable signal quenching and enhancing. Anal Chem 89(16):8282–8287
Xu L-l, Zhang W, Shang L, Ma R-n, Jia L-p, Jia W-l, Wang H-s, Niu L (2018) Perylenetetracarboxylic acid and carbon quantum dots assembled synergistic electrochemiluminescence nanomaterial for ultra-sensitive carcinoembryonic antigen detection. Biosens Bioelectron 103:6–11

Yang N, Chen X, Ren T, Zhang P, Yang D (2015) Carbon nanotube based biosensors. Sensors Actuators B Chem 207:690–715
Yang L, Li Y, Zhang Y, Fan D, Pang X, Wei Q, Du B (2017a) 3D nanostructured palladium-functionalized graphene-aerogel-supported Fe3O4 for enhanced Ru(bpy)32+-based Electrochemiluminescent Immunosensing of prostate specific antigen. ACS Appl Mater Interfaces 9(40):35260–35267
Yang L, Zhu W, Ren X, Khan MS, Zhang Y, Du B, Wei Q (2017b) Macroporous graphene capped Fe3O4 for amplified electrochemiluminescence immunosensing of carcinoembryonic antigen detection based on CeO2@TiO2. Biosens Bioelectron 91:842–848
Zaman AC, Üstündağ CB, Kaya F, Kaya C (2012) OH and COOH functionalized single walled carbon nanotubes-reinforced alumina ceramic nanocomposites. Ceram Int 38(2):1287–1293
Zhang A, Lieber CM (2015) Nano-bioelectronics. Chem Rev 116(1):215–257
Zhang X, Ke H, Wang Z, Guo W, Zhang A, Huang C, Jia N (2017) An ultrasensitive multi-walled carbon nanotube–platinum–luminol nanocomposite-based electrochemiluminescence immunosensor. Analyst 142(12):2253–2260
Zhao L, Li J, Liu Y, Wei Y, Zhang J, Zhang J, Xia Q, Zhang Q, Zhao W, Chen X (2016) A novel ECL sensor for determination of carcinoembryonic antigen using reduced graphene Oxide-BaYF5:Yb, Er upconversion nanocomposites and gold nanoparticles. Sensors Actuators B Chem 232:484–491
Zheng XT, Ananthanarayanan A, Luo KQ, Chen P (2015) Glowing graphene quantum dots and carbon dots: properties, syntheses, and biological applications. Small 11(14):1620–1636
Zhu S, Xu G (2010) Single-walled carbon nanohorns and their applications. Nanoscale 2(12):2538–2549
Zhu W, Lv X, Wang Q, Ma H, Wu D, Yan T, Hu L, Du B, Wei Q (2016) Ru (bpy) 3 2+/nanoporous silver-based electrochemiluminescence immunosensor for alpha fetoprotein enhanced by gold nanoparticles decorated black carbon intercalated reduced graphene oxide. Sci Rep 6:20348

Green Synthesis of Colloidal Metallic Nanoparticles Using Polyelectrolytes for Biomedical Applications

5

Ana M. Herrera-González, M. Caldera-Villalobos, J. García-Serrano, and M. C. Reyes-Ángeles

Abstract

Metallic nanoparticles have found applications in numerous fields, such as biosensors, antimicrobial agents, theragnostic, cancer hyperthermia, gene and drug delivery platforms, and bactericide. However, the development of these and other applications is still limited by the methods of synthesis of nanoparticles which usually use toxic substance as reducing and stabilizing agents. Because of this, there is an increasing interest in the development of methods of green synthesis. In this chapter there is a review to green synthesis of metallic nanoparticles using polyelectrolytes that act as a reducing and stabilizer agents, without toxic substances. The use of colloidal poses an easy way of administration and manipulation for metallic nanoparticles in living organisms. Colloidal nanoparticles of gold have been studied as vectors in the administration of pharmaceuticals for the treatment of tumors, as immunosensor or as DNA sequence colorimetric detector. There are reports about the synthesis and stabilization of metallic nanoparticles in colloidal solution using polyelectrolytes. Recent research has been focused on the design of new anionic polyelectrolytes and macroelectrolytes with reducing agent properties for the synthesis and stabilization of colloidal metallic nanoparticles. The use of polyelectrolytes as reducing and stabilizing agents is a viable alternative in green chemistry to avoid the use of toxic

A. M. Herrera-González (✉) · J. García-Serrano
Laboratorio de Polímeros, Instituto de Ciencias Básicas e Ingeniería, Universidad Autónoma del Estado de Hidalgo, Hidalgo, Mexico
e-mail: mherrera@uaeh.edu.mx

M. Caldera-Villalobos
Doctorado en Ciencias de los Materiales, Instituto de Ciencias Básicas e Ingeniería, Universidad Autónoma del Estado de Hidalgo, Hidalgo, Mexico

M. C. Reyes-Ángeles
Maestría en Ciencias de los Materiales, Instituto de Ciencias Básicas e Ingeniería, Universidad Autónoma del Estado de Hidalgo, Hidalgo, Mexico

P. Chandra, R. Prakash (eds.), *Nanobiomaterial Engineering*,
https://doi.org/10.1007/978-981-32-9840-8_5

substances during synthesis. The choice of natural or bio-based polyelectrolytes like proteins and polysaccharides or their derivatives allows for the obtaining of stabilized nanoparticles with molecules that confer them biocompatibility and that facilitate their function of biological reconnaissance.

Keywords
Polyelectrolytes · Colloidal nanoparticles · Green synthesis · Metallic nanoparticles

5.1 Introduction

5.1.1 Polyelectrolytes

According to IUPAC, a polyelectrolyte is a polymer in which a substantial portion of the repetitive units that make it up contain ionic, ionizable, or both kinds of groups covalently bonded. These types of groups are also called ionogenic groups. Polyelectrolytes have both the properties of electrolytes and polymers, and these distinguish themselves from nonelectrolyte polymers by their physicochemical behavior in a solution. When polyelectrolytes are dissolved in water or in a solvent with a high dielectric constant, they are electrolytically dissociated, thus leaving a large amount of electrically charged groups along all the polymeric chain. According to their dissociative capacity, polyelectrolytes are classified as weak or strong. Weak polyelectrolytes dissociate partially in a solution. In this case, the dominant intermolecular interactions are van der Waals forces. Strong polyelectrolytes are totally dissociated, and the dominant intermolecular interactions are coulombic (Fig. 5.1).

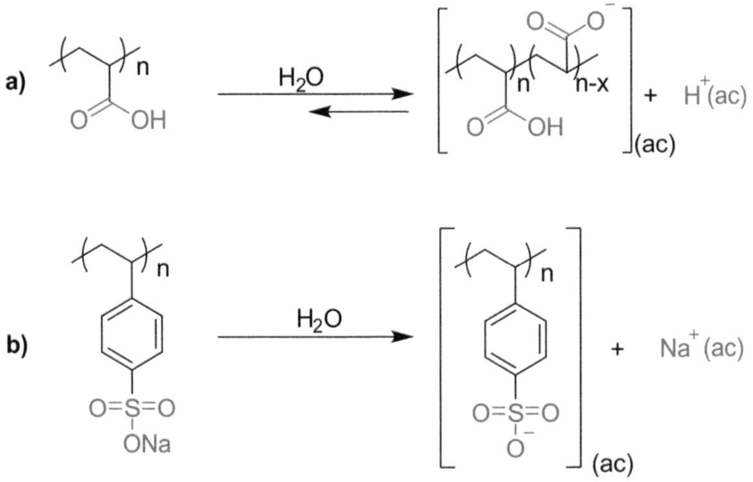

Fig. 5.1 Example of weak (**a**) and strong (**b**) polyelectrolytes

As a consequence of the dissociation, the polymeric chain of a polyelectrolyte may adopt different settings ranging from a random ball to a stretched chain depending on different factors such as the chemical nature of the solvent, the pH, the temperature, and the ionic force of the solution.

5.1.2 Classification of Polyelectrolytes

Polyelectrolytes may also be classified in accordance with the nature of the ionic group, and they may be polyacid, polybasic, polyampholytes, and polyzwitterions. Polyacids are polyelectrolytes that contain acid groups joined covalently such as carboxylic, sulfonic, phosphonic, and arsonic acids. The dissociation of these groups produces anionic groups, and, as such, they are known as polyanions. Polybasic polyelectrolytes or polycations contain amine groups which are either primary, secondary, tertiary, or quaternary. Primary, secondary, and tertiary may protonate, thus forming cations. Quaternary amine groups already possess a cationic character because of the quaternation of the nitrogen atom. Other cationic groups which can be found in polyelectrolytes are sulfonium, phosphonium, and cobaltocenium. Polyampholytes and polyzwitterions are polyelectrolytes which contain acid and basic groups within their structures. These differ in the fact that polyampholytes contain acid and basic groups in different repetitive units of the polymeric chain. Meanwhile, polyzwitterions contain an acid and a basic group within the same repetitive unit (Laschewsky 2012). Another important classification for polyelectrolytes is according to their origin. Polyelectrolytes may be natural, synthetic, and semisynthetic. Natural polyelectrolytes are found forming part of living organism, and they have different functions within them. Proteins, nucleic acids, and some polysaccharides are examples of natural polyelectrolytes. Proteins are polymers in which their monomers are amino acids which are linked to a peptide bond to form polymeric chains also known as polypeptides. Amino acids are carboxylic acids which contain an amine group, and the peptide bond is the link between a carboxyl group within an amino acid with the amine group in another amino acid to form an amide. There are 20 types of amino acids which form the proteins within living organisms, and these main contain other substitutes with an acid, basic, or neutral character (Fig. 5.2). Acid and basic substitutes give proteins the properties of a polyelectrolyte. Proteins are very important from a biological point of view because they perform several different functions within living organisms. They perform structural, hormonal, neurotransmissive, transport, restoration,

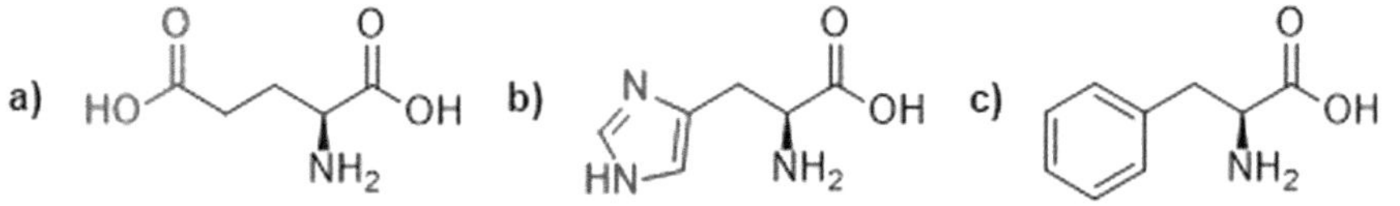

Fig. 5.2 Amino acids: (**a**) glutamic acid (acid character), (**b**) histidine (basic character), (**c**) phenylalanine (neutral character)

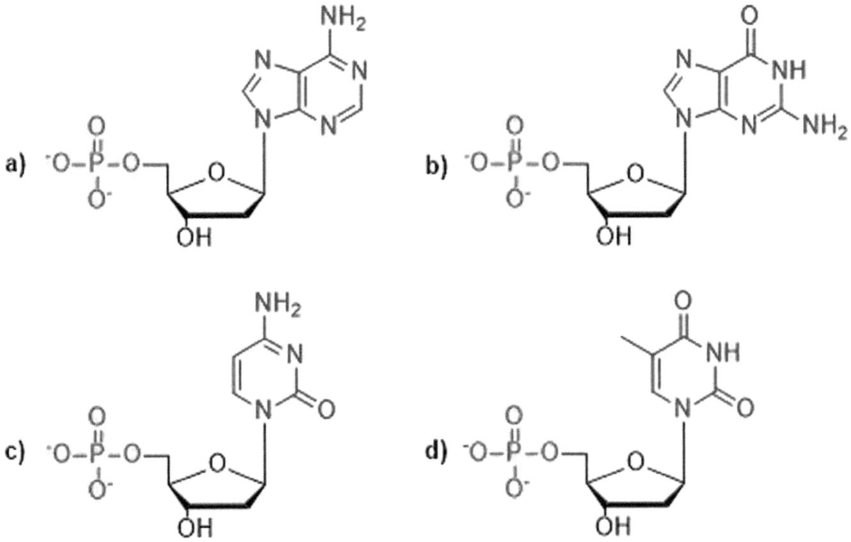

Fig. 5.3 Nucleotides: (**a**) adenine, (**b**) guanine, (**c**) cytosine, (**d**) thymine

enzymatic, or other functions. Nucleic acids such as ribonucleic (RNA) and deoxyribonucleic (DNA) acids are biopolymers responsible for storing, processing, and transmitting the genetic information within living organisms.

The monomers which form these polymers are called nucleotides, and they are made up of sugar, a heterocyclic nitrogenated base, and a phosphate group (Fig. 5.3). These acid and basic groups give polyelectrolytical properties to nucleic acids. Finally, some polysaccharides possess polyelectrolytical properties because they contain some ionogenic groups within their structure. Some examples of anionic polysaccharides are alginic acid, gum arabic, hyaluronic acid, pectin, polygalacturonic acid, agaropectin, and heparin. The majority of these natural polyelectrolytes have structural functions within plants. Because of their highly hydrophilic character, innocuity, and biodegradability, these polyelectrolytes are commonly used in the food industry as modifiers for viscosity and gelling agents. Heparin is found in mammals, and it has anticoagulant functions.

The range of synthetic polyelectrolytes available today is very wide. There is a large variety of monomers commercially available, and some synthetic polyelectrolytes such as poly(methacrylic acid), poly(acrylic acid), poly(sodium styrenesulfonate), and poly(diallyldimethylammonium chloride) are produced in great amounts. They have important applications such as superabsorbent materials, ionic exchangers, and ultrafiltration membranes. Semisynthetic polyelectrolytes are materials obtain from the chemical modification of a natural polymer. The reaction of chemical modification allows for the introduction of the ionogenic group within the structure of the natural polymer, thus conferring it with the properties of a polyelectrolyte. For example, starch can be transformed into an anionic polyelectrolyte by means of a phosphatation reaction (Dencs et al. 2004) or an esterification with an octenylsuccinic anhydride (Assaad et al. 2011). Starch may also become a polycation by making a chemical modification with substances containing quaternary amine groups (Haack et al. 2002). The applications for polyelectrolytes are very diverse. As well as those mentioned, we can cite catalysis, sensors, controlled administration of pharmaceuticals, metallic ion recovery and concentration, protonic conduction membranes for cell fuels, and formation and stabilization of metallic nanoparticles (García Serrano et al. 2008). Particularly, in the field of nanotechnology, polyelectrolytes are materials with great usefulness, and they are used as reducing agents and stabilizers in the synthesis of metallic nanoparticles.

5.2 Methods for Synthesis of Nanoparticles

Nanostructure is defined as that which measures 100 or less nm in at least one of its dimensions. Nanostructures are classified according to the number of dimensions that they have in a nanometric scale as follows: zerodimensional (0D), unidimensional (1D), and bidimensional (2D). In zerodimensional nanostructures, all three dimensions in the nanostructure are nanometric (height, width, and depth). Some of these nanostructures are nanoclusters, quantum dots, and nanoparticles. Unidimensional two dimensions of the nanostructures are nanometric (depth and width), and another dimension may have a variable length. Nanotubes and nanowires are examples. In bidimensional nanostructures, only one of the dimensions is nanometric. Superthin films and graphene are examples. Nowadays, interest in nanostructures occurs because they possess physical and chemical properties which are different from those the bulk material exhibits. The key aspect for these materials are fundamental properties such as magnetization, melting point, hardness, chemical reactivity, elasticity, optical and electrical properties, as well as others, which are modified by reducing the particle size until its nanometric without changing the chemical composition of the material. Today, there is a great interest in the synthesis and stabilization of metallic nanoparticles. These materials possess a range of very diverse applications such as biomedicine. Au nanoparticles have found applications in numerous fields, such as biosensors, antimicrobial agents, theranostics, cancer hyperthermia, and gene and drug delivery platforms (Baranwal et al. 2016) (Daniel and Astruc 2004). Ag nanoparticles have bactericide activity, and they have been used in several consumer products for imparting antimicrobial effect such as storage wares, textiles, nutritional additives, kitchen appliance surface coatings, and hospital consumables (Baranwal et al. 2018). Ag nanoparticles have also proven useful in the treatment of diseases such as cancer (Prabhu and Poulose 2012).

However, the development of these and other applications is still limited by the methods of synthesis which use toxic substance as reducing and stabilizing agents. These cause many of the materials to have low biocompatibility unfit to be applied to living systems. Because of this, there is an increasing interest in the development of methods of green synthesis that decrease the toxic effects of the substances used in the synthesis of nanoparticles, thus improving their biocompatibility. These methods of synthesis have the goal of controlling the shape and size of the nanoparticles to obtain a uniform range in all of them. There is a wide range of methods used in the synthesis of nanoparticles which may be classified in two: bottom-up and top-down. Top-down methods consist in the fragmentation of macroscopic (bulk) solids in smaller portions. The progressive reduction in size leads to nanostructures either by grinding, wear, or ablation. In general, these methods have little control over the size and shape of the nanostructures, and they require complex and expensive equipment. As such, in many cases, bottom-up methods are preferred. These consist in the fabrication of nanostructures by the ensemble of atoms or molecules in a liquid or gas phase. This method is more popular in the synthesis of nanostructures because of its ease of operation and low cost, as well as the variety of precursor substances which permits the synthesis of different nanomaterials. There are diverse methods based on the bottom-up approach. The most common are those which use chemical procedures. The most representative are sol-gel, pyrolysis, chemical precipitation, solvothermal, micellar, and colloidal. Among these, the colloidal method is one of the most commonly used, and it is described in detail in the following section.

5.2.1 Colloidal Synthesis of Nanoparticles

Colloids are mixtures that do not exhibit properties from a suspension or a true solution. They are made up of a group of fine particles dispersed throughout a continuous or dispersant phase. Colloids possess properties such as the ability to disperse a ray of light that passes through them and Brownian movement. These properties are caused by the particle size which ranges between 1 and 100 nm. Then the size of the particles is less than 1 nm; the behavior of the mixture is that of a true solution. When it is greater than 100 nm, it is that of a suspension. Many colloids are stabilized by adding a material which covers the nanoparticles suspended, and it stops their agglomeration. For example, soap yields a colloidal suspension of oil in water, thus forming an emulsion. The molecules of soap form a negatively charged layer which produces electrostatic repulsion between the particles of oil. This repulsion stops the agglomeration of the particles of oil, which, when set together, would end up separating from water. In milk, the particles of fat suspended are stabilized by a protecting layer of casein. Casein is a protein and a natural polyelectrolyte, and because of the electrostatic repulsion produced by the ionogenic groups within its structure, it can stabilize the colloidal suspension. The range of particle sizes that form colloids coincides with the range of sizes for nanomaterials. For this reason, the synthesis of nanomaterials in colloidal solutions tends to be an efficient method for their preparation and stabilization, and it is known as colloidal method. This method was systematically implemented by Turkevich and Nord. Turkevich made the synthesis of nanoparticles using sodium citrate as a reducing agent and stabilizer, thus obtaining metallic Au nanoparticles in colloidal solutions using hydrophilic polymers such as poly(vinylic alcohol) (PVA), polyethylene glycol (PEG), and polyvinylpyrrolidone (PVP) as stabilizers. The colloidal synthesis of nanoparticles consists in dissolving a precursor for the metal to be prepared, a reducer, and a stabilizer in a continuous or

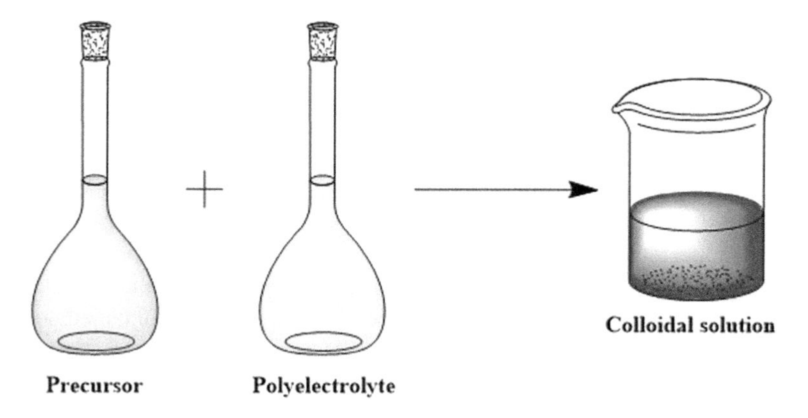

Fig. 5.4 The colloidal synthesis of nanoparticles

dispersing (liquid) phase (Fig. 5.4). With this method, the average size, the size distribution, and the shape of the nanoparticles can be controlled via the variation of the concentrations of the precursor, the reducer, or the stabilizer as well as the nature of the dispersing medium. Stable colloidal suspensions can also be obtained for very long periods of time. In a colloidal system, the nanoparticles which are small enough remain stable because of the Brownian movement which is an irregular movement attributed to the collective bombing of a multitude of particles which are thermally agitated and suspended in a liquid phase. When the nanoparticles are large enough, their dynamic behavior depending on the time is governed by gravitational forces, and sedimentation is produced. The generation of stable colloidal solutions which contain nanoparticles with controlled size and shape requires a synthetic strategy based on the perfect knowledge of the intimate relationship that exists between the surface of the particles and the stabilizing agent.

There is a great variety of substances that may be used as precursors in the synthesis of metallic nanoparticles, mainly salts and organometallic complexes. The colloidal method is very versatile and allows for the obtaining of nanoparticles of a wide range of metals. Examples are gold, silver (Angshuman Pal et al. 2007), samarium, lanthanum (Nomoto et al. 2009), copper (Dang et al. 2011), ytterbium (Pan et al. 2012), nickel (Ely et al. 1999), platinum, ruthenium, palladium, iridium (Bonet et al. 1999), and rhodium (Choukroun et al. 2001). The reduction of the precursors may be done by means of different strategies. In the conventional processes of chemical reduction, well-known reducing agents are employed such as sodium citrate ($C_{12}H_{13}O_{13}Na_3$) (Yang et al. 2006; Angshuman Pal et al. 2007), hydrazine (N_2H_4), sodium borohydride ($NaBH_4$), and ascorbic acid ($C_6H_8O_6$) (Gurav et al. 2014; Medina Ramírez et al. 2009). The reduction by means of radiochemical methods employs different kinds of radiation such as microwaves (Gupta et al. 2018), ultraviolet radiation (Mallick et al. 2005), as well as gamma rays (Dhayagude et al. 2018). Reduction may also be made using electrochemical methods (Marius et al. 2011).

Stabilizers in the reaction medium have two functions: they limit the growth of the nanoparticles (surface passivation), and they stabilize the system, thus stopping its aggregation. Stabilizers must have a high solubility in the reaction medium, and they should induce repulsions between the nanoparticles. There is a wide range of substances and protecting agents used in the stabilization of metallic nanoparticles. These have been classified in seven categories: microorganisms and bacteria, plant extracts and physiological molecules, inorganic reagent and metallic complexes organic molecules, organic acids and their salts, liposomes, and polymers. The molecules of surfactant interact with the individual atoms in the surface of the nanoparticles not unlike to what is observed between ligands and metallic atoms deficient in electrons of the metallic complexes. Hence, it is a requirement for the surfactants to have at least an atom which contains free electron pairs. The colloidal method has a certain superiority in the stabilization of metallic nanoparticles because of the use of polymers as stabilizing agents. Polymers may form a protective layer around the nanoparticles, and they provide a robust barrier against agglomeration even in extremely acid or alkaline conditions as well as high temperatures (Dumur et al. 2011). Polymers are common stabilizing agents in the colloidal synthesis of metallic nanoparticles, the main ones being hydrophilic polymers such as poly(ethylene oxide) (PEO), polyvinyl alcohol (PVA), and polyvinylpyrrolidone (PVP). As well as these polymers, there is a wide range of polymers used as stabilizers in the colloidal method. These were classified by Zhang et al. in four categories: non-electrolytic homopolymers, copolymers, dendrimers, and polyelectrolytes (Zhang et al. 2013).

Non-electrolytic homopolymers are used generally in the presence of a reducing agent such as $NaBH_4$. These are adsorbed on the surface of the nanoparticles throughout the polymeric chains, thus protecting the nanoparticles by means of steric effects. Some examples of these polymers are PVP and PVA. Copolymers used in the colloidal synthesis of nanoparticles have structures like a block copolymer, and they are composed of two or more segments formed by hydrophilic and hydrophobic monomers. The hydrophilic segment interacts with the metallic ions, and once it reduces them, the hydrophobic segment is adsorbed on the surface of the aggregates, thus acting as a protective barrier which controls the growth. For these copolymers, there is generally no use of reducing agents required. An example of this kind of copolymers is poly(ethylene oxide-co-methacrylic acid) (PEO-co-PMAA). Dendrimers control the growth of the nanoparticles through their ramified structure by means of steric effects, while the functional groups present produce selectivity effects. The large amount of polar functional groups in a dendrimer represents its main advantage against lineal polymers. An example of a dendrimer used as a nanoparticle-stabilizing agent is polyamidoamine (PAMAM).

Because of the absence of ionogenic groups, the polymers formerly described stabilize nanoparticles only through steric effects. Meanwhile, polyelectrolytes may stabilize nanoparticles by a combination of steric and electrostatic effects. Stabilization by steric effects occurs because of the high molecular mass of the polyelectrolytic chain which is found adsorbed on the surface of the nanoparticles, thus creating a physical barrier which stops agglomeration. Stabilization by electrostatic effects is caused by the presence of acid or basic ionogenic groups which, when dissociated, create an electrostatic field which induces coulombic repulsion between the nanoparticles, thus stopping their agglomeration. Stabilization through the combination of both effects poses the main advantage of using polyelectrolytes as stabilizing agents against other kinds of polymers. As well as being adequate stabilizing agents, polyelectrolytes can also act as reducing agents, thus avoiding the use of reducing agents such as NaBH4 and N2H2 during the synthesis of the nanoparticles. Polyelectrolytes have an adequate solubility in an aqueous medium because of their ability for dissociation. For this reason, the synthesis of metallic nanoparticles using polyelectrolytes is made in an aqueous medium (neutral, basic, or acid), and it avoids the use of toxic solvents such as *N*,*N*-dimethylformamide, THF, dichloromethane, and toluene. The synthesis of nanoparticles using polyelectrolytes is considered a green chemical method, and it decreases the toxicological and pharmacological hazards associated with the use of the aforementioned toxic substances. This characteristic turns them in materials with potential to be used in biomedical applications.

5.2.2 Green Synthesis of Metallic Nanoparticles Using Polyelectrolytes

There is an extensive amount of reports about the synthesis and stabilization of metallic nanoparticles using conventional polyelectrolytes. Examples of these are poly(acrylic acid), poly(methacrylic acid), poly(sodium styrenesulfonate), poly(dialyldimethylammonium chloride), and poly(sodium methacrylate), as well as some of their copolymers. Nowadays, the design and synthesis of new polyelectrolytes is an expanding research line which has the goal of developing biomedical applications of nanoparticles stabilizing using polyelectrolytes. For this, the design of useful polyelectrolytes in the synthesis of nanoparticles should consider the nature of the ionogenic groups and their dissociation capacity, the molecular mass of the polyelectrolyte, and the presence of oxidizable groups to accomplish the reduction of the metallic ions. Recent research has been focused on the design of new anionic polyelectrolytes and macroelectrolytes with reducing agent properties for the synthesis and stabilization of metallic nanoparticles. These polyelectrolytes contain imine or amide groups to give them the properties of a reducing agent. The properties of a stabilizing agent and the hydrophilic character are given by the sulfonic and arsonic acid groups (Herrera González et al. 2016; Caldera-Villalobos et al. 2018). These polyelectrolytes are useful in the green synthesis of nanoparticles, and they are a viable alternative to $NaBH_4$ and N_2H_4. Because of their high reactivity, polyelectrolytes reduce metallic ions from the first minutes of the reaction, and they form stable colloidal solutions with nanoparticles with controlled size and shape (Herrera-González et al. 2018a). The structural design of the polyelectrolytes and macroelectrolytes allows for the selective synthesis of nanoparticles with quasi-spherical shapes or anisotropic nanostructures (Fig. 5.5) such as triangular and hexagonal nanoplates as well as decahedrons and dodecahedrons (Herrera-González et al. 2018b). This way, the green synthesis of metallic nanoparticles using polyelectrolytes allows for some aspects to be solved such as the control on

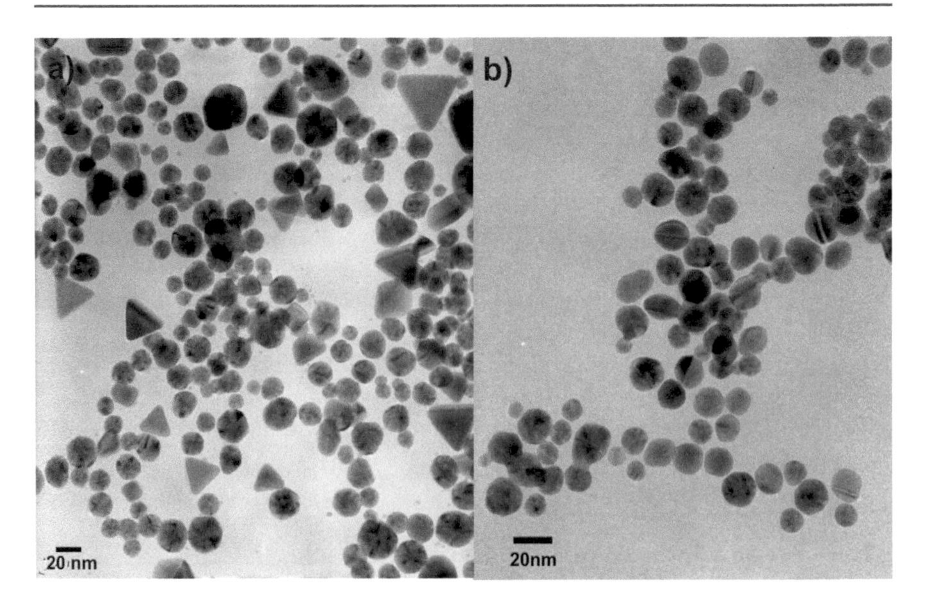

Fig. 5.5 Colloidal nanoparticles stabilized with bio-based polyelectrolytes: (**a**) anisotropic, (**b**) quasi-spherical

the shape and size of the nanoparticles and the elimination of dangerous substances during the reaction. For biomedical applications, other factors to consider are the biodegradability and biocompatibility of the polyelectrolytes as well as their dissociation in the physiological environment.

5.2.3 Synthesis of Nanoparticles Using Natural and Bio-Based Polyelectrolytes

There is a wide range of natural polyelectrolytes that may be used in the green synthesis of metallic nanoparticles (Fig. 5.6). The use of molecules of biological origin such as DNA, proteins, amino acids, polysaccharides, and vitamins for the synthesis of nanoparticles is very attractive to obtain new nanomaterials with potential biomedical and bioanalytical applications.

The modification of natural polymers may improve the performance of these polymers during the synthesis of metallic nanoparticles. This way, the quantity of nanoparticles obtained, their polydispersity, and their stability in a colloidal solution may improve. For example, the chemical modification of chitosan by means of grafting *N*-vinylpyrrolidone or ε-caprolactone helps improving its performance as a nanoparticle stabilizer (Leiva et al. 2015). The use of semisynthetic molecules or polymers may improve the interactions between the nanoparticles and the molecules or polymers used without a loss in biocompatibility and bioactivity. For example, the chemical modification of chitosan to obtain ampholytic derivatives improves its compatibility with the sanguine environment (Amiji 1998). The chemical

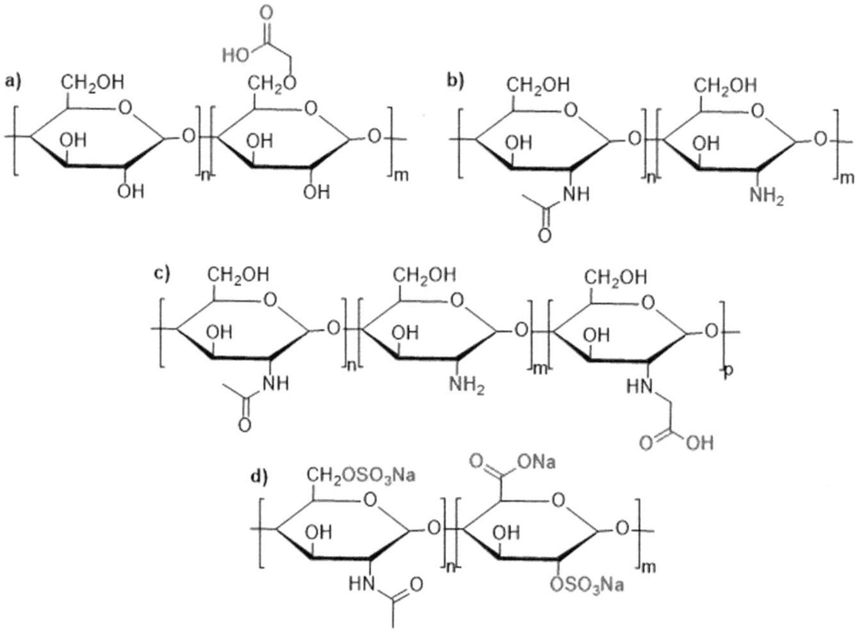

Fig. 5.6 Polysaccharides useful in green synthesis of metallic nanoparticles: (**a**) carboxymethylcellulose, (**b**) chitosan, (**c**) *N*-carboxymethylchitosan, (**d**) heparin

modification of chitosan by carboxylmethylation improves the function of this polyelectrolyte as a stabilizing agent of Au nanoparticles. By controlling the degree of modification, it is possible to conserve -OH and -NH_2 groups which are reducing agents for Au^{3+} ions. This way, the polyelectrolyte might perform the functions of the reducing or stabilizing agents (Sun et al. 2018). The chemical modification of chitosan using formylbenzenesulfonates improves the properties as a reducing agent of this polyelectrolyte, and it confers the properties of a polyampholyte. This allows it to interact with anionic and cationic species and to dissociate in a wider range of pH values (Caldera-Villalobos et al. 2017). As mentioned, one of the characteristics of a polyelectrolyte to be applied in the green synthesis of nanoparticles is the ability to reduce metallic ions. For this, the polyelectrolyte must contain oxidizable groups (Park et al. 2011). For example, some polysaccharides such as heparin, hyaluronic acid (Kemp et al. 2009b), alginic acid (Pal et al. 2005), and carboxymethylcellulose (Hebeish et al. 2010) have a large amount of -OH oxidizable groups and may be employed as reducing agents and stabilizers in the synthesis of nanoparticles. These polysaccharides also have polyelectrolytic properties because of the presence of anionic groups.

Hence, it is possible to apply natural and semisynthetic polyelectrolytes in the synthesis of metallic nanoparticles to be used in biomedical applications. For examples, the nanoparticles of Au and Ag stabilized with heparin derivatives have antiangiogenic properties, and they have a potential application in the treatment of diseases such as cancer and inflammatory disorders (Kemp et al. 2009a). Travan et al. reported the stabilization of Ag nanoparticles with a bioactive polysaccharide derived from chitosan. The nanocomposite obtained exhibits an efficient bactericide activity without the cytotoxic effects produced on some eukaryotic cells (Travan et al. 2009). Nucleic acids (DNA and RNA) may interact with the surface of metallic nanoparticles through nitrogenated heterocyclic bases or phosphodiester. Nucleic acids are capable of reducing metallic ions and stabilizing metallic nanoparticles by means of steric and electrostatic effects. An interesting aspect of nucleic acids is that properties of the colloidal nanoparticles may be modified according to the sequence of nucleotides employed. Nanoparticles stabilized with nucleic acids are biocompatible and apt for biological reconnaissance functions (Berti and Burley 2008). Different nucleotides such as adenosine triphosphate may also be employed as stabilizing agents for metallic nanoparticles (Zhao et al. 2007). Nanoparticles stabilized with oligonucleotides may interact between themselves forming supramolecular systems which self-assemble reversibly. The distance between particles within the system may be modulated by varying the sequence and length of the oligonucleotide chain as well as the temperature, the solvent, or the concentration of the supporting electrolyte. These properties are useful in the design of sensing systems (Mirkin et al. 1996).

5.3 Biomedical Applications of Colloidal Nanoparticles Stabilized Using Polyelectrolytes

Metallic nanoparticles have awoken interest also because of their potential for biomedical use. The use of colloidal materials poses an easy way of administration and manipulation for these materials in living organisms. Colloidal nanoparticles of gold have been studied as vectors in the administration of pharmaceuticals for the treatment of tumors (Paciotti et al. 2006), as immunosensors (Sadowski and Maliszewska 2011), or as DNA sequence colorimetric detectors (Li and Rothberg 2004). Transferrin is an animal protein responsible for the transport of iron in plasma. This protein is capable of reducing Au^{3+} ions and forming colloidal Au nanoparticles with a quasi-spherical shape. The interaction between Au nanoparticles and transferrin increases the exposition of hydrophobic groups in the protein towards the surface, and it changes its conformation. As a result of this conformational change, transferrin gains the capacity of interacting with monolayers of lipids or with cell walls (McDonagh et al. 2015). Conformational changes in proteins induced by Au nanoparticles have also been observed in bovine seroalbumin, bovine α-lactalbumin, and lysozyme. Au nanoparticles stabilized with these proteins also possess properties such as fluorescence and localized plasmon resonance, which are useful in biosensing applications (Lystvet et al. 2013). Wu et al. reported the construction of nanoprobes formed by Au nanoparticles and Au nanocumuli used as substrates to produce localized surface plasmon resonance. These nanoprobes exhibit fluorescence and enhanced Raman scattering, and they may be used as fluorescent markers in the early detection of cancer. These nanoprobes are stabilized using bovine seroalbumin to improve their biocompatibility, and their surface is functionalized by an oligopeptide which improves selectivity during detection. This kind of probes serves in the detection of tumors formed by a very small number of cells, thus aiding in the early diagnosis of the disease (Wu et al. 2018). The antimicrobial properties of Au nanoparticles and some peptides are well known. Au nanoparticles stabilized with antibacterial peptides produces a synergic effect in its antibacterial properties. This represents a viable alternative to fight bacteria which exhibit resistance to antibiotics (Rajchakit and Sarojini 2017).

It is important to remember that the properties of the nanoparticles do not depend solely on their size but also on their shape. As such, when the shape of the nanoparticle changes, these may become useful in different applications. Anisotropic Au nanoparticles are interesting in biosensor design because of their optical properties. Different anisotropic nanostructures such as Au nanorods, nanourchins, and nanocages have been applied in the detection of tumor cells, quantitative determination of proteins such as kinase A, detection of enterovirus 71 and influenza virus, and determination of poly(ADP-ribose) polymerase-1. These sensing systems are based on phenomena such as fluorescence quenching, surface plasmon resonance, or surface-enhanced Raman scattering produced by the anisotropic nanostructures, or they make use of colorimetric methods. The development of these applications is closely linked to the shape of the nanoparticles and their particular optical properties (Kohout et al. 2018). Silver nanoparticles are also studied because of their antibacterial activity (Panacek et al. 2006; Sondi and Salopek Sondi 2004; Sharma et al. 2009; Choi et al. 2008). Casolaro et al. reported the stabilization of silver nanoparticles using hydrogels prepared with polyelectrolytes derived from proteic amino acids L-phenylalanine, L-histidine, and L-valine. Hydrogels loaded with Ag nanoparticles exhibited antibacterial activity against Gram-positive and Gram-negative bacteria as well as antifungal against *S. cerevisiae*. The use of bio-based monomers to obtain hydrogels allows for the obtaining of materials with greater biocompatibility which could be used in the treatment of contact lenses to prevent fungal keratitis (Casolaro et al. 2018). Phycoerythrin is a protein responsible for the catching of light in cyanobacteria which has been isolated and used in the synthesis of silver nanoparticles. This protein is capable of reducing Ag^+ and forming and stabilizing spherical Ag nanoparticles in colloidal solutions. The nanoparticles which were obtained exhibit cytotoxic activity against tumor cells according to in vitro experiments made. The use of Ag nanoparticles stabilized using phycoerythrin allows to decrease the size of tumor cells as well as to reduce the size of tumors by means of photothermal therapy. These nanoparticles also possess antihemolytic properties (El Naggar et al. 2018). Similarly, the protein pigment known as phycocyanin is capable of reducing Ag^+ ions and forming stable colloidal solutions of quasi-spherical Ag nanoparticles. These nanoparticles exhibit anticarcinogenic against MCF-7 cell lines, and in vivo research shows that they may inhibit tumor growth (El-naggar et al. 2017). Galactomannan is a polysaccharide isolated from the skin of *Punica granatum* fruit, and it possesses antioxidant, immunomodulatory, and anticarcinogenic properties. This polysaccharide has been used in the colloidal synthesis of Ag nanoparticles acting as a reducing and stabilizing agent. Quasi-spherical nanoparticles stabilized using this polysaccharide exhibited selectivity against human adenocarcinoma cells, colorectal carcinoma, and hepatocellular carcinoma. Because of the nontoxic nature of this polysaccharide, nanoparticles stabilized with it possess high biocompatibility with blood cells (Padinjarathil et al. 2018). Antibacterial activity has also been observed in Ag nanoparticles stabilized with a semisynthetic polysaccharide obtained from tamarind. The polysaccharide possesses properties like an anionic polyelectrolyte because of the presence of carboxymethyl, and it may be used in the green synthesis of Ag nanoparticles. These nanoparticles exhibit antibacterial activity against Gram-positive and Gram-negative bacteria which are resistant to antibiotics (Sanyasi et al. 2016). This property may also be observed in Ag nanoparticles stabilized using sodium alginate against *Staphylococcus aureus* and *Escherichia coli*. The antibacterial properties of these nanoparticles were attributed to the cellular death induced by the increase in the permeability of the cell membrane and the disruption of the cell wall (Shao et al. 2018).

An interesting property of silver nanoparticles is their larvicide activity. Mane et al. reported the colloidal synthesis of Ag nanoparticles using fibroin as reducing and stabilizing agent. The films formed by evaporation of the colloidal solution showed a larvicide activity against mosquito *Aedes aegypti* larvae. The observed mortality rate was 100%, and it contributes in fighting the transmission vector of the Zika virus (Mane et al. 2017). Lee et al. reported the functionalization of silver nanoparticles with oligonucleotides using disulfide groups. The chains of oligonucleotides contain sequences which are complimentary to DNA that, when combined, allow for the reversible agglomeration of nanoparticles, thus giving them optical properties which depend on their distance (Lee et al. 2007). This way, these nanoparticles may be applied as a programmable material used in diagnosis. Although we have focused on Au and Ag colloidal nanoparticles, other metallic elements may also have biomedical applications. Antibacterial activity is a property which has also been observed in copper nanoparticles (Ruparelia et al. 2008). Colloidal iron particles also have bactericide properties against Gram-positive and Gram-negative bacteria. They also have antifungal activity against some species of Candida (Da'na et al. 2018). Pan et al. have reported the synthesis of colloidal ytterbium nanoparticles stabilized using biotin derivatives. Yb nanoparticles showed great potential as a contrast agent in computer tomography (Pan et al. 2012). Summing up, colloidal synthesis of metallic nanoparticles is a method which offers many advantages obtaining nanomaterials with biomedical applications. The use of polyelectrolytes as reducing and stabilizing agents is a viable alternative in green chemistry to avoid the use of toxic substances during synthesis. The choice of natural or bio-based polyelectrolytes like proteins, polysaccharides, and nucleic acids or their derivatives allows for the obtaining of stabilized nanoparticles with molecules that confer them with biocompatibility and that facilitate their function of biological reconnaissance. This allows for the use of metallic nanoparticles in biomedical applications such as diagnosis and genome therapy. The combination of the properties of the nanoparticles and the polyelectrolytes also allows for an improvement in their antibacterial, antifungal, and antilarval properties of the nanoparticles.

References

Amiji MM (1998) Platelet adhesion and activation on an amphoteric chitosan derivative bearing sulfonate groups. Colloids Surf B: Biointerfaces 10(5):263–271. https://doi.org/10.1016/S0927-7765(98)00005-8

Assaad E, Wang YJ, Zhu XX, Mateescu MA (2011) Polyelectrolyte complex of carboxymethyl starch and chitosan as drug carrier for oral administration. Carbohydr Polym 84(4):1399–1407. https://doi.org/10.1016/j.carbpol.2011.01.048

Baranwal A, Mahato K, Srivastava A, Maurya PK, Chandra P (2016) Phytofabricated metallic nanoparticles and their clinical applications. RSC Adv 6(107):105996–106010. https://doi.org/10.1039/C6RA23411A

Baranwal A, Srivastava A, Kumar P, Bajpai VK, Maurya PK, Chandra P (2018) Prospects of nanostructure materials and their composites as antimicrobial agents. Front Microbiol 9:422. https://doi.org/10.3389/fmicb.2018.00422

Berti L, Burley GA (2008) Nucleic acid and nucleotide-mediated synthesis of inorganic nanoparticles. Nat Nanotechnol 3(2):81–87. https://doi.org/10.1038/nnano.2007.460

Bonet F, Delmas V, Grugeon S, Herrera Urbina R, Silvert PY, Elhesissen T (1999) Synthesis of monodisperse Au, Pt, Pd, Ru and Ir nanoparticles in ethylene glycol. Nanostruct Mater 11(8):1277–1284. https://doi.org/10.1016/S0965-9773(99)00419-5

Caldera-Villalobos M, García-Serrano J, Peláez-Cid AA, Herrera-González AM (2017) Polyelectrolytes with sulfonate groups obtained by chemical modification of chitosan useful in green synthesis of Au and Ag nanoparticles. J Appl Polym Sci 134(38):45240. https://doi.org/10.1002/app.45240

Caldera-Villalobos M, Herrera-González AM, García-Serrano J (2018) Polyelectrolytes based on poly(p-acryloyloxybenzaldehyde) with arsonic acid groups useful in the colloidal synthesis of silver nanoparticles. J Polym Res 25(8):190. https://doi.org/10.1007/s10965-018-1582-7

Casolaro M, Casolaro I, Akimoto J, Ueda M, Ueki M, Ito Y (2018) Antibacterial properties of silver nanoparticles embedded on polyelectrolyte hydrogels based on α-amino acid residues. Gels 4(2):42. https://doi.org/10.3390/gels4020042

Choi O, Deng KK, Kim NJ, Ross L, Surampalli RY, Hu Z (2008) The inhibitory effects of silver nanoparticles, silver ions, and silver chloride colloids on microbial growth. Water Res 42(12):3066–3074. https://doi.org/10.1016/j.watres.2008.02.021

Choukroun R, de Caro D, Chaudret B, Lecante P, Snoeck E (2001) H2-induced structural evolution in non-crystalline rhodium nanoparticles. New J Chem 25(4):525–527. https://doi.org/10.1039/B009192H

Da'na E, Taha A, Afkar E (2018) Synthesis of iron nanoparticles by Acacia nilotica pods extract and its catalytic, adsorption, and antibacterial activities. Appl Sci 8(10):1922. https://doi.org/10.3390/app8101922

Dang TMD, Le TTT, Fribourg-Blanc E, Dang MC (2011) Synthesis and optical properties of copper nanoparticles prepared by a chemical reduction method. Adv Nat Sci Nanosci Nanotechnol 2(1):015009. https://doi.org/10.1088/2043-6262/2/1/015009

Daniel MC, Astruc D (2004) Gold nanoparticles: assembly , supramolecular chemistry , quantum-size-related properties, and applications toward biology, catalysis, and nanotechnology. Chem Rev 104(1):293–346

Dencs J, Nos G, Bencs B, Marton G (2004) Investigation of solid-phase starch modification reactions. Chem Eng Res Des 82(2):215–219. https://doi.org/10.1205/026387604772992792

Dhayagude AC, Das A, Joshi SS, Kapoor S (2018) γ-Radiation induced synthesis of silver nanoparticles in aqueous poly(N-vinylpyrrolidone) solution. Colloids Surf A 556:148–156. https://doi.org/10.1016/j.colsurfa.2018.08.028

Dumur F, Guerlin A, Dumas E, Bertin D, Gigmes D, Mayer CR (2011) Controlled spontaneous generation of gold nanoparticles assisted by dual reducing and capping agents. Gold Bull 44(2):119–137. https://doi.org/10.1007/s13404-011-0018-5

El Naggar NEA, Hussein MH, El Ssawah AA (2018) Phycobiliprotein-mediated synthesis of biogenic silver anticancer activities. Sci Rep 8(1):8925. https://doi.org/10.1038/s41598-018-27276-6

El-naggar NE, Hussein MH, El-sawah AA (2017) Bio-fabrication of silver nanoparticles by phycocyanin , characterization, in vitro anticancer activity against breast cancer cell line and in vivo cytotxicity. Sci Rep 7(1):10844. https://doi.org/10.1038/s41598-017-11121-3

Ely TO, Amiens C, Chaudret B, Snoeck E, Verelst M, Cedex T et al (1999) Synthesis of nickel nanoparticles. Influence of aggregation induced by modification of poly(vinylpyrrolidone) chain length on their magnetic properties. Chem Mater 11(3):526–529. https://doi.org/10.1021/cm980675p

García Serrano J, Pal U, Herrera AM, Salas P, Ángeles Chávez C (2008) One-step "green" synthesis and stabilization of Au and Ag nanoparticles using ionic polymers. Chem Mater 20(13):5146–5153. https://doi.org/10.1021/cm703201d

Gupta SSR, Kantam ML, Bhanage BM (2018) Shape-selective synthesis of gold nanoparticles and their catalytic activity towards reduction of p-nitroaniline. Nano Struct Nano Objects 14:125–130. https://doi.org/10.1016/j.nanoso.2018.01.017

Gurav P, Naik SS, Ansari K, Srinath S, Kishore KA, Setty YP, Sonawane S (2014) Stable colloidal copper nanoparticles for a nanofluid: production and application. Colloids Surf A Physicochem Eng Asp 441:589–597. https://doi.org/10.1016/j.colsurfa.2013.10.026

Haack V, Heinze T, Oelmeyer G, Kulicke WM (2002) Starch derivatives of high degree of functionalization, 8 synthesis and flocculation behavior of cationic starch polyelectrolytes. Macromol Mater Eng 287(8):495–502. https://doi.org/10.1002/1439-2054(20020801)287:8<495::AID-MAME495>3.0.CO;2-K

Hebeish AA, El Rafie MH, Abdel Mohdy FA, Abdel Halim ES, Emam HE (2010) Carboxymethyl cellulose for green synthesis and stabilization of silver nanoparticles. Carbohydr Polym 82(3):933–941. https://doi.org/10.1016/j.carbpol.2010.06.020

Herrera González AM, Caldera Villalobos M, García-Serrano J, Peláez Cid AA (2016) Polyelectrolytes with sulfonic acid groups useful in the synthesis and stabilization of Au and Ag nanoparticles. Des Monomers Polym 19(4):330–339. https://doi.org/10.1080/15685551.2016.1152543

Herrera-González AM, Caldera-Villalobos M, Bocardo-Tovar PB, García-Serrano J (2018a) Synthesis of gold colloids using polyelectrolytes and macroelectrolytes containing arsonic moieties. Colloid Polym Sci 296(5):961–969. https://doi.org/10.1007/s00396-018-4309-8

Herrera-González AM, García-Serrano J, Caldera-Villalobos M (2018b) Synthesis and stabilization of Au nanoparticles in colloidal solution using macroelectrolytes with sulfonic acid groups. J Appl Polym Sci 135(8):45888. https://doi.org/10.1002/app.45888

Kemp MM, Kumar A, Mousa S, Dyskin E, Yalcin M, Ajayan P et al (2009a) Gold and silver nanoparticles conjugated with heparin derivative possess anti-angiogenesis properties. Nanotechnology 20(45):455104. https://doi.org/10.1088/0957-4484/20/45/455104

Kemp MM, Kumar A, Mousa S, Park T, Ajayan P, Kubotera N et al (2009b) Synthesis of gold and silver nanoparticles stabilized with glycosaminoglycans having distinctive biological activities. Biomacromolecules 10(3):589–595. https://doi.org/10.1021/bm801266t

Kohout C, Santi C, Polito L (2018) Anisotropic gold nanoparticles in biomedical applications. Int J Mol Sci 19(11):3385. https://doi.org/10.3390/ijms19113385

Laschewsky A (2012) Recent trends in the synthesis of polyelectrolytes. Curr Opin Colloid Interf Sci 17(2):56–63. https://doi.org/10.1016/j.cocis.2011.08.001

Lee JS, Lytton-Jean AKR, Hurst SJ, Mirkin CA (2007) Silver nanoparticle-oligonucleotide conjugates based on DNA with triple cyclic disulfide moieties. Nano Lett 7(7):2112–2115. https://doi.org/10.1021/nl071108g

Leiva A, Bonardd S, Pino M, Saldías C, Kortaberria G, Radic D (2015) Improving the performance of chitosan in the synthesis and stabilization of gold nanoparticles. Eur Polym J 68:419–431. https://doi.org/10.1016/j.eurpolymj.2015.04.032

Li H, Rothberg L (2004) Colorimetric detection of DNA sequences based on electrostatic interactions with unmodified gold nanoparticles. Proc Natl Acad Sci 101(39):14036–14039. https://doi.org/10.1073/pnas.0406115101

Lystvet SM, Volden S, Singh G, Yasuda M, Halskau O, Glomm WR (2013) Tunable photophysical properties, conformation and function of nanosized protein-gold constructs. RSC Adv 3(2):482–495. https://doi.org/10.1039/c2ra22479h

Mallick K, Witcomb MJ, Scurrell MS (2005) Polymer-stabilized colloidal gold: a convenient method for the synthesis of nanoparticles by a UV-irradiation approach. Appl Phys A 80:395–398. https://doi.org/10.1007/s00339-003-2298-y

Mane PC, Chaudhari RD, Shinde MD, Kadam DD, Song CK, Amalnerkar DP, Lee H (2017) Designing ecofriendly bionanocomposite assembly with improved antimicrobial and potent on-site zika virus vector larvicidal activities with its mode of action. Sci Rep 7(1):15531. https://doi.org/10.1038/s41598-017-15537-9

Marius S, Lucian H, Marius M, Daniela P, Irina G, Simona D, Viorel M (2011) Enhanced antibacterial effect of silver nanoparticles obtained by electrochemical synthesis in poly (amide-hydroxyurethane) media. J Mater Sci Mater Med 22:789–796. https://doi.org/10.1007/s10856-011-4281-z

McDonagh BH, Volden S, Lystvet SM, Singh G, Ese MHG, Ryan JA et al (2015) Self-assembly and characterization of Transferrin-gold nanoconstructs and their interaction with bio-interfaces. Nanoscale 7(17):8062–8070. https://doi.org/10.1039/C5NR01284H

Medina Ramírez I, González García M, Liu JL (2009) Nanostructure characterization of polymer-stabilized gold nanoparticles and nanofilms derived from green synthesis. J Mater Sci 44(23):6325–6332. https://doi.org/10.1007/s10853-009-3871-3

Mirkin CA, Letsinger RL, Mucic RC, Storhoff JJ (1996) A DNA-based method for rationally assembling nanoparticles into macroscopic materials. Nature 382(6592):607. https://doi.org/10.1038/382607a0

Nomoto A, Kido A, Kakiuchi K, Mitani I, Tatsumi M, Ogawa A (2009) Hydrophilic polymer supported nanoparticles prepared from samarium trichloride and lanthanum trichloride. Research on Chemical Intermerdiates 35:1027–1032. https://doi.org/10.1007/s11164-009-0084-y

Paciotti GF, Kingston DGI, Tamarkin L (2006) Colloidal gold nanoparticles: a novel nanoparticle platform for developing multifunctional tumor-targeted drug delivery vectors. Drug Dev Res 67(1):47–54. https://doi.org/10.1002/ddr.20066

Padinjarathil H, Joseph MM, Unnikrishnan BS, Preethi GU, Shiji R, Archana MG et al (2018) Galactomannan endowed biogenic silver nanoparticles exposed enhanced cancer cytotoxicity with excellent biocompatibility. Int J Biol Macromol 118:1174–1182. https://doi.org/10.1016/j.ijbiomac.2018.06.194

Pal A, Esumi K, Pal T (2005) Preparation of nanosized gold particles in a biopolymer using UV photoactivation. J Colloid Interface Sci 288:396–401. https://doi.org/10.1016/j.jcis.2005.03.048

Pal A, Shah S, Devi S (2007) Synthesis of Au, Ag and Au – Ag alloy nanoparticles in aqueous polymer solution. Colloid Surf A Physicochem Eng Asp 302(1–3):51–57. https://doi.org/10.1016/j.colsurfa.2007.01.054

Pan D, Schirra CO, Senpan A, Schmieder AH, Stacy AJ, Roessl E et al (2012) An early investigation of ytterbium nanocolloids for selective quantitative "multicolor" spectral CT and imaging. ACS Nano 6(4):3364–3370. https://doi.org/10.1021/nn300392x. C2012

Panacek A, Kvitek L, Prucek R, Kolar M, Vecerova R, Pizurova N et al (2006) Silver colloid nanoparticles: synthesis, characterization, and their antibacterial activity. J Phys Chem B 110(33):16248–16253. https://doi.org/10.1021/jp063826h

Park Y, Hong YN, Weyers A, Kim YS, Linhardt RJ (2011) Polysaccharides and phytochemicals: a natural reservoir for the green synthesis of gold and silver nanoparticles. IET Nanobiotechnol 5(3):69–78. https://doi.org/10.1049/iet-nbt.2010.0033

Prabhu S, Poulose EK (2012) Silver nanoparticles: mechanism of antimicrobial action, synthesis, medical applications, and toxicity effects. Int Nano Lett 2(1):32. https://doi.org/10.1186/2228-5326-2-32

Rajchakit U, Sarojini V (2017) Recent developments in antimicrobial peptide conjugated gold nanoparticles. Bioconjug Chem 28(11):2673–2686. https://doi.org/10.1021/acs.bioconjchem.7b00368

Ruparelia JP, Chatterjee AK, Duttagupta SP, Mukherji S (2008) Strain specificity in antimicrobial activity of silver and copper nanoparticles. Acta Biomater 4(3):707–716. https://doi.org/10.1016/j.actbio.2007.11.006

Sadowski Z, Maliszewska IH (2011) Applications of gold nanoparticles : current trends and future prospects. In: Rai M, Duran N (eds) Metal nanoparticles in microbiology. Springer, Berlin, pp 225–248. https://doi.org/10.1007/978-3-642-18312-6

Sanyasi S, Majhi RK, Kumar S, Mishra M, Ghosh A, Suar M et al (2016) Polysaccharide-capped silver nanoparticles inhibit biofilm formation and eliminate multi- drug-resistant bacteria by disrupting bacterial cytoskeleton with reduced cytotoxicity towards mammalian cells. Sci Rep 6:24929. https://doi.org/10.1038/srep24929

Shao Y, Wu C, Wu T, Yuan C, Chen S, Ding T et al (2018) Green synthesis of sodium alginate-silver nanoparticles and their antibacterial activity. Int J Biol Macromol 111:1281–1292. https://doi.org/10.1016/j.ijbiomac.2018.01.012

Sharma VK, Yngard RA, Lin Y (2009) Silver nanoparticles: green synthesis and their antimicrobial activities. Adv Colloid Interf Sci 145(1–2):83–96. https://doi.org/10.1016/j.cis.2008.09.002

Sondi I, Salopek Sondi B (2004) Silver nanoparticles as antimicrobial agent: a case study on E. coli as a model for gram-negative bacteria. J Colloid Interface Sci 275(1):177–182. https://doi.org/10.1016/j.jcis.2004.02.012

Sun L, Pu S, Li J, Cai J, Zhou B, Ren G et al (2018) Size controllable one step synthesis of gold nanoparticles using carboxymethyl chitosan. Int J Biol Macromol 122:770–783. https://doi.org/10.1016/j.ijbiomac.2018.11.006

Travan A, Pelillo C, Donati I, Marsich E, Benincasa M, Scarpa T et al (2009) Non-cytotoxic silver nanoparticle-polysaccharide nanocomposites with antimicrobial activity. Biomacromolecules 10(6):1429–1435. https://doi.org/10.1021/bm900039x

Wu X, Peng Y, Duan X, Yang L, Lan J, Wang F (2018) Homologous gold nanoparticles and nanoclusters composites with enhanced surface Raman scattering and metal fluorescence for cancer imaging. Nano 8(10):819. https://doi.org/10.3390/nano8100819

Yang J, Lee JY, Too HP (2006) Phase-transfer identification of core-shell structures in bimetallic nanoparticles. Plasmonics 1(1):67–78. https://doi.org/10.1007/s11468-005-9003-2

Zhang AQ, Cai LJ, Sui L, Qian DJ, Chenn M (2013) Reducing properties of polymers in the synthesis of noble metal nanoparticles. Polym Rev 53(2):240–276. https://doi.org/10.1080/15583724.2013.776587

Zhao W, Gonzaga F, Li Y, Brook MA (2007) Highly stabilized nucleotide-capped small gold nanoparticles with tunable size. Adv Mater 19(13):1766–1771. https://doi.org/10.1002/adma.200602449

Peroxidase-Like Activity of Metal Nanoparticles for Biomedical Applications

6

Swachhatoa Ghosh and Amit Jaiswal

Abstract

Nanoparticles are versatile proponents in modern-day research. Upon scaling down to the nano-range, materials exhibit a host of interesting properties, for use in imaging, sensing, and therapeutic approaches. Enzymes as biocatalysts require optimal thermodynamic conditions for maximal activity. Their purification protocols are both labor intensive and uneconomical, paving the need for development of simpler alternatives. Metal nanoparticles act as redox enzymes due to the electron exchange escalated by their superficial atoms. Also, monolayer-protected metal nanoparticles electrostatically interact with different substrates, promoting catalysis. This enables their use in industries, in detection of environmental pollution, and in biomedical applications and other clinical approaches. Peroxidases are an essential family of enzymes, capable of removing harmful metabolic by-products from the cellular environment and involved in the maintenance of cellular defense and integrity. This chapter lays its focus on the peroxidase-like activity of metal nanoparticles and their role in development of biosensors and immunoassays and detection of tumor cells while discussing its present catalytic limitations and future outlook.

Keywords

Nanoparticle · Peroxidase · Immunosensors · Biosensors · Catalysis

S. Ghosh · A. Jaiswal (✉)
School of Basic Sciences, Indian Institute of Technology Mandi,
Mandi, Himachal Pradesh, India
e-mail: j.amit@iitmandi.ac.in

P. Chandra, R. Prakash (eds.), *Nanobiomaterial Engineering*,
https://doi.org/10.1007/978-981-32-9840-8_6

Abbreviations

ABTS	2,2′-Azino-bis(3-ethylbenzothiazoline-6-sulphonic acid)
ADP	Adenosine diphosphate
AUR	Amplex ultra Red
BSA	Bovine serum albumin
CSF	Cerebrospinal fluid
DS	Dermatan sulfate
GAG	Glycosaminoglycan
HS	Heparin sulfate
SERS	Surface-enhanced Raman scattering
TMB	3,3′,5,5′-Tetramethylbenzidine
MS	Mass spectroscopy
GC-MS	Gas chromatography/mass spectroscopy
MALDI-MS	Matrix-assisted laser desorption/ionization
CGM	Continuous glucose monitoring
CTAB	Cetyltrimethylammonium bromide
TnI	Cardiac troponin I
GOx	Glucose oxidase
FAD	Flavin adenosine dinucleotide
$FADH_2$	Reduced flavin adenosine dinucleotide
hMSC	Mesenchymal stem cells
MNP	Magnetic nanoparticle
PBS	Phosphate buffer saline
MRI	Magnetic resonance imaging
ELISA	Enzyme-linked immunosorbent assay
ROS	Reactive oxygen species
HER2	Human epidermal growth factor receptor 2
OPD	o-Phenylenediamine dihydrochloride

6.1 Introduction

The fundamental role of enzymes is to act as biological catalysts. With the evolution of science over the years, enzymes are no longer restricted to proteins. Ribozymes, which are RNA molecules with enzymatic activities, can catalyze specific reactions, as can be done by certain DNA molecules. Enzymes may enhance the rate of a biochemical reaction by several folds. Cells contain many different enzymes, and their catalytic activity decides the regular events taking place within a living system (Cooper and Hausman 2004). Natural enzymes exhibit high sensitivity and selectivity towards the reactions they catalyze but involve intensive purification protocols, making the production costly. They also undergo rapid degradation when subjected to harsh temperature and pH changes in the reaction system. Therefore, the hunger for developing alternate enzyme-like systems with improved robustness and easier production methods has been persistent in the scientific arena. Bionics is an emerging field dedicated to the study and design of engineering systems and modern technology, using naturally found biological systems (Bonser and Vincent 2007). Biomimicking is a concept that was popularized by Charles Darwin, as early as the nineteenth century. Artificial enzymes developed primarily to mimic the natural enzymes were found to be highly stable, robust, and easy to manufacture (Dramou and Tarannum 2016). Materials upon reduction to the nanoscale undergo drastic change in their physicochemical behavior, promoting their use in imaging (Han et al. 2019; Ahlawat et al. 2019), catalysis (Singh et al. 2017), sensing (Pallela et al. 2016; Skrabalak et al. 2007; Majarikar et al. 2016), diagnostics (Banerjee and Jaiswal 2018), and therapy (Yadav et al. 2019; Roy and Jaiswal 2017a). Catalytically active nanomaterials emerged as promising enzyme mimics, due to the attributes they shared with natural enzymes (Lin et al. 2014a). Metals like gold, silver, and platinum, considered to be inert under ordinary conditions, become efficient catalysts in nano-dimension (Lin et al. 2014b). This is primarily due to the increased surface area to volume ratio of nanoparticles, exposing a greater number of superficial metal atoms, leading to increased catalytic activity. Nanoparticles solely or with surface functionalization and their small size and varying morphology largely resemble natural enzymes and further boost their ability to replicate these catalysts (Wei and Wang 2013). Functionalization with polymers, ligands, drugs, proteins, surfactants, and other small molecules not only renders stability and regulates optical and physiochemical properties of nanoparticles but also aids in the interaction of the substrate molecules with its surface (Mahato et al. 2019). This enhances the catalytic output. However, unmodified nanoparticles are proven to be better catalysts. This is largely because the enzymatic activity stems from the nanoparticle itself and does not rely on the functional groups on its surface (Wang et al. 2012). The superficial atoms on the surface of metal nanoparticles contribute in electron exchange with the reaction system, spurring their redox-like properties (El-Sayed et al. 2017). Enhanced catalytic properties of porous nanoparticles are due to greater number of surfaces available for interaction. The inner and outer walls of the outer shell and the inner gold core of gold nanorattles exhibit varied catalytic profiles. The lack of stabilizers inside the gold nanorattle is responsible for increased catalytic activity of the inner core. Pasquato et al. termed their thiol-protected nanogold "nanozymes," in analogy with synzymes, which were enzyme-like polymers (Manea et al. 2004). Today, this term holds meaning in defining nanosystems which can show enzymatic activity in vitro and is already an intensely explored field (Lin et al. 2014b), primarily in sensing (Howes et al. 2014). The distinctive optical properties of metal nanoparticles, in combination with functionalized bioreceptors, impart sensing of miniscule analyte proportions in biological fluids (Chandra et al. 2011; Akhtar et al. 2018). Point-of-care devices and nanoparticles are together being used for rapid immunosensing applications. These multifaceted nanoparticles are increasingly being explored for simple healthcare (Roy and Jaiswal 2017b; Khandelia et al. 2013, 2014; Singh et al. 2019) and diagnosis (Chandra et al. 2017). In this chapter, the focus will primarily lie on the mechanism of peroxidase-like action, the properties of metal nanoparticles that contribute towards their activity, and their biomedical applications with focus on biosensing, immunoassay platforms, and cancer cell detection. Lastly, their pitfalls and scope for further research will be discussed.

6.2 Metal Nanoparticles as Catalysts

The ability of metal nanoparticles to act as catalysts was discovered by Haruta when he demonstrated room temperature oxidation of carbon monoxide by gold nanostructures (Haruta et al. 1987). This led to countless experiments on the catalytic property of metal nanoparticles. The properties of materials transform upon reduction from bulk to nanometer range. Noble metals, considered to be inert under ordinary conditions, show unparalleled catalytic properties on attaining nanometer dimensions. Bulk metals tend to reflect light falling on them. Electron clouds on the surface of metal nanoparticles, however, resonate with different wavelengths of light depending upon their frequency. The variation in size and shape is responsible for absorbing different wavelengths of light and hence most metal nanoparticles with different sizes and shapes differ in color. The melting temperature also drops with decrease in size (Noguez 2007a). The number of coordinatively unsaturated atoms on the surfaces and edges of the nanoparticles is far less than in the bulk form and thus becomes highly reactive. Atoms at the corner, steps, and edges of the nanoparticles have the lowest coordination number and tend to be a lot more interactive with substrates and reagents (Noguez 2007b). These, therefore, show the highest catalytic potential (Navalón and García 2016a). As nanoparticles increase in size, they may undergo aggregation, limiting their reactivity. Adsorbing nanoparticles on the surfaces of insoluble solids provides stability against sintering and growth (Navalón and García 2016b).

Considered to be hard, crystalline and diametrically opposite of proteins, nanoparticles in reality, resemble them quite remarkably. Their overall size, charge, and shape along with the organic functional groups on their surface make them protein-like catalysts. Interaction with the substrate is dependent on the media parameters for both nanoparticles and natural enzymes, as even the former may have bulky functionalities on its surface (Kotov 2010). Thus, the superficial atoms on the metal nanoparticles, the size of the nanoparticles, and their thermodynamic properties decide their catalytic potential.

6.3 Peroxidase-Like Activity of Metal Nanoparticles

Peroxidases are a group of enzymes, capable of catalyzing the oxidation of hydrogen peroxide into hydroxyl radicals. These radicals further participate in electron exchange with substrates producing color on oxidation. Found in a range of organisms, peroxidases remove the toxic hydrogen peroxide, released as a by-product of respiration. The standard reactions are demonstrated using colorimetric substrates like TMB, ABTS, and so on. The redox reactions catalyzed by peroxidases can be replicated by the superficial metal atoms of metal nanoparticles, thereby promoting catalysis. Fe_3O_4 nanoparticles were first reported to be potent peroxidase mimics by Gao et al. (2007a). The reduced specificity of horseradish peroxidase makes it a model enzyme for studying peroxidase-catalyzed reactions. Initially, peroxidases were mostly involved in detoxification of polluted water. Nobel metal nanoparticles show peroxidase-like action. CTAB-reduced PtNCs exhibited peroxidase-mimicking activity, while galvanic replacement of tellurium produced highly porous Pt nanotubes (Ma et al. 2011; Cai et al. 2013). Unmodified AuNPs show the highest catalytic activity, when compared to charged nanoparticles (Wang et al. 2012). Bimetallic nanoparticles of iron and platinum, in a cage of apoferritin (Aft-FePt), were reported as peroxidase-mimicking platforms and were more efficient as compared to single metal counterparts Aft-Fe and Aft-Pt, as Pt localized over Fe (Zhang et al. 2012; Xie et al. 2012). FeCo nanoparticles also demonstrated increased peroxidase activity as compared to their singular forms (Xie et al. 2012). Heavy metals like Hg^{2+} and antioxidants result in catalytic inhibition. The loss in catalytic activity in presence of heavy metals was regained in their absence and promoted heavy metal detection. BSA-stabilized AuNPs were used as sensing platforms for dopamine, xanthine, and uric acid (Zhao et al. 2012; Wang et al. 2011; Tao et al. 2013a).

6.4 Biomedical Applications of Peroxidase-Like Metal Nanoparticles

6.4.1 Biosensing

6.4.1.1 Detection of Heparin in Blood

Heparin or unfractionated heparin (UFH) is an anticoagulant drug used for prevention and treatment of thrombosis. The activity of antithrombin, a natural anticoagulant, is instantaneously upregulated upon intravenous administration of heparin. Inside the body, heparin non-specifically binds to plasma proteins and interacts with the cell surfaces of leukocytes and endothelial cells resulting in rapid clearance and reduced therapeutic activity (McRae Simon and Jeffrey 2004). Patients are generally subjected to different doses of the drug to study its response (Baughman Robert et al. 1998). You et al. devised a highly sensitive and selective fluorescent probe for detection of heparin from human serum (You et al. 2018). The probe consisted of adenosine-analogue functionalized gold nanoparticles which aggregated in presence of surfen, a small molecule antagonist of heparin sulfate, and lost their peroxidase activity. On adding negatively charged heparin, the surfen molecule detached from the AuNPs dismantling the aggregates (Fig. 6.1). The increase in their peroxidase action on administration of heparin was monitored spectroscopically and used for estimating a dose as low as 30 nM in spiked blood serums.

Hu et al. developed a heparinase sensing system by conjugating heparin to boost the enzymatic activity of BSA-stabilized AuNCs. Heparinase is a heparin-degrading

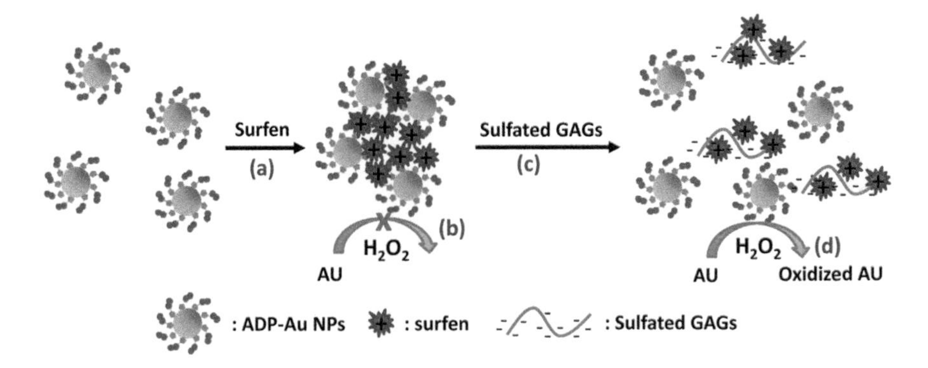

Fig. 6.1 Schematic demontration of sensing mechanism for detecting heparin- (**a–b**) surfen-induced aggregation of ADP-AuNPs, inhibiting the oxidation of AUR (**c–d**) in presence of sulfated GAGs like heparin, the surfen binding is disrupted allowing oxidation of AUR. (Reproduced with permission from You et al. (2018). Copyright 2018 American Chemical Society)

enzyme with pathological impact in tumor metastasis and membrane vascularization (Vlodavsky et al. 2011). It breaks heparin into small fragments inhibiting the peroxidase activity of the nanocomposites. The limit of detection achieved for heparinase was 0.06 μg/ml with a high signal to noise ratio, proving to be a sensitive and reliable system. Detection of both heparin and heparinase can be useful for detection of thrombosis and other cerebral conditions (Hu et al. 2018).

6.4.1.2 Detection of Glucose

Blood-Glucose Monitoring Sensor

Glucose is an essential fuel responsible for the major metabolic activities taking place in the human body. In India and globally, diabetes mellitus is considered to be a silent killer. This clinical condition is characterized by insulin insensitivity leading to hyperglycemia. Different types of glucose sensors have been flocking the market since the past 40 years (Chen et al. 2013). Glucose oxidase-based enzyme electrodes are applied for easy-to-use glucose monitoring, due to their cost-effectiveness and increased stability (Wang 2001; Wilson and Turner 1992).

$$GOx\left(\mathrm{FAD}\right) + \mathrm{glucose} \rightarrow GOx\left(\mathrm{FADH_2}\right) + \mathrm{gluconic\ acid}$$

$$GOx\left(\mathrm{FADH_2}\right) + Med_{\mathrm{ox}} \rightarrow GOx\left(\mathrm{FAD}\right) + Med_{\mathrm{red}}$$

The amperometric glucose biosensors involve the following three steps: (a) electron (and protons) transfer from glucose causing reduction of FAD to $FADH_2$ in the reaction centers of GOx; (b) electron transfer from $FADH_2$ centers to the mediator, transforming the mediators from Medox to their reduced state Medred; and (c) the final shuttle of electrons through the mediators to the electrode (Chen et al. 2013). Three generations of amperometric glucose sensors are known, based on the

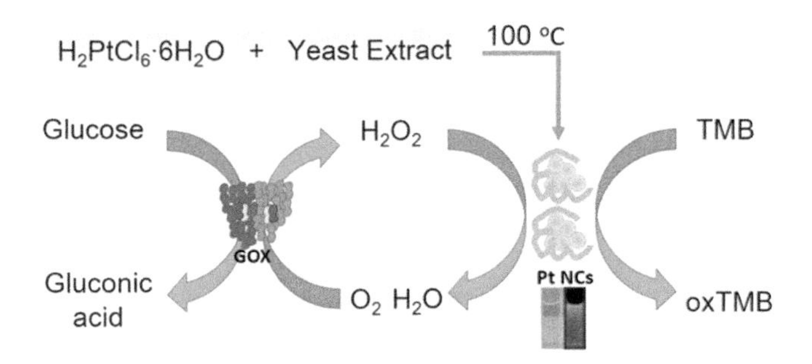

Fig. 6.2 Synthesis procedure of yeast-stabilized PtNCs and its mechanism of catalysis for glucose detection. (Reproduced with permission from Jin et al. (2017) Copyright 2017 American Chemical Society)

mediator (Med). The first generation of amperometric glucose biosensors uses O_2 for regeneration of GOx (FAD) and detects glucose based on either the amount of O_2 consumed or the amount of H_2O_2 generated in the process. These generations of sensors were simple and easy to use but were sensitive to oxygen for detection. The reliability on oxygen was addressed in the second-generation glucose sensors by using artificial electron shuttlers like ferrocene derivatives and conducing organic salts, transferring electrons into and out of the enzyme active site. The need to stabilize these artificial electron shuttlers resulted in development of third-generation glucose sensors. Electrochemical potential of GOx was used for detection of glucose in this generation of sensors.

Metal nanoparticles can be conjugated to glucose oxidase for triggering their peroxidase-like activity and promoting glucose detection. A unique yeast-stabilized PtNC system with peroxidase-mimicking ability was developed by Jin et al. for simultaneous oxidation of glucose and reduction of H_2O_2 (Jin et al. 2017) (Fig. 6.2). This system was then extended to human serum for colorimetric estimation of glucose. The limit of detection of this colorimetric system was 0.28 μM, which is significantly less when compared to other colorimetric sensors.

AuNPs, with intrinsic peroxidase-like activity, were used for colorimetric detection of H_2O_2 and glucose (Jv et al. 2010). A colorimetric glucose detection platform was developed by Wang et al. by combining the catalytic activities of GOx and MNPs. This sensor was fabricated based on the electrochemical depletion of electroactive species in the diffusion layer. Electroactive species like ascorbic acid and their role in glucose detection was tested by the interference-free micro-circumstance of the substrate electrode (GOD). The limit of detection for glucose was 0.005 mM with an overall range of 0.01–1 mM (Wang et al. 2005).

6.4.1.3 Detection of Cholesterol and Galactose Using Peroxidase-Like MNPs

Colorimetric detection of cholesterol was achieved by developing a nanostructured multicatalyst system consisting of MNPs and cholesterol oxidase, immobilized in large pore-sized mesoporous silica (Kim et al. 2011). This multicatalyst system consisted of MNPs embedded in the wall of mesocellular silica pores, resulting in magnetic mesoporous silica (MMS) and cholesterol oxidases. Cholesterol oxidase immobilized in the MMS reacted with cholesterol to generate H_2O_2, subsequently activating MNPs in the mesocellular silica pores for colorimetric conversion of its substrate. The limit of detection for cholesterol was as low as 5 μM. Detection of galactose using a nanostructured multicatalyst system consisting of MNPs and galactose oxidase was reported by Kim et al. This multicatalytic system was tested as a promising analytical tool for galactosemia diagnosis, by determining the galactose concentration from the dried blood specimens obtained from clinical hospitals (Kim et al. 2012).

6.4.1.4 Plasmonic and Catalytic Activity of Gold Nanoparticles for Sensing Glucose and Lactate in Living Tissues

Lin et al. investigated the cross-talk between glucose, lactate, and ascorbate in the ascorbate modulating neuronal metabolism for better understanding of brain ischemia (Lin et al. 2014c). A microfluidic chip-based sensor was fabricated using three surface-modified indium-tin oxide electrodes as working electrodes. A stainless steel tube was used as a counter electrode and an Ag/AgCl wire as reference electrode. Electrochemical oxidation of ascorbic acid was achieved by using single-walled carbon nanotubes, while for glucose and lactate, a dehydrogenase-based mechanism was implemented. Fluctuation in the level of these metabolites may help in early diagnosis and treatment of neuronal conditions.

A peroxidase-like gold nanoparticle-impregnated metal-organic framework AuNP@MIL-101 was used for SERS activation of leucomalachite green (LMG) and subsequent detection of glucose and lactate in living tissues (Hu et al. 2017) (Fig. 6.3). This oxidase-integrated nanoparticle system for detecting glucose resulted in reduction of oxygen to hydrogen peroxide. This H_2O_2 facilitated the conversion of LMG, and the SERS signal intensity was used for quantification.

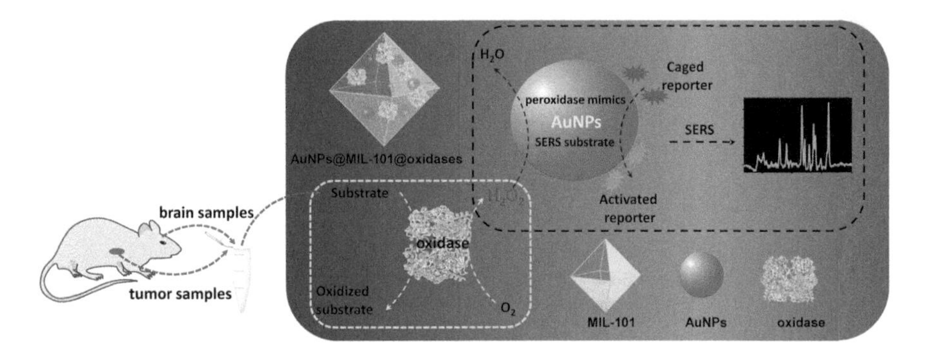

Fig. 6.3 Schematic showing AuNP@MIL-101@oxidase nanozyme. The mechanism of oxidation of substrate for H_2O_2 production and simultaneous peroxidase activity for obtaining activated SERS reporter (malachite green). The peroxidase activity of the integrated system enables enhancement in SERS signal. (Reproduced with permission from Hu et al. (2017) Copyright 2017 American Chemical Society)

A similar protocol was followed for detection of lactate from living tissues. The combination of plasmonic properties and SERS activity of gold nanoparticles show promise in successful real-time probing of analytes and can be used for designing novel immunoassays. In tumor tissues, the glucose metabolism is enhanced leading to low glucose levels. Hypoxia or low oxygen levels may enhance anaerobic glycolysis leading to lactate production, making these integrated nanosystems highly significant in clinical diagnosis.

6.4.1.5 Using Bare Nanoparticles for Detection of Food Contaminants

To fulfill the increasing demands for different food products, artificial taste enhancers are often added. The melamine debacle of China in 2008 is one such example (Pei et al. 2011). Melamine is a nitrogenous compound added to increase the protein amount in food artificially. This organic compound is reportedly toxic beyond a daily intake of 0.5 mg/kg of body mass and leads to reproductive damage and bladder or kidney stones (Ingelfinger 2008). Melamine detection is critical to evaluate the quality of food, and the intensive experimental techniques were replaced by nanozymes for ease of access. Nanoparticles have also been used for detection of harmful reactive oxygen species in food samples (Bajpai et al. 2018).

A simple colorimetric detection strategy consisting of bare AuNPs was developed by Ni et al. for sensitive detection of melamine in food (Ni et al. 2014). On exposure to melamine, the AuNPs form aggregates with melamine, resulting in better oxidation of the substrate and increased colorimetric detection. Melamine and H_2O_2 combine to form an addition compound which was used for detection of melamine according to Ding et al. The consumption of H_2O_2 was used as a measure of melamine levels in spiked raw milk and milk powder samples (Ding et al. 2010).

6.4.1.6 Detection of Nucleic Acids

A label-free, colorimetric sensor consisting of MNPs was developed for detection of known DNA samples. The DNA was first amplified using polymerase chain reaction and then conjugated with MNPs. The peroxidase-like activity shown by unconjugated MNPs was inhibited by the electrostatically adsorbed DNA. The electrostatic interaction between positively charged substrate OPD and negatively charged DNA and the adsorption of DNA on the surface of the enzyme inhibited the peroxidase activity of MNPs, reducing the signal output. This decrease in the colorimetric product led to the detection of *Chlamydia trachomatis*, a common bacterium found in sexually transmitted diseases, in human urine. Other nanoparticles like CeO_2 were also used for detection of nucleic acids (Park et al. 2011).

6.4.1.7 Detection of Thrombin

Zhang et al. developed an aptamer-based biosensor for detection of thrombin in blood samples. Aptamers are ssDNA or ssRNA and can be designed to specifically bind to a target molecule, replacing antibodies for detection. A sandwich-type assay developed using chitosan-conjugated MNPs and two thrombin aptamers could detect a 1 nM level of thrombin (Zhang et al. 2010a). Following the principle of

ELISA, streptavidin was subjected onto a 96 well plate, and the non-specific binding sites were blocked using BSA. A biotinylated 29mer aptamer I adsorbed onto the 96 well plate due to biotin-streptavidin interaction. On adding the sample and a 15mer aptamer 2 modified MNPs to the wells, a colorimetric change was observed, in presence of thrombin. This was due to the distinct aptamer binding sites present on opposite faces of thrombin. A 1 nM concentration of thrombin was successfully detected using this biosensor.

6.4.2 Detection of Tumor Cells

The hallmark of a typical cancer cell is the overexpression of certain proteins when compared to a healthy cell. Folate receptors, for example, are most commonly expressed on tumor cells. Antibodies against these receptors or folic acid functionalized onto nanohybrids help in easy detection of certain tumor types. Successful ultrasensitive detection of HER2-positive SKBR-3 cells was done by anti-HER2-fabricated AuNPs (Zhu et al. 2013). Amperometric detection of metastatic tumor cells was also achieved using gold nanoparticles (Pallela et al. 2016). Colorimetric detection, by peroxidase-mimicking AuNPs, has also been reported for easy and sensitive detection of tumors.

6.4.2.1 Hybrid AuNCs for Rapid Colorimetric Detection of Cancer Cells

AuNCs show peroxidase-like action under acidic pH. Graphene oxide can undergo easy surface functionalization due to the presence of a large number of surface functionalities. A GO-AuNC hybrid was prepared which was capable of showing peroxidase-like activity at physiological pH. Tao et al. proposed a simple, cost-effective, folate conjugated GO-AuNC nanohybrid which electrostatically interacted with positively charged TMB to produce a colorimetric product (Tao et al. 2013b) (Fig. 6.4). The folate receptor-overexpressing MCF-7 cells were targeted using this nanohybrid system, and the catalytic activity increased with increase in the number of cancer cells. Also, the developed composite was found to be highly selective for MCF-7 cells when compared with both healthy and cancerous cell lines. This strategy could be employed for further detection of other folate overexpressing cell lines.

6.4.2.2 Stem Cell Proliferation and Imaging

Biosafety concerns limit the use of nanoparticles in stem cell imaging. Huang et al. reported ferucarbotran, an ionic superparamagnetic iron oxide nanoparticle, to be safe for mesenchymal stem cell imaging and growth. Upon internalization into hMSCs, the intrinsic peroxidase-like activity of ferucarbotran resulted in the quenching of intracellular H_2O_2, thereby helping in the growth of stem cells. The SPIO nanoparticles underwent lysosomal degradation resulting in leaching of Fe from their surface. However, this could not block the stem cell progression, and an exact reason for the above observation was not known. However, the

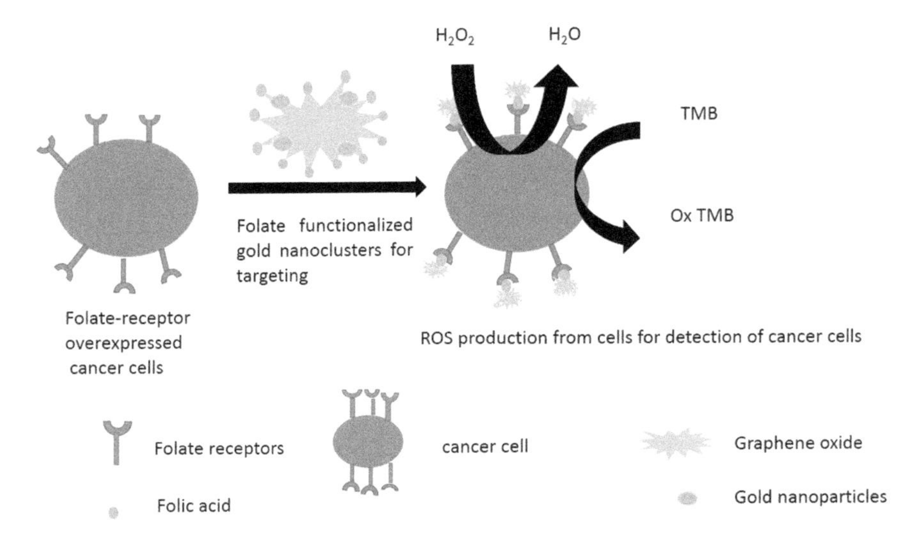

Fig. 6.4 Schematic showing folate-functionalized hybrid nanoclusters for detection of cancer cells. The ROS produced by the cells facilitates colorimetric detection of folate-overexpressed cancer cells

ferucarbotran-labelled hMSCs were very easily visualized using MRI. Thus, a simple, biocompatible cell-imaging platform was reported by Huang et al. and could be effectively used in regenerative medicine(Huang et al. 2009).

6.4.3 Immunoassays

In biological systems, immunoassays are used for tracking different hormones, proteins, and antibodies. Enzymes with colorimetric substrates like horseradish peroxidase and alkaline phosphatase are extensively used for conjugation with secondary antibodies and amplify the detection signal manyfold (Micheli et al. 2002). However, these enzymes have a short shelf life and undergo easy denaturation on long-term storage (Gao et al. 2008). Thus, nanozyme systems with peroxidase-mimicking properties have been used for developing immunoassays. A typical immunoassay platform developed using the enzymatic metal nanoparticles is shown in Fig. 6.5.

6.4.3.1 Immunoassay for Detection Cardiac Troponin I

Gao et al. developed a sandwich-type immunoassay, similar to ELISA, using dextran-functionalized MNPs for suitable detection of cardiac troponin I, a well-known biomarker of cardiac myopathy (Gao et al. 2007b). An antibody of TnI antigen was conjugated with MNPs and then mixed with serum to capture the cardiac troponin antigen. On binding with the antigen, the bound MNPs were separated using a magnetic field and then loaded along with the reaction buffer, into a microtiter plate, for measuring the absorbance at 652 nm. Both the catalytic and magnetic properties of MNPs were demonstrated using this immunoassay platform.

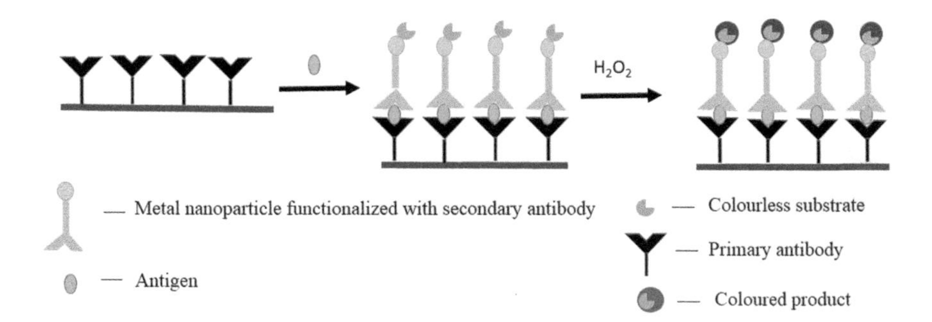

Fig. 6.5 Schematic showing a colorimetric immunoassay platform developed using gold nanoparticles

6.4.3.2 Immunoassay for Detection of Cancer

An ultrafast, specific immunoassay platform was designed for the colorimetric detection of HER2, a common biomarker for breast cancer. Kim et al. developed a nanocomposite of MNPs and PtNPs and immobilized it on mesoporous carbon for utilizing its synergistic catalytic potential in detection of HER2 and rotavirus like model pathogenic antigens (Kim et al. 2014). An antibody against HER2 was immobilized on the nanocomposite MMC-10/Pt-10, and cell lysates from human breast cancer cell lines SKBR-3 and MCF-7 along with human cell melanoma WM-266-4 were employed in different wells along with the nanocomposite. A typical ELISA type reaction demonstrated a dense blue color in the well containing SKBR-3 cell lysate, indicating the selectivity of the immunoassay developed. Among SKBR-3 and MCF-7, the level of HER2 expression is much higher in SKBR-3, justifying the observation reported by Kim et al. Rotavirus detection reported using the same nanocomposite was equally selective, and the magnetic properties of the MNPs were also demonstrated for the separation of the antigens. This nanocomposite with distinct catalytic and magnetic properties could be extended for point-of-care detection in clinical applications.

6.4.3.3 Chitosan-Modified MNPs for Detection of Mouse IgG and CEA

The use of MNPs for immunoassay designing is reliant on four major properties: easy dispersion in aqueous solution at physiological pH, proper surface functionalization for linking with antibodies, large enough saturation magnetization for separation using a moderate magnetic field, and easy separation of aggregated nanoparticles upon removal of applied magnetic field.

Chitosan-modified MNPs, used by Gao et al. for the development of a sandwich antigen-down type immunoassay, possessed all the above features and were used for detection of mouse IgG and carcinoembryonic antigen (CEA) detection (Gao et al. 2008) (Fig. 6.6). The wells of a 96 well microtiter plate were loaded with the antigen (mouse IgG) in bicarbonate buffer at 4 °C. The unbound antigen was washed thrice using PBS followed by addition of non-specific protein BSA to the wells. After 2 hours of incubation, the wells were again washed with PBS. Antimouse IgG

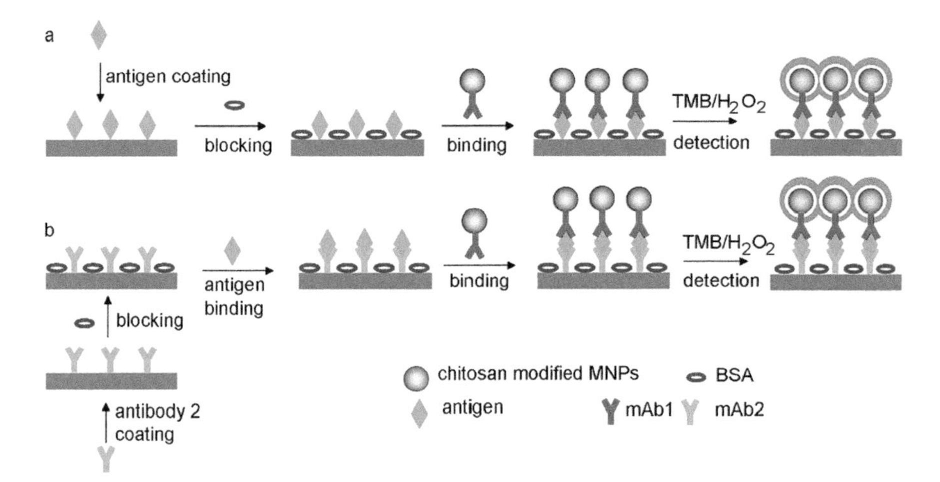

Fig. 6.6 Schematic showing the sandwich-type immunoassay developed by Gao et al. The binding of the antigen to the well was followed by addition of the antibody functionalized MNPs for detection of the chromogenic product. (Reproduced with permission from Gao et al. (2008) Copyright 2008 American Chemical Society)

antibody functionalized MNPs were then loaded into the wells and incubated for 1 h at room temperature. The unbound nanoparticles were washed out using PBS, and the reaction buffer was added into the wells. The MNPs, acting as peroxidases, oxidize the substrate producing a color change that can be detected at 652 nm for estimation of the amount of antigen.

For CEA detection, the wells were first coated with anti-CEA antibody. The unbound antibodies were washed with PBS and a blocking agent BSA was added. This was followed by further washing of the wells and addition of the antigen CEA into it. After incubation, the unbound CEA was washed using PBS, and the anti-CEA antibody-functionalized MNPs were added into the well. Upon washing the wells with PBS, the reaction buffer containing the peroxidase substrate TMB and H_2O_2 was added, and absorbance of the oxidized substrate at 652 nm was measured spectroscopically. Gao et al. also demonstrated the magnetic concentration and separation of the antigen from the wells. This immunoassay platform reported the sensing of 1 ng/mL CEA.

6.4.4 Detection of IgG Using Prussian Blue-Modified Iron Oxide Nanoparticles

The peroxidase-mimicking activities of Prussian blue-modified iron oxide nanoparticles were used in the development of an immunoassay platform for detection of IgG. Zhang et al. used a staphylococcal protein A(SPA) as the antibody for conjugating with the modified nanoparticles (Zhang et al. 2010b). This protein-nanoparticle composite was used for the detection of IgG immobilized on 96 well plates. The catalytic and magnetic properties of the nanoparticles were intact and even increased on functionalizing with Prussian blue, confirming its usage for further designing of improved biosensing platforms.

6.5 Limitations

Nanozymes as artificial enzyme systems demonstrate high operational stability, low cost, and easy bioconjugation. Their catalytic potential is supported by the unsatiated atoms on their surface and further boosted by functionalization. Their applications in biosensing, immunoassay development, and tumor detection platforms is therefore a highly investigated field. In-depth studies of their catalytic behavior have revealed some limitations. Of the different catalytic behaviors manifested by metal nanoparticles, redox enzyme mimics seem to dominate. Docking and other forms of simulations can help discover other enzymatic properties of nanometals (Wang et al. 2016). Despite being functionalized by different organic groups, nanozymes cannot compete with the selectivity and catalytic properties of natural enzymes. New nanozymes with improved catalytic features need to be designed in the future. Better cascade mimics can be developed by focusing on the natural action of enzymes. Studying the stability of metal nanoparticles in response to functionalization may lead to detailed investigation of their theranostic properties. Their enzymatic properties have been scarcely explored for therapeutic purposes. A major reason for this is the cytotoxicity associated with the unshielded metal surface (Vlamidis and Voliani 2018). In-depth evaluation of their biocompatibility may promote their use in clinical medicine and therapeutics.

6.6 Summary and Outlook

- Metal nanoparticles are suitable alternatives to natural enzymes. A summary of the biomedical application of different metal nanoparticles arising from its peroxidase-like activity is represented in Table 6.1.
- Their small size, easy surface functionalization and reduced manufacturing cost enhance their applicability in scientific research.
- This chapter is a brief account of their peroxidase-like action in several biological applications.
- Biosensing of small molecules, clinically significant biomarkers and food adulterants can be successfully achieved through these enzyme mimics. Their sensitivity is enhanced by colorimetric oxidation of different peroxidase substrates.
- Nanozymes play a significant role in tumor cell detection and imaging, making their clinical applicability remarkable.
- Further efforts made into designing biocompatible nanocomposites with improved catalytic potential will lead to distinct therapeutic applications and approval for clinical use.
- Bionics and other fields which are exploring these mimics ambitiously can achieve a lot in the future.

Table 6.1 Biomedical application of different metal nanoparticles arising from its peroxidase-like activity

Application	Nanozyme	Detection method	Details	Reference
Biosensing	MNPs	Colorimetric	Glucose biosensor	Wang et al. (2005)
	AuNPs	Colorimetric	H_2O_2 and glucose biosensor	Jv et al. (2010)
	AuNP (impregnated on metal-organic framework with oxidase)	Colorimetric, SERS based	Glucose, lactate, and ascorbate detection	Hu et al. (2017)
	MNPs (with oxidase in mesoporous silica)	Colorimetric	Glucose and cholesterol biosensor	Kim et al. (2011)
	MNPs (with oxidase in mesoporous silica)	Colorimetric	Galactose biosensor	Kim et al. (2012)
	MNPs	Colorimetric	Nucleic acid detection	Park et al. (2011)
	Chitosan-modified MNPs with thrombin aptamers	Colorimetric	Thrombin detection	Zhang et al. (2010a)
	MNPs	Colorimetric	Melamine detection	Ding et al. (2010)
	AuNPs	Colorimetric	Melamine detection	Ni et al. (2014)
Immunoassay	MNPs	Colorimetric	Cardiac Troponin I (TnI)	Gao et al. (2007b)
	MNPs-PtNPs in mesoporous carbon	Colorimetric	Detection of cancer biomarkers	Kim et al. (2014)
	Chitosan-modified MNPs	Colorimetric	Mouse IgG and carcinoembryonic antigen (CEA)	Gao et al. (2008)
	Prussian blue-modified γ-Fe_2O_3 NPs	Colorimetric	IgG detection	Zhang et al. (2010b)
Cancer cell detection	Superparamagnetic iron oxide NPs	Colorimetric	Promotion of stem cell growth	Huang et al. (2009)
	GO-AuNCs	Colorimetric	MCF-7 cell detection in mice	Tao et al. (2013b)

Acknowledgment AJ gratefully acknowledge the support from BioX Centre, Advanced Materials Research Centre, Indian Institute of Technology Mandi. Support from Department of Science and Technology (DST) under project number SERB/F/5627/2015-16 and Department of Biotechnology (DBT), Government of India, under project number BT/PR14749/NNT/28/954/2015 is also acknowledged.

References

Ahlawat M et al (2019) Gold nanorattles with intense Raman in silica nanoparticles (Nano-IRIS) as multimodal system for imaging and therapy. Chem Nano Mat 5(5):625–633
Akhtar MH et al (2018) Ultrasensitive dual probe immunosensor for the monitoring of nicotine induced-brain derived neurotrophic factor released from cancer cells. Biosens Bioelectron 116:108–115
Bajpai VK et al (2018) Prospects of using nanotechnology for food preservation, safety, and security. J Food Drug Anal 26(4):1201–1214
Banerjee R, Jaiswal A (2018) Recent advances in nanoparticle-based lateral flow immunoassay as a point-of-care diagnostic tool for infectious agents and diseases. Analyst 143(9):1970–1996
Baughman Robert A et al (1998) Oral delivery of anticoagulant doses of heparin. Circulation 98(16):1610–1615
Bonser RHC, Vincent JFV (2007) Technology trajectories, innovation, and the growth of biomimetics. Proc Inst Mech Eng C J Mech Eng Sci 221(10):1177–1180
Cai K et al (2013) Aqueous synthesis of porous platinum nanotubes at room temperature and their intrinsic peroxidase-like activity. Chem Commun 49(54):6024–6026
Chandra P et al (2011) Detection of daunomycin using phosphatidylserine and aptamer co-immobilized on Au nanoparticles deposited conducting polymer. Biosens Bioelectron 26(11):4442–4449
Chandra P, Tan YN, Singh SP (2017) Next generation point-of-care biomedical sensors technologies for cancer diagnosis, vol 10. Springer, Singapore
Chen C et al (2013) Recent advances in electrochemical glucose biosensors: a review. RSC Adv 3(14):4473–4491
Cooper GM, Hausman RE (2004) The cell: molecular approach, Medicinska naklada
Ding N et al (2010) Colorimetric determination of melamine in dairy products by Fe3O4 magnetic nanoparticles−H2O2−ABTS detection system. Anal Chem 82(13):5897–5899
Dramou P, Tarannum N (2016) 3 – Molecularly imprinted catalysts: synthesis and applications. In: Li S et al (eds) Molecularly imprinted catalysts. Elsevier, Amsterdam, pp 35–53
El-Sayed R et al (2017) Importance of the surface chemistry of nanoparticles on peroxidase-like activity. Biochem Biophys Res Commun 491(1):15–18
Gao L et al (2007a) Intrinsic peroxidase-like activity of ferromagnetic nanoparticles. Nat Nanotechnol 2:577
Gao L et al (2007b) Intrinsic peroxidase-like activity of ferromagnetic nanoparticles. Nat Nanotechnol 2(9):577
Gao L et al (2008) Magnetite nanoparticle-linked immunosorbent assay. J Phys Chem C 112(44):17357–17361
Han X et al (2019) Applications of nanoparticles in biomedical imaging. Nanoscale 11(3):799–819
Haruta M et al (1987) Novel gold catalysts for the oxidation of carbon monoxide at a temperature far below 0 C. Chem Lett 16(2):405–408
Howes PD, Chandrawati R, Stevens MM (2014) Colloidal nanoparticles as advanced biological sensors. Science 346(6205):1247390
Hu Y et al (2017) Surface-enhanced Raman scattering active gold nanoparticles with enzyme-mimicking activities for measuring glucose and lactate in living tissues. ACS Nano 11(6):5558–5566
Hu L et al (2018) Accelerating the peroxidase-like activity of gold nanoclusters at neutral pH for colorimetric detection of heparin and heparinase activity. Anal Chem 90(10):6247–6252
Huang D-M et al (2009) The promotion of human mesenchymal stem cell proliferation by super-paramagnetic iron oxide nanoparticles. Biomaterials 30(22):3645–3651
Ingelfinger JR (2008) Melamine and the global implications of food contamination. N Engl J Med 359(26):2745–2748
Jin L et al (2017) Ultrasmall Pt nanoclusters as robust peroxidase mimics for colorimetric detection of glucose in human serum. ACS Appl Mater Interfaces 9(11):10027–10033

Jv Y, Li B, Cao R (2010) Positively-charged gold nanoparticles as peroxidiase mimic and their application in hydrogen peroxide and glucose detection. Chem Commun 46(42):8017–8019
Khandelia R et al (2013) Gold nanoparticle–protein agglomerates as versatile nanocarriers for drug delivery. Small 9(20):3494–3505
Khandelia R et al (2014) Polymer coated gold nanoparticle–protein agglomerates as nanocarriers for hydrophobic drug delivery. J Mater Chem B 2(38):6472–6477
Kim MI et al (2011) Fabrication of nanoporous nanocomposites entrapping Fe3O4 magnetic nanoparticles and oxidases for colorimetric biosensing. Chem Eur J 17(38):10700–10707
Kim MI et al (2012) Colorimetric quantification of galactose using a nanostructured multi-catalyst system entrapping galactose oxidase and magnetic nanoparticles as peroxidase mimetics. Analyst 137(5):1137–1143
Kim MI et al (2014) A highly efficient colorimetric immunoassay using a nanocomposite entrapping magnetic and platinum nanoparticles in ordered mesoporous carbon. Adv Healthc Mater 3(1):36–41
Kotov NA (2010) Chemistry. Inorganic nanoparticles as protein mimics. Science 330(6001): 188–189
Lin Y, Ren J, Qu X (2014a) Catalytically active nanomaterials: a promising candidate for artificial enzymes. Acc Chem Res 47(4):1097–1105
Lin Y, Ren J, Qu X (2014b) Nano-gold as artificial enzymes: hidden talents. Adv Mater 26(25):4200–4217
Lin Y et al (2014c) Continuous and simultaneous electrochemical measurements of glucose, lactate, and ascorbate in rat brain following brain ischemia. Anal Chem 86(8):3895–3901
Ma M, Zhang Y, Gu N (2011) Peroxidase-like catalytic activity of cubic Pt nanocrystals. Colloids Surf A Physicochem Eng Asp 373(1):6–10
Mahato K et al (2019) Gold nanoparticle surface engineering strategies and their applications in biomedicine and diagnostics. 3 Biotech 9(2):57
Majarikar V, Ranjan A, Jaiswal A (2016) Nanoparticle-based sensing of oligonucleotides and proteins. In: Nanobiosensors for personalized and onsite biomedical diagnosis. Institution of Engineering and Technology, pp 583–592
Manea F et al (2004) Nanozymes: gold-nanoparticle-based transphosphorylation catalysts. Angew Chem Int Ed 43(45):6165–6169
McRae Simon J, Jeffrey SG (2004) Initial treatment of venous thromboembolism. Circulation 110(9_Suppl_1):I-3-I-9
Micheli L et al (2002) Production of antibodies and development of highly sensitive formats of enzyme immunoassay for saxitoxin analysis. Anal Bioanal Chem 373(8):678–684
Navalón S, García H (2016a) Nanoparticles for catalysis. Multidisciplinary Digital Publishing Institute
Navalón S, García H (2016b) Nanoparticles for catalysis. Nano 6(7):123
Ni P et al (2014) Visual detection of melamine based on the peroxidase-like activity enhancement of bare gold nanoparticles. Biosens Bioelectron 60:286–291
Noguez C (2007a) Surface plasmons on metal nanoparticles: the influence of shape and physical environment. J Phys Chem C 111(10):3806–3819
Noguez C (2007b) Surface plasmons on metal nanoparticles: the influence of shape and physical environment. J Phys Chem C 111(10):3806–3819
Pallela R et al (2016) An amperometric nanobiosensor using a biocompatible conjugate for early detection of metastatic cancer cells in biological fluid. Biosens Bioelectron 85:883–890
Park KS et al (2011) Label-free colorimetric detection of nucleic acids based on target-induced shielding against the peroxidase-mimicking activity of magnetic nanoparticles. Small 7(11):1521–1525
Pei X et al (2011) The China melamine milk scandal and its implications for food safety regulation. Food Policy 36(3):412–420
Roy S, Jaiswal A (2017a) Graphene-based nanomaterials for theranostic applications. Rep Adv Phys Sci 1(04):1750011

Roy S, Jaiswal A (2017b) SERS-based biosensors as potential next-generation point-of-care cancer diagnostic platforms. In: Next generation point-of-care biomedical sensors technologies for cancer diagnosis. Springer, Singapore, pp 173–204

Singh P, Roy S, Jaiswal A (2017) Cubic gold nanorattles with a solid octahedral core and porous shell as efficient catalyst: immobilization and kinetic analysis. J Phys Chem C 121(41):22914–22925

Singh P et al (2019) Gold nanostructures for photothermal therapy. In: Nanotechnology in modern animal biotechnology. Springer, Singapore, pp 29–65

Skrabalak SE et al (2007) Gold nanocages for biomedical applications. Adv Mater 19(20): 3177–3184

Tao Y et al (2013a) A dual fluorometric and colorimetric sensor for dopamine based on BSA-stabilized Aunanoclusters. Biosens Bioelectron 42:41–46

Tao Y et al (2013b) Incorporating graphene oxide and gold nanoclusters: a synergistic catalyst with surprisingly high peroxidase-like activity over a broad pH range and its application for cancer cell detection. Adv Mater 25(18):2594–2599

Vlamidis Y, Voliani V (2018) Bringing again noble metal nanoparticles to the forefront of cancer therapy. Front Bioeng Biotechnol 6:143–143

Vlodavsky I, Elkin M, Ilan N (2011) Impact of heparanase and the tumor microenvironment on cancer metastasis and angiogenesis: basic aspects and clinical applications. Rambam Maimonides Med J 2(1):e0019–e0019

Wang J (2001) Glucose biosensors: 40 years of advances and challenges. Electroanalysis 13(12):983–988

Wang K et al (2005) Selective glucose detection based on the concept of electrochemical depletion of electroactive species in diffusion layer. Biosens Bioelectron 20(7):1366–1372

Wang X-X et al (2011) BSA-stabilized au clusters as peroxidase mimetics for use in xanthine detection. Biosens Bioelectron 26(8):3614–3619

Wang S et al (2012) Comparison of the peroxidase-like activity of unmodified, amino-modified, and citrate-capped gold nanoparticles. Chem Phys Chem 13(5):1199–1204

Wang X, Hu Y, Wei H (2016) Nanozymes in bionanotechnology: from sensing to therapeutics and beyond. Inorganic Chem Front 3(1):41–60

Wei H, Wang E (2013) Nanomaterials with enzyme-like characteristics (nanozymes): next-generation artificial enzymes. Chem Soc Rev 42(14):6060–6093

Wilson R, Turner APF (1992) Glucose oxidase: an ideal enzyme. Biosens Bioelectron 7(3):165–185

Xie J et al (2012) Analytical and environmental applications of nanoparticles as enzyme mimetics. TrAC Trends Anal Chem 39:114–129

Yadav V et al (2019) 2D MoS2-based nanomaterials for therapeutic, bioimaging, and biosensing applications. Small 15(1):1803706

You J-G, Wang Y-T, Tseng W-L (2018) Adenosine-related compounds as an enhancer for peroxidase-mimicking activity of nanomaterials: application to sensing of heparin level in human plasma and total sulfate glycosaminoglycan content in synthetic cerebrospinal fluid. ACS Appl Mater Interfaces 10(44):37846–37854

Zhang Z et al (2010a) Magnetic nanoparticle-linked colorimetric aptasensor for the detection of thrombin. Sensors Actuators B Chem 147(2):428–433

Zhang X-Q et al (2010b) Prussian blue modified iron oxide magnetic nanoparticles and their high peroxidase-like activity. J Mater Chem 20(24):5110–5116

Zhang W et al (2012) Caged-protein-confined bimetallic structural assemblies with mimetic peroxidase activity. Small 8(19):2948–2953

Zhao H et al (2012) Uricase-based highly sensitive and selective spectrophotometric determination of uric acid using BSA-stabilized Au nanoclusters as artificial enzyme. Spectrosc Lett 45(7):511–519

Zhu Y, Chandra P, Shim Y-B (2013) Ultrasensitive and selective electrochemical diagnosis of breast cancer based on a hydrazine–Au nanoparticle–aptamer bioconjugate. Anal Chem 85(2):1058–1064

7 Biomedical Applications of Lipid Membrane-Based Biosensing Devices

Georgia-Paraskevi Nikoleli, Marianna-Thalia Nikolelis, Spyridoula Bratakou, and Vasillios N. Psychoyios

Abstract

The investigation of lipid films for the construction of biosensors has recently given the opportunity to manufacture devices to selectively detect a wide range of compounds of clinical interest. Biosensor miniaturization using nanotechnological tools has provided novel routes to immobilize various "receptors" within the lipid film. This chapter reviews and exploits platforms in biosensors based on lipid membrane technology that are used to detect various analytes of clinical interest. Examples of applications are described with an emphasis on novel systems, new sensing techniques, and nanotechnology-based transduction schemes. The compounds that can be monitored are urea, cholesterol, glucose, toxins, antibiotics, microorganisms, hormones, etc. Finally, limitations and future prospects are presented herein on the evaluation/validation and eventually commercialization of the proposed sensors.

Keywords

Lipid film-based biosensors · Nanotechnological platforms · Graphene electrodes · ZnO nanowalls and nanowires · Clinical analysis

7.1 Introduction

Nanobiosensors have been widely used for the monitoring and determination of biomolecules (Chandra 2016; Nikolelis and Nikoleli 2018), and it is expected that this technology in combination with the recent nanotechnological advances will

G.-P. Nikoleli (✉) · M.-T. Nikolelis · S. Bratakou · V. N. Psychoyios
Laboratory of Inorganic & Analytical Chemistry, School of Chemical Engineering, Department 1, Chemical Sciences, National Technical University of Athens, Athens, Greece
e-mail: dnikolel@chem.uoa.gr

P. Chandra, R. Prakash (eds.), *Nanobiomaterial Engineering*,
https://doi.org/10.1007/978-981-32-9840-8_7

further promote the therapeutic applications in personalized diagnostics. Biosensor nanotechnologies include a wide range of devices that were developed to monitor biomolecules of clinical interest such as glucose, urea, uric acid, cholesterol, calcium ions, etc. The importance of nanomaterials in biosensing mechanism and the various physicochemical techniques that were used to exploit the mechanism of signal generation were described in previous reports (Nikolelis and Nikoleli 2018; Mahato et al. 2018; Prasad 2014). In accordance probe fabrication techniques, analytical characteristics of the biosensing devices such as selectivity, sensitivity, interferences, and evaluation; and validation for commercialization of these biosensors were extensively discussed in these books and reports (Nikolelis and Nikoleli 2018; Mahato et al. 2018; Prasad 2014; Chandra et al. 2011, 2017).

Since Mueller et al.'s works on bilayer lipid membranes (BLMs) (Mueller et al. 1962), devices that were based on lipid membranes to monitor food toxicants, environmental pollutants, and compounds of clinical interest have increased tremendously. However, the "black" lipid membranes that were based on the technique of Mueller et al. were very unstable and broke under an electrical and mechanical field and were unstable outside a KCl electrolyte solution, and this has influenced their practicality. During the last decade, a number of advances to prepare stabilized lipid-based devices were explored, and this has given the opportunity to construct biosensors to detect food toxicants and environmental pollutants in real samples and in the field. The advantages of lipid film devices are summarized as follows: the membranes are biocompatible and therefore can be used to be implanted in the human body, they are fast-responding devices with response times of seconds, they have high sensitivity and can detect biomolecules in the mM concentration range, they also have high selectivity with minor interferences which allow them to be used in real samples such as human serum, etc. Their size is small due to nanotechnological advances; they can have portability, hold a large number of advantages toward the liquid and gas chromatographic units which are bulky, and have a high cost to buy them; and their analysis times are too long on the order of sometimes of days.

This work describes the platforms of nanosensors based on lipid membranes that were investigated to detect compounds of clinical interest. The chapter provides novel routes for the design and nanofabrication of lipid film-based biosensors for the rapid detection and monitoring of compounds of clinical interest such as urea, cholesterol, glucose, antibiotics, hormones, toxins, etc.

7.2 The Preparation of Lipid Membranes

The methods of the preparation of lipid membranes have been extensively described in the literature and mainly of biosensors that are based on lipid films and are not prone to electrical or mechanical interferences and breakage (Nikolelis et al. 2006, 2008a).

These techniques can prepare lipid film biosensors that are stable in air and outside an electrolyte solution for period of times of more than a month; therefore these devices can be extensively used for practical applications. These biosensing devices have been used in electrochemical experimentation. An exception is the development of stabilized polymerized lipid films on a filter paper that switch on and off their fluorescence and therefore belong to optical biosensors. Tien and Salamon (1989) in the past have suggested a simple and reliable technique for the construction of bilayer lipid membrane (sBLM) that was stable outside an electrolyte solution and was not prone to electrical or mechanical breakage; these sensors were prepared at a tip of a freshly cut tip of a Teflon-coated metallic wire, and their stability is due to the nascent metallic surface. The method has used a Teflon-coated stainless steel metal wire (0.1–0.5 mm in diameter) which was cut to provide a nascent surface while it was inside in lipid solution of phosphatidylcholine in a solvent of chloroform using a miniature guillotine. The tip of the wire is covered with the lipid solution that turns into a lipid membrane; when is placed in an electrolyte (i.e., 0.1 M KCl), the lipid film spontaneously thins into a self-assembled lipid bilayer membrane (sBLM) is formed.

7.3 Methods for the Preparation of Two Most Important Platforms of Stabilized Devices Based on Lipid Films

During the last decade, the construction of stabilized lipid film-based biosensors that do not break when electrical or mechanical shock is applied and are stable outside an electrolyte solution has been the investigation of a number of reports; these investigations will provide devices that can be commercialized due to their practical applications. Nanotechnological advances have provided a route to construct devices that their size is less than 1 μm size and therefore belong to the class of nanosensors. Techniques for the construction of this class of biosensors based on lipid films are further described and have a number of advantages such as ease of construction, rapid response times, small size high selectivity and sensitivity, and most importantly are stable outside an electrolyte solution that will allow them to be eventually commercialized.

7.3.1 Stabilized Lipid Films Formed on a Glass Fiber Filter

A route of the construction of stabilized in electrolyte lipid membranes was first reported by Nikolelis et al. group, and these films were prepared on glass Whatman filter disks (Nikolelis et al. 2008a); this has permitted a large number of evaluation and validation in real samples, i.e., the continuous monitoring of aflatoxin M_1 in milk and cheese products (Andreou and Nikolelis 1998). The lipid membranes were constructed on a GF/F glass microfiber, which has 0.9 cm of diameter and 0.7 μm nominal pore size (Andreou and Nikolelis 1998; Nikolelis et al. 1995).

Two Plexiglas blocks that were divided by a Saran-Wrap partition film (10 μm thick) folded in half (the Saran-Wrap film had a hole having 0.16 mm of radius) were used as the electroanalytical unit to construct these stabilized lipid film devices; a glass GF/F microporous fiber disk (Whatman™, UK) was centered in the hole

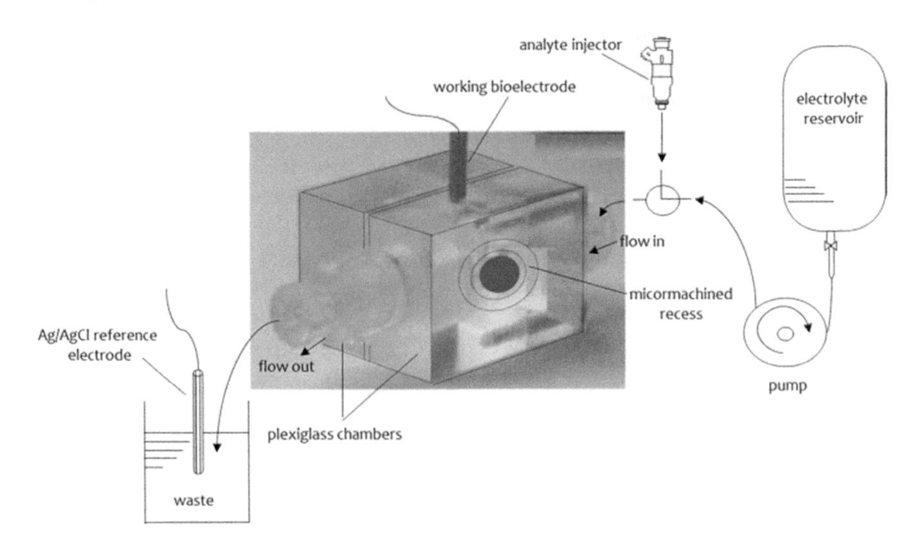

Fig. 7.1 A diagram of the setup used for the construction of lipid membranes that were stable in electrolyte. (From Nikoleli et al. 2018)

between the two plastic Saran-Wrap films. This plastic Saran-Wrap film was clamped between the two Plexiglas block units. The shape and dimensions of these chamber blocks were as follows: one of them was circular with a radius of 0.5 cm and depth 0.5 cm, and the other one was cylindrical. An Ag/AgCl electrode which acted as a reference was positioned in the wastes. This chamber was connected with a carrier electrolyte flow system. The cylindrical chamber had a circular upper hole (diameter 0.2 cm^2) and an elliptical lower hole. An Ag/AgCl electrode was positioned at the center of the cylindrical cell, and a 25 mV was applied between the two reference electrodes; at the same time, the current was measured using a Keithley instrument. A volume of 75 microliters samples were injected that contained the analyte. This experimental setup was inside a grounded Faraday cage. A simplified diagram of the instrumentation used is provided in Fig. 7.1. The stabilized in electrolyte solution lipid membranes were formed by established procedure (Andreou and Nikolelis 1998; Nikolelis et al. 1995): Once the lipid films were formed, the current was brought down at the pA levels, and gramicidin D is used to provide the bimolecular structure of these bilayers.

7.3.2 Polymer-Supported Bilayer Lipid Membranes

The polymeric stable in air lipid membranes were constructed as previously reported (Nikolelis et al. 2006, 2008a). UV irradiation and not heating at 60 °C is preferable, because the latter deactivates protein molecules. Differential scanning calorimetry and Raman spectrophotometry have shown that it is required 4 h to finish the polymerization.

7.3.3 Polymeric Lipid Membranes Supported on Graphene Microelectrodes

Graphene nanomaterials have been extensively utilized in biosensors because of their advantages such as enhanced physicochemical properties (mechanical, electrical, and thermal), high biocompatibility, and low toxicity. The large surface-to-volume ratio minimizes the device size and provides rapid response times and lower sensitivity without biofouling. Our group has constructed a device that was composed from a stabilized lipid membrane on graphene electrodes (Nikoleli et al. 2012; Bratakou et al. 2015). These nanodevices were used for the rapid determination of food toxicants, environmental pollutants, and compounds of clinical interest (Bratakou et al. 2016, 2017; Karapetis et al. 2016).

The preparation of graphene electrodes was extensively reported (Bratakou et al. 2016, 2017; Karapetis et al. 2016). *N*-methyl-pyrrolidone (NMP) was mildly sonicated for about 7.5 days and centrifuged at 700 rpm for 120 min. This dispersion was poured onto a copper wire (0.25 mm in diameter) which was positioned on a GF/F microfiber disk, and the solvent was slowly converted into vapor. The copper acted as the connection for the electrochemical experiments.

The method of preparation of the lipid membrane biosensors was previously described in detail (Bratakou et al. 2016, 2017; Karapetis et al. 2016): The stable lipid membranes were constructed following the polymerization stage as above.

"Receptor" molecules were placed in the lipid film devices before they were polymerized by injecting 15 μL of the "receptor" solution on the filter. The filter-supported polymeric BLMs were finally mounted onto the electrode.

7.3.4 Polymerized Lipid Membranes on ZnO Electrodes

Nanostructured ZnO is a promising material for the construction of nanoelectrodes for food, environmental, and clinical applications because it has a large number of advantages such as low cost, ease of preparation, biocompatibility, and catalytic surface activity. Other advantages include high isoelectric point (IEP) and nanostructured ZnO electrodes have high sensitivity and small size. IEP of ZnO is 9.5 which is higher than the IEP of a large number of biomolecules, and therefore it can be used as a matrix to immobilize these compounds through electrostatic bonding. ZnO nanoelectrodes has widely been used for the construction of devices to detect medical important compounds such as cholesterol, glucose, L-lactic acid, uric acid, metal ions, and pH.

7.3.4.1 Potentiometric Biosensors Based on ZnO Nanowalls and Stabilized Polymerized Lipid Film

The unmodified ZnO nanowall electrodes on an aluminum foil can be constructed by the sonochemical technique of Nayak et al. (2012) which briefly is as follows: Unimolar solutions of zinc nitrate hexahydrate, with almost 100% purity, and $C_6H_{12}N_4$ are mixed using 350 rpm for 5 min. An Al coated wire was then placed in this solution and sonicated for about 10 m. The electrode was then washed with distilled water and placed in a N_2 atmosphere.

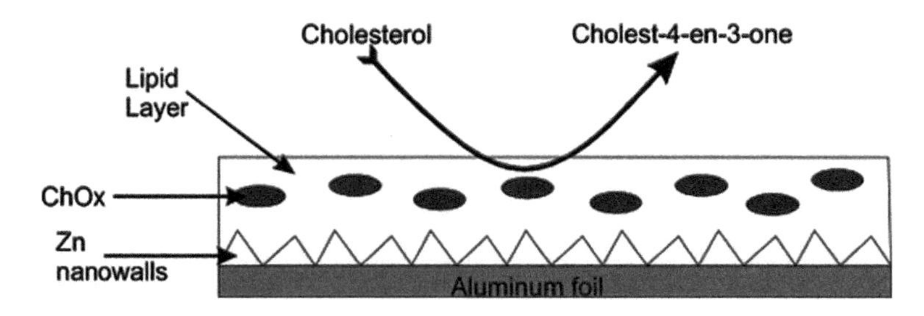

Fig. 7.2 A simple diagram of the device design. (From Psychoyios et al. (2013) with permission)

The preparation of the polymeric stable lipid membranes for the determination of cholesterol has been previously described (Psychoyios et al. 2013), and an extensive procedure was previously has been provided (Psychoyios et al. 2013).

Cholesterol oxidase was placed in these lipid membranes before these films were polymerized by placing 15 μL of the enzyme suspension on these filter disks using a microliter syringe.

The final stage to construct the device was to encapsulate the polymeric lipid membranes onto the wire that contained the electrode. Figure 7.2 shows a simplified diagram of the device.

7.3.4.2 Potentiometric Biosensors Based on Lipid Stabilized Membranes ZnO Nanowires

The biosensor was constructed it was previously reported (Usman Ali et al. 2011). This paper describes in detail the preparation of these electrodes (Usman Ali et al. 2011; Vaface and Youzbashizade 2007).

The enzyme (uricase) was placed in the lipid membranes before the membranes were made polymeric as it was previously described (Tzamtzis et al. 2012). These sensors were used in flow injection analysis (FIA) experiments. The FIA system used was previously described in detail (Andreou and Nikolelis 1998; Tzamtzis et al. 2012). The design of the experimental apparatus was reported previously in literature (Andreou and Nikolelis 1998; Nikoleli et al. 2018; Tzamtzis et al. 2012). The Ag/AgCl reference electrode was immersed in the waste of the carrier electrolyte solution, whereas the ZnO electrode was placed into the cylindrical cell.

7.4 Practical Applications of Lipid Membrane Devices in Biomedical and Clinical Analysis

7.4.1 Applications of Lipid Film Devices Based on Polymeric Lipid Membranes

A synthetic "receptor" (calixarene) was prepared and immobilized on lipid membranes on glass microfiber filters. Calixarene was inserted into the lipid structure and provided a signal which was adequate to rapidly determine insecticides rapidly, with a sensitive and selective response, and was used to determine these compounds in real samples of fruits and vegetables (Nikolelis et al. 2008b). Similar devices were constructed to selectively and rapidly determine food hormones (i.e., naphthalene acetic acid) in fruits and vegetables (Nikolelis et al. 2008c) and a zinc in water (Nikolelis et al. 2009).

A report that exploits the construction of a receptor for the fast FIA monitoring of zinc and used stabilized lipid films on a methacrylate polymer on a fiber filter disk with an incorporated receptor (Nikolelis et al. 2009). This receptor was prepared by replacing the hydroxyl groups of resorcin into phosphoryl groups. This nanosensor was prepared specifically for the FIA monitoring of zinc and was based on these air stabilized lipid membranes that were polymeric. The nanobiosensor can determine zinc in a drop (75 μL) of the sample. The analyte (i.e., zinc) was injected into the flowing electrolyte streams of 0.1 M KCl electrolyte. A complex formation between the phosphoryl receptor and Zn^{2+} takes place and causes pre-concentration of the analyte at the lipid film/solution interface which in turn causes changes in the electrostatic fields and phase structure of the lipid films; as a result we have obtained ion current transients, and the peak height of these signals was correlated to the analyte concentration. The response times were on the order of ca 5 s, and Zn^{2+} could be determined at very low levels of concentration (i.e., nM detection levels). The analytical curve was linear in the concentration range $1.00 \times 10^{-7} - 1.20 \times 10^{-6}$ M with detection limits of 5.00×10^{-8} M and a rsd of less than 4%. Potent interferences were examined including a wide range of other metals, lipids, and proteins. As an analytical demonstration and evaluation of this technique, trace concentrations of Zn(II) were successfully determined in real samples of waters.

7.4.2 Applications of Graphene-Based Devices

A potentiometric urea lipid membrane-based minisensor on graphene has been appeared recently (Nikoleli et al. 2012). A potentiometric urea device based on lipid film technology on graphene nanosheets has been constructed; a simplified setup of this biosensor is shown in Fig. 7.3. The main characteristics of this biosensor are excellent reproducibility, sensitivity, selectivity, reusability, and rapid response times; the slope of the electrode is ca. 70 mV/decade over the urea logarithmic concentration range which can be determined from 1×10^{-6} M to 1×10^{-3} M.

The interactions of cholera toxin with polymeric lipid membranes with incorporated ganglioside GM1 were reported in the literature (Nikoleli et al. 2011). An injection of cholera toxin in the flowing streams of a KCl 0.1 M carrier solution provided a current signal. The peak height of the ion current signal was correlated with the concentration of cholera toxin in the sample solution and had detection limits of 0.06 μM.

The response times and detection limits were improved using polymerized lipid membranes on graphene nanosheets (i.e., response times of 5 min and detection limits of 1 nM) (Karapetis et al. 2016). The construction of this sensor was easy and has shown excellent reproducibility, reusability, selectivity, long shelf life, and a

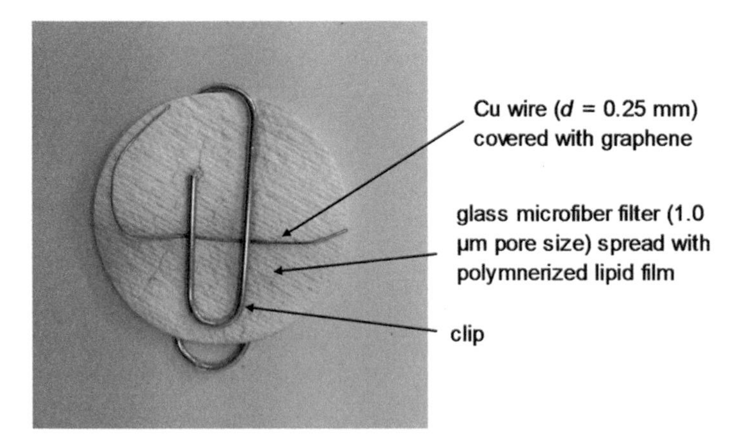

Fig. 7.3 Picture of the lipid film device on graphene minielectrode which was used for the potentiometric detection of urea. (Reprinted from Nikoleli et al. 2017)

slope of 60 mV/decade of toxin concentration. The method was evaluated/validated in lake water samples.

An electrochemical biosensor for the determination of saxitoxin based on graphene nanosheets with stable in air lipid membranes and immobilized anti-STX was provided in the literature (Bratakou et al. 2017). An excellent selectivity, sensitivity, and detection limits (1 nM) for the determination of saxitoxin with rapid response times (i.e., 5–20 min) were noticed. The sensor was easily constructed, lasted long periods of time with a slope of 60 mV/decade over saxitoxin concentration. The method was evaluated/validated for the determination of STX in lake waters and shellfish.

7.4.3 Applications of the ZnO Nanoelectrode-Based Devices

A potentiometric cholesterol device was constructed by immobilizing cholesterol oxidase into polymerized lipid membrane on ZnO nanowalls (Psychoyios et al. 2013). The enzyme was codeposited into the lipid membrane prior to polymerization on the ZnO nanowalls surface and provided a sensitive, selective, stable, and reproducible cholesterol device. The electrode slope was 57 mV/decade of cholesterol. No interferences were noticed by ascorbic acid, glucose, urea, proteins, and lipids. The nanosensor device has shown biocompatibility and could be implanted in the human body.

A uric acid electrochemical device was reported in the literature by immobilizing the enzyme uricase into polymerized lipid membranes on Zn nanowires (Usman Ali et al. 2011). The enzyme was codeposited with the lipid membrane prior to polymerization on the surface of the electrode. The biosensor was sensitive, selective, stable, and reproducible. The presence of a cationic lipid in membranes has increased the electrode slope by twofold. No interferences were observed by the presence of ascorbic acid, glucose, urea, proteins, and lipids.

ZnO nanowires (NW) were tailored for the immobilization of glucose oxidase in order to fabricate a glucose sensor (Zang et al. 2007). The high specific surface area and isoelectric point provide the electrode efficient immobilization of high concentration of acidic enzymes. The apparent Michaelis constants were adjusted by tailoring the thickness of the GOD/ZnO nanowire layer and the enzyme loading in the nanowires. Through this route, linear region of sensitivity and reaction rates could be obtained. The long-term stability of this biosensor was high due to the inorganic ZnO NW.

Well-aligned ZnO nanowires were constructed on gold-coated plastic substrates using a low-temperature aqueous chemical growth method (Hsu et al. 2017). These arrays had 50–130 nm diameters and were applied to construct a urea biosensor using urease within the concentration range 0.1 mM to 100 mM with logarithmic response. The electrode slope was 52.8 mV/decade for 0.1–40 mM of urea, and response times were less than 4 s; this urea biosensor had excellent selectivity and reproducibility and shown no response to interferents such as ascorbic acid and uric acid, glucose, and K(+) and Na(+) ions.

Well-aligned ZnO nanowires decorated with Pt nanoparticles (NPs) were recently used to construct a nonenzymatic glucose biosensor (Miao et al. 2016). The use of Pt NPs decoration increased the sensitivity by tenfold. The high specific surface area and isoelectric point (IEP) of ZnO have provided the electrode biocompatibility. A similar glucose biosensor on silicon NWs (ZnO/Si NWs) was also reported in the literature (Fung et al. 2017). These nanowire nanocomposites have shown an excellent amperometric sensitivity to glucose (129 $\mu A \cdot mM^{-1}$), low detection limits (12 μM), and good stability, reproducibility, and selectivity in the presence of common interferents.

A ZnO NWs/Au electrode was constructed by immobilizing DNA for the fast detection of breast cancer 1 (BRCA1) gene (Mansor et al. 2014). This DNA biosensor was able to detect the target sequence in the concentration range between 10.0 and 100.0 μM with a detection limit of 3.32 μM. A sensitive and selective label-free DNA ZnO NW device which was based on a Schottky contacted was also reported in the literature (Cao et al. 2016). The performance of this device was greatly increased by the use of piezotronic effect (Cao et al. 2016).

7.5 Conclusions and Future Prospects

The present paper describes the recent platforms which are based on lipid membranes and used for biomedical applications for the rapid detection of analytes of clinical interest. These technologies include the construction of stable in solution and in air and are supported on microfiber glass filters and are polymerized on graphene and ZnO microelectrodes. The polymeric lipid film devices can be portable and used in the field. These biosensors have detection limits in the nM concentrations. It is expected soon to commercially prepare units for market production.

The results have exhibited that these lipid membrane-based detectors can be stored and used after remaining in the air for periods of 1 month and can be easily constructed at low cost. The response times of these nanosensors are on the order of s and are not bulky and much cheaper than chromatographic units; these detectors can be complimentary to LC and gas chromatographic instruments for biomedical applications in clinical analysis. These toxicants include toxins, metals, hormones, urea, glucose, cholesterol, etc. with high sensitivity and selectivity, rapid response times, portability, etc.

References

Andreou VG, Nikolelis DP (1998) Flow injection monitoring of aflatoxin M_1 in milk and milk preparations using filter-supported bilayer lipid membranes. Anal Chem 70:2366–2371

Bratakou S, Nikoleli G-P, Nikolelis DP, Psaroudakis N (2015) Development of a potentiometric chemical sensor for the rapid detection of carbofuran based on air stable lipid films with incorporated calix[4]arene phosphoryl receptor using graphene electrodes. Electroanalysis 27:2608–2613

Bratakou S, Nikoleli G-P, Siontorou CG, Nikolelis DP, Tzamtzis N (2016) Electrochemical biosensor for naphthalene acetic acid in fruits and vegetables based on lipid films with incorporated auxin-binding protein receptor using graphene electrodes. Electroanalysis 28:2171–2177

Bratakou S, Nikoleli G-P, Siontorou GC, Nikolelis DP, Karapetis S, Tzamtzis N (2017) Development of an electrochemical biosensor for the rapid detection of saxitoxin based on air stable lipid films with incorporated Anti-STX using graphene electrodes. Electroanalysis 29:990–997

Cao X, Cao X, Guo H, Li T, Jie Y, Wang N, Wang ZL (2016) Piezotronic effect enhanced label-free detection of DNA using a Schottky-contacted ZnO nanowire biosensor. ACS Nano 10:8038–8044

Chandra P (ed) (2016) Nanobiosensors for personalized and onsite biomedical diagnosis, IET Digital Library, July 2016.

Chandra P, Noh H-B, Won M-S, Shim Y-B (2011) Detection of daunomycin using phosphatidylserine and aptamer co-immobilized on Au nanoparticles deposited conducting polymer. Biosens Bioelectron (*11*):4442–4449

Chandra P, Tan Y-N, Singh S (2017) Next generation point-of-care biomedical sensors technologies for cancer diagnosis. In: Springer

Fung CM, Lloyd JS, Samavat S, Deganello D, Teng KS (2017) Facile fabrication of electrochemical ZnO nanowire glucose biosensor using roll to roll printing technique. Sensors Actuators B Chem 247:807–813

Hsu CL, Lin JH, Hsu DX, Wang SH, Lin SY, Hsueh TJ (2017) Enhanced non-enzymatic glucose biosensor of ZnO nanowires via decorated Pt nanoparticles and illuminated with UV/green light emitting diodes. Sensors Actuators B 238:150–159

Karapetis S, Nikoleli G-P, Siontorou CG, Nikolelis DP, Tzamtzis N, Psaroudakis N (2016) Development of an electrochemical biosensor for the rapid detection of cholera toxin based on air stable lipid films with incorporated ganglioside GM1 using graphene electrodes. Electroanalysis 28:1584–1590

Mahato K, Maurya PK, Chandra P (2018) Fundamentals and commercial aspects of nanobiosensors in point-of-care clinical diagnostics. Biotech *8*:149. https://doi.org/10.1007/s13205-018-1148-8

Mansor NA, Zain ZM, Hamzah HH, Noorden MSA, Jaapar SS, Beni V, Ibupoto ZH (2014) Detection of Breast Cancer 1 (BRCA1) gene using an electrochemical DNA biosensor based on immobilized ZnO nanowires. Open J Appl Biosens 3:9–17

Miao F, Lu X, Tao B, Li R, Chu PK (2016) Glucose oxidase immobilization platform based on ZnO nanowires supported by silicon nanowires for glucose biosensing. Microelectron Eng 149:153–158

Mueller P, Rudin DO, Tien HT, Wescott WC (1962) Reconstitution of cell membrane structure in vitro and its transformation into an excitable system. Nature 194:979–980

Naval, A.P., . Katzenmeyer, A.M., Gosho, Y., Tekin, B., Islam, M.S., Sonochemical approach for rapid growth of zinc oxide nanowalls, Appl Phys A, 2012, 107 (3), 661–667.

Nikoleli G-P, Nikolelis DP, Tzamtzis N (2011) Development of an electrochemical biosensor for the rapid detection of cholera toxin using air stable lipid films with incorporated ganglioside GM1. Electroanalysis 23(9):2182–2189

Nikoleli G-P, Israr MQ, Tzamtzis N, Nikolelis DP, Willander M, Psaroudakis N (2012) Structural characterization of graphene nanosheets for miniaturization of potentiometric urea lipid film based biosensors. Electroanalysis 24:1285–1295

Nikoleli G-P, Siontorou CG, Nikolelis DP, Bratakou S, Karapetis S, Tzamtzis N (2017) Biosensors based on lipid modified graphene microelectrodes. Carbon 3(1):9. https://doi.org/10.3390/c3010009

Nikoleli G-P, Nikolelis D, Siontorou CG, Karapetis S (2018) Lipid membrane nanosensors for environmental monitoring: The art, the opportunities, and the challenges. Sensors 18(1):284

Nikolelis DP, Nikoleli G-P (2018) Nanotechnology and biosensors, 1st edn, Elsevier

Nikolelis DP, Siontorou CG, Andreou VG, Krull UJ (1995) Stabilized bilayer-lipid membranes for flow-through experiments. Electroanalysis 7:531–536

Nikolelis DP, Raftopoulou G, Nikoleli GP, Simantiraki M (2006) Stabilized lipid membrane based biosensors with incorporated enzyme for repetitive uses. Electroanalysis *18*:2467–2474

Nikolelis DP, Raftopoulou G, Chatzigeorgiou P, Nikoleli GP, Viras K (2008a) Optical portable biosensors based on stabilized lipid membrane for the rapid detection of doping materials in human urine. Sensors Actuators B Chem 130:577–582

Nikolelis DP, Raftopoulou G, Simantiraki M, Psaroudakis N, Nikoleli G-P, Hianik T (2008b) Preparation of a selective receptor for carbofuran for the development of a simple optical spot test for its rapid detection using stabilized in air lipid films with incorporated receptor. Anal Chim Acta 620:134–141

Nikolelis DP, Ntanos N, Nikoleli G-P, Tampouris K (2008c) Development of an electrochemical biosensor for the rapid detection of naphthalene acetic acid in fruits by using air stable lipid films with incorporated auxin-binding protein 1 receptor. Protein Pept Lett 15:789–794

Nikolelis DP, Raftopoulou G, Psaroudakis N, Nikoleli G-P (2009) Development of an electrochemical chemosensor for the rapid detection of zinc based on air stable lipid films with incorporated calix4arene phosphoryl receptor. Int J Environ Anal Chem 89:211–222

Prasad S (2014) Nanobiosensors: the future for diagnosis of disease? Dovepress *3*:1–10

Psychoyios VN, Nikoleli G-P, Tzamtzis N, Nikolelis DP, Psaroudakis N, Danielsson B, Israr MQ, Willander M (2013) Potentiometric cholesterol biosensor based on ZnO nanowalls and stabilized polymerized lipid film. Electroanalysis 25(2):367–372

Ti Tien H, Salamon Z (1989) Formation of self-assembled lipid bilayers on solid substrates. J Electroanal Chem 276:211–218

Tzamtzis N, Psychoyios VN, Nikoleli G-P, Nikolelis DP, Psaroudakis N, Willander M, Israr MQ (2012) Flow potentiometric injection analysis of uric acid using lipid stabilized films with incorporated uricase on ZnO nanowires. Electroanalysis 24(8):1719–1725

Usman Ali SM, Alvi NH, Ibupoto Z, Nur O, Willander MB, Danielsson B (2011) Selective potentiometric determination of uric acid with uricase immobilized on ZnO nanowires. Sensors Actuators B *152*:241–247

Vaface M, Youzbashizade H (2007) Production of zinc oxide nanoparticles by liquid phase processing: an investigation on optical properties. Mater Sci Forum *553*:252–256

Zang J, Li CM, Cui X, Wang J, Sun X, Dong H, Sun CQ (2007) Tailoring Zinc Oxide Nanowires for High Performance Amperometric Glucose Sensor. Electroanalysis 19(9):1008–1014

Dextran-based Hydrogel Layers for Biosensors

8

Andras Saftics, Barbara Türk, Attila Sulyok, Norbert Nagy, Emil Agócs, Benjámin Kalas, Péter Petrik, Miklós Fried, Nguyen Quoc Khánh, Aurél Prósz, Katalin Kamarás, Inna Szekacs, Robert Horvath, and Sándor Kurunczi

Abstract

Biofunctional coatings are key elements of biosensors regulating interactions between the sensing surface and analytes as well as matrix components of the sample. These coatings can improve sensing capabilities both by amplifying the target signal and attenuating interfering signals originating from surface fouling (non-specific binding). Considering the tested materials so far, hydrogel-based layers have been verified to be among the most effective layers in improving biochip performance. The polysaccharide dextran can be efficiently used to form hydrogel layers displaying extended three-dimensional structure on biosensor surfaces. Owing to their high water content and flexible structure, dextran coatings present advanced antifouling abilities, which can be exploited in classic bioanalytical measurements as well as in the development of cell-on-a-chip

A. Saftics · B. Türk · A. Prósz · I. Szekacs · R. Horvath (✉) · S. Kurunczi
Nanobiosensorics Laboratory, Centre for Energy Research, Hungarian Academy of Sciences, Budapest, Hungary

A. Sulyok
Thin Film Physics Department, Centre for Energy Research, Hungarian Academy of Sciences, Budapest, Hungary

N. Nagy · E. Agócs · B. Kalas · P. Petrik · M. Fried
Photonics Department, Centre for Energy Research, Hungarian Academy of Sciences, Budapest, Hungary

N. Q. Khánh
Microtechnology Department, Centre for Energy Research, Hungarian Academy of Sciences, Budapest, Hungary

K. Kamarás
Institute for Solid State Physics and Optics, Wigner Research Centre for Physics, Hungarian Academy of Sciences, Budapest, Hungary

P. Chandra, R. Prakash (eds.), *Nanobiomaterial Engineering*,
https://doi.org/10.1007/978-981-32-9840-8_8

type biosensors. However, in spite of the numerous applications, the deep characterization of dextran layers has been missing from the literature. This phenomenon can be attributed to the challenging analysis of few nanometer-thick layers with high water content. The lack of available data is more pronounced regarding the layer behaviors under aqueous conditions. In this chapter we present various surface analytical methods (including biosensor-type techniques) suitable for the complex characterization of hydrogel coatings whose thickness ranges from few to several ten nanometers. As a case study, we focus on the analysis of carboxymethyl dextran (CMD) layers developed for waveguide-based label-free optical biosensor applications. Examination methodologies both under dry and aqueous conditions as well as testing of antifouling abilities are also presented.

Keywords
Dextran · Hydrogel · Label-free biosensor · OWLS · QCM · Ellipsometry · Non-specific binding

8.1 Introduction

The role of biofunctional coatings is the modification of solid supports (substrates) by thin films (thickness is usually below 1 μm) of synthetic or naturally derived materials using various types of surface chemistries. Generally, the aim of modification is to maintain the desired and controlled interaction between the support and the biological system (Knoll 2013). In the field of biosensors, although biofunctional coatings are usually not defined as special sensor elements, they are of high importance in exploiting specific and detectable signals from biosensor response (Chandra 2016; Chandra et al. 2017). These coatings or so-called interface chemistries make the sensitization of the transducer surface by biorecognition elements possible. The role of biosensor coatings is more significant when the detection is based on a label-free method. Label-free biosensors measure signals generated only by the physical presence of analytes (targets). Due to this detection principle, all kinds of sample components present in the detectable field (mainly on the sensor surface) can produce signal which is therefore interfered by the non-targets as well. This type of interference is called biofouling or more specifically non-specific binding (NSB). As a result of the fact that protein molecules are main components of biological samples and they can adsorb to various surfaces, the NSB signal originates primarily from proteins. Additionally, as the adhesion of mammalian cells is mediated by surface-protein interactions, adsorbed proteins promote the biofouling of surfaces by cells and cell clusters. The essential aim of using well-designed biofunctional coatings on biosensor surfaces is the minimization of biofouling by a protein- and cell-repellent layer and furthermore the amplification of the target-related specific signal. It has been shown that the thermodynamically unfavorable water exclusion in case of heavily hydrated coatings made of hydrogels significantly contributes to the protein-repellent abilities (Tanaka et al. 2013).

Hydrogels are hydrophilic polymers which are able to absorb large amount of water. Hydrogels have found applications in a number of fields including drug delivery systems (Sivakumaran et al. 2011; Secret et al. 2014; Liu et al. 2016), tissue engineering (Balakrishnan and Banerjee 2011; Zhang and Khademhosseini 2017; Liu et al. 2017; Xu et al. 2018) or biosensors (Mateescu et al. 2012; Lian et al. 2016; Peppas and Van Blarcom 2016; Tavakoli and Tang 2017). Hydrogel-based coatings are key elements of biosensors. The widespread application of hydrogels in biosensor design originates from surface plasmon resonance (SPR)-based biochip developments. The layers are designed to modify the physicochemical properties of the sensor transducer surface, which is originally not biocompatible, promotes the non-specific adsorption of proteins, and has reduced specific surface area (low immobilization capacity). The large number of conjugable sites provided by the 3D polymer network allows to accommodate huge amount of various biomolecules. Hydrogels provide a natural-like wet microenvironment for biomolecules enabling to maintain their stability. The structural similarity to the extracellular matrix (ECM), the porous framework, and – mainly in case of naturally derived polymers – the biomimetic characteristic all provide great benefits for cell survival, proliferation, and migration (Balakrishnan and Banerjee 2011; Liu et al. 2017).

The naturally derived polysaccharide dextran can be used to form high capacity hydrogel layers on biosensor chip surfaces improving measurement sensitivity. Also, conjugated with suitable anchor molecules, dextran can provide a mechanical support and ECM-mimicking microenvironment for living cells. Dextran is constructed of α-(1,6) linked anhydroglucose units in its linear chains. Its flexible branches (providing ca. 5% branching degree) are linked by α-(1,3) glycoside bonds. The dextran chemical structure can be seen in Fig. 8.1A. Using dextran layers on biosensor surfaces has been proven to be an efficient strategy to limit the NSB, increase the number of immobilized bioreceptors, and achieve higher signal-to-noise ratio, resulting in better sensitivity. The advanced antifouling ability of dextran mainly originates from its hydrophilic nature and flexible chains. The free motion of mobile chains plays a key role in generating high interfacial entropy and thus steric repulsion that has a determinate contribution (steric-entropic effect) against protein adsorption (Löfås and Johnsson 1990; McArthur et al. 2000; Massia et al. 2000).

Dextran layers present remarkable water absorption and swelling, resulting in even four times of increase in their thickness (Elender et al. 1996; Piehler et al. 1999). Originally, the term hydrogel is used for crosslinked polymers; however, without sticking to the pure definition, one can point that even layers made of uncrosslinked dextran present hydrogel properties or at least hydrogel-like behavior (Elam et al. 1984; Löfås and Johnsson 1990).

Owing to the biocompatibility, biomimetic characteristic, and ECM-like viscoelastic properties, hydrogels made of dextran derivatives are effectively used in tissue engineering as scaffolds and ECM model components (Ferreira et al. 2004; Lee et al. 2008; Cutiongco et al. 2014).

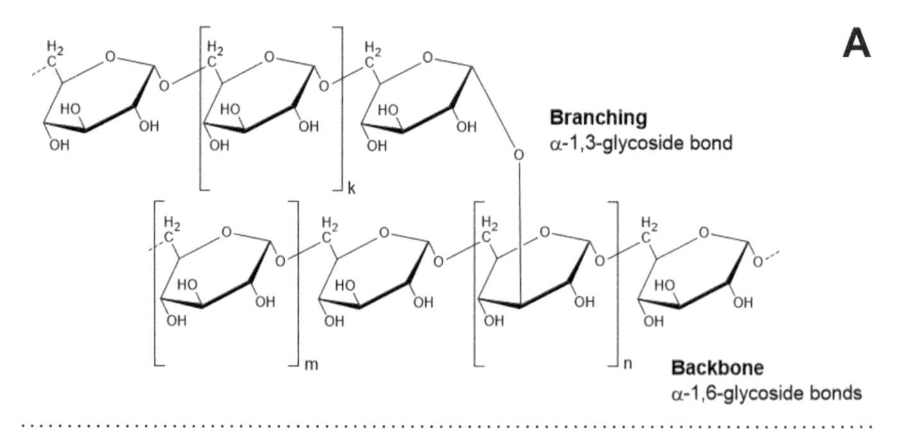

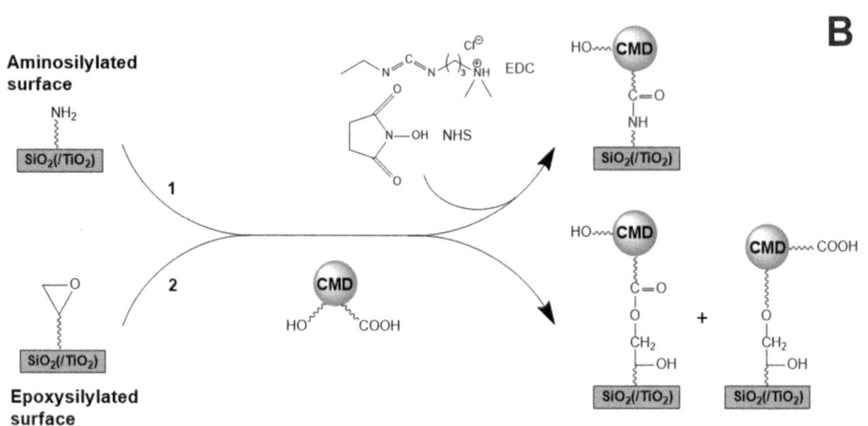

Fig. 8.1 (**A**) Chemical structure of dextran. The subscriptions *m*, *n*, and *k* indicate the number of anhydroglucose units in the backbone (*m*), branches (*n*), and branched chains (*k*). [Reprinted with permission from (Saftics 2018)]. (**B**) Reaction schemes of covalent grafting to amino- and epoxysilylated surfaces. While using aminosilane coating (**1**), EDC and NHS reagents are required, and the grafting chemistry results in amide bonds between the surface amino groups and CMD molecules. The grafting chemistry of using epoxide groups (**2**) can be performed without additional reagents, and it can form both ester and ether bonds between CMD and the surface. [Adapted with permission from (Saftics et al. 2016), copyright 2016 Elsevier]

Combining the achievements of biosensors and tissue engineering, label-free biosensors endowed with dextran-based hydrogel coatings are unique tools for revealing the surface-related events of cell clusters or even single cells, such as cell-cell or substrate-cell adhesion, spreading, migration, proliferation, and signaling. In spite of the number of advantageous properties and verified applications, dextran layers – mainly their structure – are poorly characterized, and descriptions about their fabrication methods are insufficient. The few available details about these coatings can be derived from the fact that the characterization of ultrathin, nanometers thick, and heavily hydrated polymer layers is still difficult and challenging, even with using the currently available modern surface analytical techniques.

To reveal the properties and dynamic behavior of such coatings, the deployment of various measurement techniques is in demand. Besides the classical bioanalytical applications, label-free biosensors and their unique surface sensitivity can be efficiently used to explore the structure and hydration of deposited polymer chains on sensing surfaces.

In this book chapter, we focus on the characterization of CMD hydrogel layers developed for various waveguide-based optical biosensor applications, such as for the analysis of controlled adhesion of living single cells and cell clusters. Due to the proposed future applications, the detailed analysis of the developed layers was highly demanded. We present our analytical methodologies applied for the characterization of layers both under dry and hydrated conditions. Regarding the different layer properties examined, our work covers the composition, topography, thickness, and wettability of the coatings. To reveal the hydration behavior and nanostructure in aqueous environment, both optical and mechanical biosensor techniques were employed. The analysis of the layers antifouling abilities is also presented.

8.2 Fabrication of Carboxymethyl Dextran Layers

Methods providing the surface grafting of stable dextran-based hydrogel layers with sufficient surface amount and thickness as well as with great antifouling ability and immobilization capacity are still under intensive research (Zhang and Horváth 2003). Generally, the fabrication of dextran layers is primarily empirically optimized, the published methods do not describe specific technical details, and the shared procedures are confusing in several cases.

Regarding the first biosensor applications, dextran layers were developed for SPR type sensors, therefore the original grafting methods apply thiol-based surface chemistries to attach dextran chains onto the gold surface of SPR chips (Löfås and Johnsson 1990; Löfås et al. 1993, 1995; Monchaux and Vermette 2007). Grafting methods to silica (glass) type surfaces are also available and can be used to functionalize waveguide-based optical transducers with dextran layers. These chemistries are mainly based on the silylation of the substrate resulting in a covalently bound silane film. This silane self-assembled monolayer (SAM) provides active functions for the covalent attachment of polysaccharide molecules. The most common silane reagents are epoxy- and aminosilanes (e.g., 3-glycidoxypropyltriethoxy silane (GOPS) and 3-aminopropyltriethoxysilane (APTES)). While CMD can directly react with epoxide groups both through its hydroxyl and carboxyl functions, the coupling of CMD via surface amino groups needs additional reagents, 1-ethyl-3-(3-dimethylaminopropyl)-carbodiimide hydrochloride (EDC) and N-hydroxysuccinimide (NHS) (Akkoyun and Bilitewski 2002). The EDC/NHS-based linking method is widely used, originally when protein molecules are intended to be coupled to other amine- or carboxyl-bearing surfaces or biomolecules. These grafting techniques are generally result in ultrathin layers whose thickness is typically in the range of 0.2–3.0 nm under dry conditions (Elender et al. 1996; Kuhner and Sackmann 1996). Thicker layers of polymers (thickness over 10–100 nm) can be achieved by the spin-coating method that is based on the spreading of a polymer solution on flat surfaces while the substrate is rotated. However, there are only very few available methods on the fabrication of dextran-based coatings using spin-coating, and they avoid detailing the stability of the layers. Specifically, procedures based on the combination of spin-coating and crosslinking in dextran layer fabrication have not been published so far, only examples for simple spin-coating of dextran can be found (Piehler et al. 1999; Linder et al. 2005).

In this study, we present the characterization of two different types of CMD layers varying in their thickness and presence of chemical crosslinks. We developed ultrathin (hereafter CMD-ut layers) as well as thicker (10–100 nm) CMD layers, in the latter case using spin-coating technique with adding crosslinking agent to the coating solution (hereafter CMD-sc layers). For the fabrication of CMD-ut layers, aminosilane- (CMD-ut-Am layers) and epoxysilane-based (CMD-ut-Ep layer) coupling chemistries were both applied (see Fig. 8.1B). The layers were prepared on SiO_2-TiO_2 waveguide type substrates and Si model wafers. The deposition of CMD layers onto the silylated substrate was performed using batch and flow-cell methods. The fabrication of CMD-sc coatings involved the grafting of a CMD-ut layer onto an aminosilylated substrate and then the spin-coating of CMD solution that contained sodium trimetaphosphate (STMP) as a crosslinker. Regarding both the prepared CMD-ut and CMD-sc layers, the samples were intensively washed to gain their stable form. Detailed fabrication methods of these layers can be found in our previous publications (Saftics et al. 2016, 2017, 2018).

8.3 CMD Layer Characterization

8.3.1 Surface Analytical Techniques

Various surface analytical techniques were used to characterize the developed CMD layers, including attenuated total reflection Fourier-transform infrared spectroscopy (ATR-FTIR), x-ray photoelectron spectroscopy (XPS), atomic force microscopy (AFM), spectroscopic ellipsometry (SE), contact angle (CA) measurements, optical waveguide lightmode spectroscopy (OWLS), quartz crystal microbalance with impedance measurement (QCM-I), as well as phase contrast microscopy. A list of the used analytical techniques as well as specific layer parameters offered by the measurement methods are summarized in Table 8.1. In the following sections, the applied characterization methods and their results are presented.

8.3.2 Composition, Thickness, and Topography of Ultrathin CMD Layers in Dry State: Characterization by ATR-FTIR, XPS, and AFM Measurements

Under dry conditions, the CMD-ut layers were characterized using ATR-FTIR, XPS, and AFM techniques. ATR-FTIR spectroscopy can be used to measure the infrared spectrum of thin layers deposited on the surface of an internal reflection

Table 8.1 Summary of the applied characterization techniques and the specific layer parameters obtained as measurement results

Analytical technique	Condition	Offered results
ATR-FTIR	Dry	Detection of carboxyl groups
XPS	Dry	Elemental composition, chemical states, thickness
SE	Dry	Lateral inhomogeneity, thickness, refractive index
AFM	Dry	Topography, surface roughness, thickness
CA meas.	Hydrated	Wettability, CA
OWLS	Hydrated	Refractive index, thickness, optical anisotropy, dry surface mass, protein-repellent ability
QCM-I	Hydrated	Viscoelastic parameters, hydrated surface mass, and thickness
Phase contrast microscopy	Hydrated	Cell-repellent ability

Adapted with permission from (Saftics 2018)

element the so-called ATR crystal. The sample layer absorbs infrared photons at certain wavelengths and attenuates the incident radiation through the evanescent field that is generated by the series of reflections on the crystal-layer interface. Graph **A** in Fig. 8.2 shows an ATR-FTIR spectrum of a CMD-ut-Am layer. The peaks at 1720–1740 cm^{-1} correspond to the characteristic stretching vibration of the carboxylic C=O groups of the grafted CMD (Zhang et al. 2005). This peak is also present in the bulk CMD at 1740 cm^{-1} (spectrum is not shown but can be found in our related paper (Saftics et al. 2016)).

XPS is used to measure the elemental composition of surfaces and thin layers by detecting the kinetic energy of photoelectrons ejected from the sample as a result of x-ray irradiation.

The goal of XPS measurements was the detection of CMD layers (by measuring changes in the elemental compositions and chemical states of elements) and determination of the thickness of CMD and silane layers under dry conditions. In contrast to aminosilane, epoxysilane does not contain distinctive heteroatom like N, and therefore the heteroatom-based thickness calculations could not be performed in case of CMD layers prepared with epoxysilane undercoating. Four types of sample wafers were measured to characterize the CMD-ut-Am layers, including bare Si, aminosilylated Si (Si/aminosilane), Si covered with CMD without aminosilane undercoating (Si/CMD, control sample), as well as aminosilylated Si covered with CMD (Si/aminosilane/CMD). Elemental compositions obtained on the examined samples can be seen in Fig. 8.2 B. In case of samples which were modified with aminosilane, a significant N signal could be observed. The Si/aminosilane/CMD samples presented particularly weakened signal of N, which was attributed to the covering effect of the CMD overlayer. Using the model that assumes an exponentially decaying probability of escaping a photoelectron with depth, the N signal provided the basis of determining the thickness of aminosilane and CMD layers (Rivière and Myhra 1998).

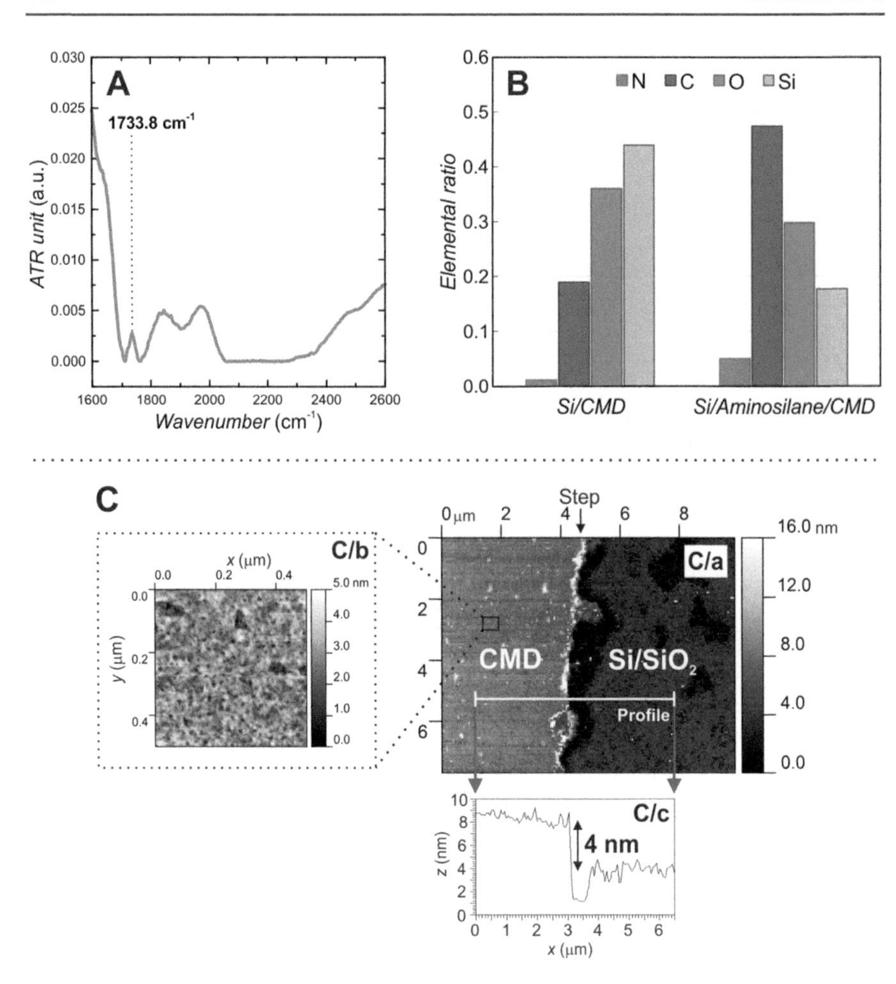

Fig. 8.2 Results on the characterization of CMD-ut layers under dry conditions. [Adapted with permission from (Saftics et al. 2016), copyright 2016 Elsevier]
(**A**) ATR-FTIR spectrum recorded on CMD-coated ATR crystal. The CMD-ut layer was grafted to the aminosilylated surface
(**B**) Elemental compositions of a CMD layer prepared without (sample Si/CMD) and with (sample Si/aminosilane/CMD) aminosilane undercoating
(**C**) AFM images about the surface of a CMD-ut-Am layer. **C/a.** The sample was prepared by gold lithography, resulting in a surface partially covered by the aminosilane/CMD coating, allowing to have a bare SiO_2 surface on the other half of the sample (10 × 10 μm image). **C/b**. 0.5 × 0.5 μm image representing the topography of CMD surface. Surface roughness RMS (root mean square) value of 1.1 nm was obtained, larger than the roughness value of native SiO_2 surface (0.2 nm), as expected. **C/c.** Height (*z*) profile of the aminosilane/CMD coating presented in image **C/a**

The thickness of the aminosilane layer was determined to be 2.30 nm. In case of control CMD samples, when the CMD was deposited onto a bare Si wafer without aminosilane, 0.07 nm was determined as the average thickness of CMD layer that means a weak partial coverage. This low value usually relates to slight contaminations representing only some physically adsorbed CMD molecules. Nevertheless, grafting the CMD to aminosilane precoating provided 0.73 ± 0.10 nm CMD layer thickness (average ± std., number of examined samples: six). According to this result, the aminosilane contributed to the efficient coupling of CMD to the surface that verifies the covalent grafting of CMD layer (Saftics et al. 2016).

AFM is a scanning probe microscopy technique that can be used for imaging surfaces at the nano- and micrometer level and for semiquantitative characterization of surfaces (e.g., measurement of surface roughness). Figure 8.2C presents AFM images captured on a CMD-ut-Am layer. The partly covered CMD sample was prepared using gold lithography. As a result, only the one-half of the Si wafer was coated with gold. After aminosilane and CMD grafting, the gold was removed allowing to achieve a sharp step between the aminosilane/CMD layer and uncoated Si (image **C**/**a**-**b**). Measuring the step heights, 3.2–5.8 nm (around 4 nm on the average) was determined for the total thickness of the covering layer (graph **C**/**c**). The resulting thickness of epoxysilane/CMD layers was in the same range. As the determined thickness corresponds to the combined thickness of the CMD and silane layer, the thickness of dry CMD coating itself should be below 4 nm. These findings are in good accordance with XPS measurements, which resulted in ca. 3.0 nm for the total thickness (2.30 nm aminosilane + 0.73 nm CMD) (Saftics et al. 2016).

It is worth highlighting that during the XPS and AFM measurements, the samples were kept in dry conditions. In this state, the CMD chains collapse and lie down on the surface forming a dense, very thin film. However, considering their real applications, CMD layers are applied in aqueous environments where they are in hydrated state. It is noted that in hydrating environments, CMD is able to swell up to even its multiple extent. Therefore, the characterization of CMD layers in their hydrated state is of high demand. For these examinations, in situ OWLS and QCM-I techniques were applied.

8.3.3 Nanostructure of Ultrathin CMD Layers Under Aqueous Conditions: Characterization by OWLS

Waveguide-based label-free optical biosensors detect refractive index changes occurring over the transducer surface. The sensing is performed by the evanescent field that is generated when the light inside the waveguide layer (that has the higher refractive index) is propagating by total internal reflections. The intensity of the reflected light, when meeting with the lower refractive index medium at the interface, extends over the interface into the medium generating an exponentially decaying (evanescent) electromagnetic field. Optical waveguide lightmode spectroscopy (OWLS) is a traditional setup of waveguide type biosensors. OWLS employs planar waveguides made of SiO_2-TiO_2 waveguide material where a zeroth-order waveguide mode (i.e., propagating standing electromagnetic wave) with two polarization states (transverse electric (TE) and transverse magnetic (TM) polarizations) is excited. The measurement of effective refractive index corresponding to the TE and TM waveguide modes enables one to determine the refractive index (n_A) and thickness (d_A) of the examined adlayer using the 4-layer mode equations as evaluation model. The adlayer surface mass density (M_A, with a common dimension of ng/cm^2) can be calculated from the previously determined n_A and d_A utilizing the de Feijter's formula (De Feijter et al. 1978):

$$M_A = \frac{d_A(n_A - n_C)}{dn/dc} \tag{8.1}$$

where dn/dc is the refractive index increment of the analyte in its solution. In case of, e.g., proteins, a value of 0.182 mL/g is used (De Feijter et al. 1978). Regarding the calculation of n_A and d_A by the 4-layer mode equations, the classical and most commonly used model assumes a homogeneous and isotropic adlayer. In case of ordered adlayer structures, the optical anisotropy results in optical birefringence. When a material is birefringent, its apparent average refractive index can be decomposed into ordinary ($n_{A,o}$) and extraordinary refractive indices ($n_{A,e}$), and their ratio refers to the sign of birefringence. With the application of two waveguide modes, only d_A and the averaged n_A can be determined. However, it is still possible to characterize the ordered structure of the adlayer using the homogenous and isotropic model (Kovacs et al. 2013). Horvath and Ramsden pointed out (Horvath and Ramsden 2007) that the values of n_A and d_A can be used to decide whether the layer is optically isotropic or it has an ordered and anisotropic structure. Applying the 4-layer homogenous and isotropic model in case of an adlayer structure that is anisotropic (quasi-isotropic analysis), unrealistic n_A is observed meaning that the resulting n_A over- or underestimates the realistic (expected) adlayer n_A (Horvath and Ramsden 2007) (see Fig. 8.3A). When the adlayer is negatively birefringent (conformation of oriented molecules parallel with the surface) overestimated n_A ($n_{A,o} > n_{A,e}$), whereas the adlayer is positively birefringent (conformation of oriented molecules perpendicular to the surface) underestimated n_A can be detected ($n_{A,o} < n_{A,e}$) (Horvath and Ramsden 2007). As a result, the observed n_A is an indicator of adlayer birefringence. Also, the value of under- or overestimation is in relationship with the extent of anisotropy. Even in case of unrealistic n_A and d_A, the surface mass density can be still precisely determined owing to the error compensation of the de Feijter formula (Horvath and Ramsden 2007).

The in situ OWLS measurements were performed using a microfluidic assembly (flow-cell) that was mounted into the measurement head of the OWLS setup, and it was applied to maintain a continuous flow of solutions over the inserted sensor chip surface (Orgovan et al. 2014). The CMD grafting experiments monitored by OWLS were carried out on amino- and epoxysilylated SiO_2-TiO_2 type sensor surfaces using different grafting solution pHs. The measured raw optical data (effective refractive indices of TE and TM modes (N_{TE}, N_{TM}) were evaluated by the homogenous isotropic 4-layer mode equations, providing the optical thickness (d_A) and apparent average refractive index (n_A) of the CMD-ut layer (De Feijter et al. 1978; Horvath and Ramsden 2007) and enabling to calculate the deposited surface mass density by the de Feijter formula.

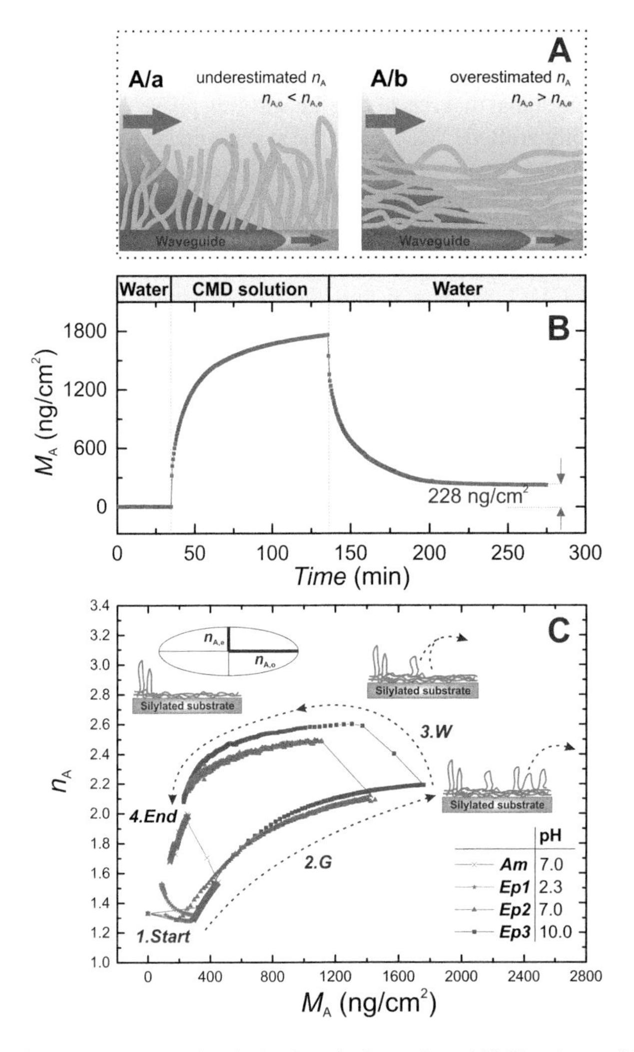

Fig. 8.3 OWLS measurement results obtained on the formation of CMD-ut layers. [Adapted with permission from (Saftics 2018) and from (Saftics et al. 2016), copyright 2016 Elsevier]
(**A**) Dependence of adlayer refractive index on different layer structures. In case of polymer chains perpendicular to the surface (**C/a**), positive birefringence, in case of chains parallel with the surface (**C/b**), negative birefringence can be observed by OWLS. The light propagating in the waveguide layer and its generated evanescent field over the surface are also shown
(**B**) Surface mass density (M_A) sensogram recorded during the deposition of CMD layer on an epoxysilylated OWLS sensor surface. The headers indicate the solutions flowed over the sample surface using a flow-cell
(**C**) Refractive index of CMD-ut layer as a function of deposited surface mass density. The curves represent different experimental conditions revealing the dependence of CMD layer structure during its formation on the applied silane precoating and pH of the grafting solution. The time-related direction of the measurements is indicated by the dashed arrows (*G* indicates the grafting, *W* the washing section of experiments). The inset table represents the different experimental conditions (*Am*, aminosilylated; *Ep*, epoxysilylated surfaces). The schemes above the curves illustrate the alteration of CMD layer nanostructure at the different experimental phases

Figure 8.3**B** The resulting mass curve of an in situ OWLS experiment revealing the kinetics of the formation of CMD-ut layer on an epoxysilylated OWLS chip surface. The stable signal that reached in the washing phase demonstrates the stability of the covalently grafted CMD layer. The surface mass of the CMD layer as well as the grafting efficiency was measured as the difference of mass values between the baseline and the end of the washing section

Our results revealing the alterations in the layers nanostructure throughout the grafting experiment as a function of deposited mass are presented in Fig. 8.3C. The conformation of CMD chains and thus the nanostructure was correlated with n_A values. As the realistic refractive index of hydrated dextran layers should be in the range of 1.36–1.52 (Piehler et al. 1999) (according to literature values), the obtained n_A = 1.52–2.21 values are clearly overestimated indicating a significant negative birefringence. Based on the quasi-isotropic analysis, the unrealistically high n_A values originate from mainly lying down CMD chains parallel with the surface. The n_A vs. M_A hysteresis curves of Fig. 8.3**C** help to understand the nanostructural alterations and layer formation mechanism by tracking the change of n_A originating from each added or removed CMD mass unit (Horvath et al. 2008, 2015; Escorihuela et al. 2015; Lee et al. 2015). The shown curves represent both the grafting (*G*) and washing (*W*) sections. The direction of the experiment is indicated by the dashed arrows, and the illustrations correspond to the supposed layer nanostructure at the given experimental section. We found that the nanostructure and amount (mass) of CMD layers significantly varied depending on the applied grafting conditions: the pH of grafting solution and grafting chemistry (type of silane coating).

Based on the high n_A values, it was observed that under neutral and alkalic conditions (pH 7.0 and 10.0), the conformation of CMD chains was dominantly parallel with the surface. Regarding the formation process, n_A was increasing during the grafting section and when the washing section started, n_A further increased indicating that first the weakly bound brush-like molecules were removed from the surface. Subsequently, in the washing section, CMD chains with lain down orientation were also desorbed; however the rate of their removal and the removed amount was smaller. The mechanism of CMD layer formation was significantly different at pH 2.3. In this case, the relative amount of lain down molecules was smaller which was supported by the n_A values (1.52) close to the realistic CMD refractive index. In the grafting section, the deposition of loops was predominant, and as a result, $n_A \approx 1.30$ was calculated at the end of the grafting phase. This refers to a layer composed of randomly oriented chains.

Nevertheless, the chains extending toward the solution were easily removed in the subsequent washing process, and finally, chains mainly with lain down conformations remained on the surface. The dependence of the CMD layer structure upon the pH on epoxysilylated surfaces is supposed to derive from the pH-dependent yield of the grafting reaction (reaction between epoxide and carboxyl/hydroxyl functions).

In case of the experiments performed on aminosilylated surfaces (pH 7.0), the shape of n_A vs. M_A curves was similar to those measured on epoxylated surfaces under neutral and basic conditions. However, as the increase of n_A was smaller, a more extended layer was assumed. This result obtained on the aminated surface suggested that compared to layers prepared on epoxylated surfaces at pH 7.0 and 10.0, the CMD was attached via smaller number of surface grafting points.

8.3.4 Viscoelastic Properties of Ultrathin CMD Layers: Characterization by QCM

We applied the QCM technique to reveal the viscoelastic properties of hydrated CMD-ut layers throughout the whole layer formation event. QCM is a label-free biosensor that applies mechanical transduction principle for sensing and for characterizing thin coatings. A QCM sensor chip is composed of a resonant piezoelectric quartz crystal disk contacted on its both sides with planar gold electrodes. The AC voltage applied on the electrodes generates shear oscillation in the piezoelectric crystal at its fundamental resonance frequency (5 MHz) and also at overtone frequencies. The oscillation results in a standing plane wave[1] (also known as acoustic wave), which vertically propagates through the crystal and penetrates into the medium over the crystal surface (Ferreira et al. 2009; Johannsmann 2015). As added mass on the sensor surface detunes the resonance frequencies (f_n, where n is the overtone number), QCM can measure the mass of deposited layers by recording the shift in the quartz crystal' resonance frequencies (Δf_n). When the added mass is not rigid and it has significant viscoelasticity (which is common in case of soft and hydrated biomaterial films (Ismail et al. 1996; Marx 2003; Laos et al. 2006; Kittle et al. 2011)), it also changes the decay characteristic of the penetrating wave. The decay characteristic is in connection with the energy dissipation of the oscillation that can be quantified by the dissipation factor (D_n or ΔD_n, measured for fundamental and each overtone frequency as well), and it can provide results on the adlayer viscoelastic properties. Moreover, QCM is also sensitive to the mass of solvent molecules coupled to the adlayer enabling to measure the hydrated mass. As viscoelastic properties and hydration cannot be measured by optical methods, these features make QCM unique among the surface analytical techniques.

Two models are mostly used for evaluating the measured Δf_n and ΔD_n data. The simplest model, the Sauerbrey equation does not take dissipation (and viscoelasticity) into account and presumes that the surface mass density (M_A) is a linear function of the measured normalized frequency shift ($\Delta f_n/n$) (Sauerbrey 1959):

$$\Delta M = -C\frac{\Delta f_n}{n} \tag{8.2}$$

[1]This principle is analogous to the measurement principle of evanescent field applying sensors, where a standing electromagnetic field over the waveguide layer is generated allowing to sense analytes.

where C refers to the mass sensitivity constant that depends on the quartz crystal physical properties.

In case of highly solvated adlayers presenting viscoelastic properties, the Sauerbrey equation is not valid. The quantitative evaluation requires to consider the layer viscoelastic behavior and continuum mechanical models, most commonly the Voigt-Kelvin model should be used. The most commonly applied implementation of this viscoelastic model was published by Voinova and co-workers, and it is usually referred as Voinova's equations (Voinova et al. 1999). The data evaluation is based on the simultaneous error-minimizing fit of the measured Δf_n and ΔD_n data with the model equations. The fit results in the thickness (d_A), shear viscosity (η_A), as well as shear elastic modulus (μ_A) of the formed adlayer, supposing a known layer mass density (ρ_A) (Voinova et al. 1999; Höök et al. 2001; Stengel et al. 2005). In case of layers made of heavily hydrated polymers, such as CMD, an approximation of 1000 kg/m^3 can be used for the value of ρ_A (Müller et al. 2005).

We used the QCM-I method (MicroVacuum Ltd., Budapest, Hungary) that employs impedance analysis for reading out frequency and dissipation data (QCM-I denotes QCM with impedance analysis). Similarly to OWLS, the in situ setup applies a flow-cell assembled in the measurement head and mounted on the sensor surface. The continuous (real-time) monitoring of surface events is achieved by the sequential measurement of impedance spectra enabling to obtain f_n and D_n data simultaneously at each measurement time. The in situ QCM experiments were performed with the same solution flow sections as in case of OWLS. The data analysis (model fit) was carried out using a home-developed evaluation program.

Our results obtained from in situ QCM-I measurements on the formation of CMD-ut-Am layers are shown in Fig. 8.4. The analysis of $\Delta f_n/n$ and ΔD_n data (graph **A** and **B**) can provide useful qualitative information about the layer. While the environmental noise can significantly disturb the penetrating acoustic wave at the fundamental resonance frequency (Dutta and Belfort 2007) ($n = 1$), only data for overtones $n > 1$ ($n = 3, 5, 7$) were considered. As it can be seen from the significant shifts in $\Delta f_n/n$ and ΔD_n data, large amount of CMD was deposited in the grafting section; however, the washing strongly affected the layer and a large amount of CMD was washed off. Besides the CMD amount, the viscoelastic properties also changed in the washing section indicated by the dissipation, which turned from a strongly dissipative state ($\Delta D_3 = 60 \times 10^{-6}$) to less dissipative ($4.8 \times 10^{-6}$). It can be assumed that a soft CMD layer formed during the grafting section; however, the loosely bound chains desorbed, and a thin CMD coverage remained. The adequacy of the application of Sauerbrey equation for evaluation was checked based on the magnitude of ΔD_n data. If ΔD_n exceeds the critical value of 2×10^{-6}, the layer cannot be treated to be rigid, and the Sauerbrey relationship is not valid (Vogt et al. 2004; Liu et al. 2011). The large values of ΔD_n ($\Delta D_3 = 4.8 \times 10^{-6}$) indicated a viscoelastic case; therefore Voinova's equations implemented in our home-developed code were used for the evaluation and quantification of layer properties. Δf_n and ΔD_n data obtained from fitting the measured curves are shown in Fig. 8.4, and as it can be seen, good fit quality could be achieved. The thickness ($d_A^{QCM,V}$), surface

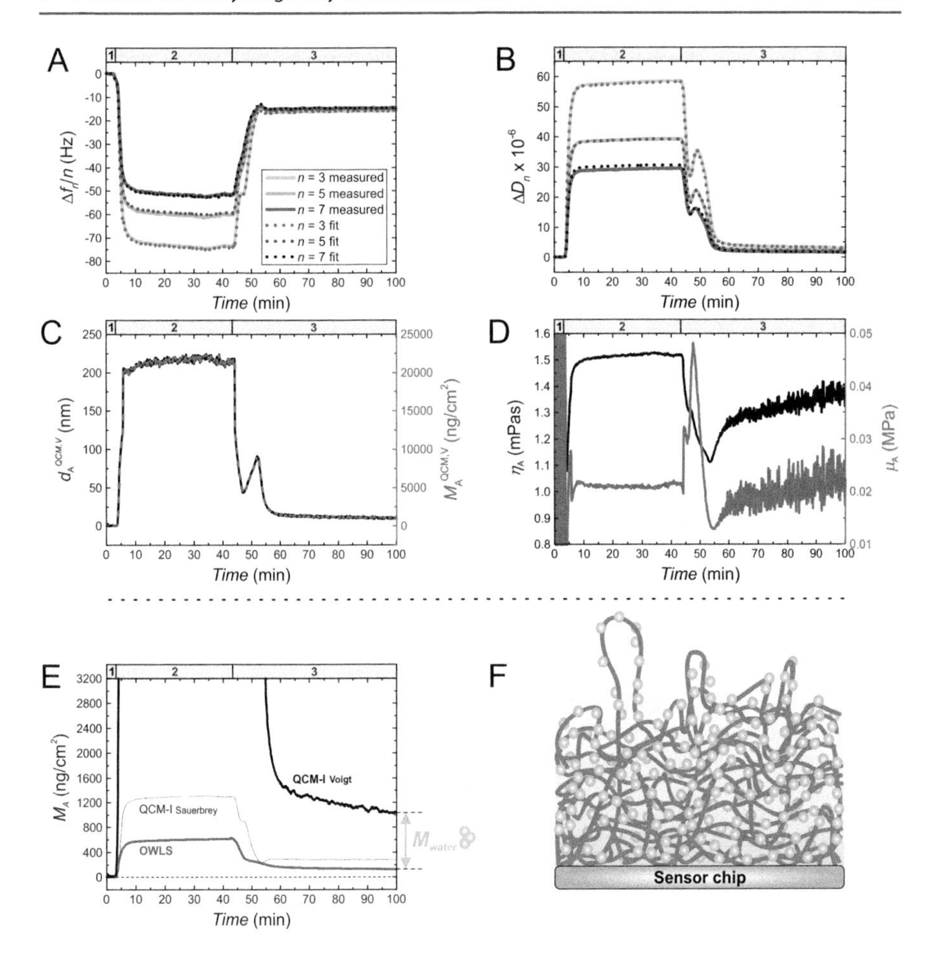

Fig. 8.4 QCM-I results measured on the formation of CMD-ut-Am nanolayers. [Adapted with permission from (Saftics et al. 2018), copyright 2018 Springer Nature Publishing AG, used under CC BY 4.0]

(**A, B**) Normalized frequency shift (**A**) and dissipation shift (**B**) data measured by QCM-I over the time of CMD-ut-Am layer formation on SiO_2-TiO_2 surfaces. The data correspond to the overtones $n = 3, 5, 7$. The in situ monitored experiments were carried out in three consecutive phases of solution flows as follows (shown by the header numbers): flow of polymer-free water to reach a stable baseline (1), polymer solution flow (2), and water flow applied to wash the weakly adsorbed molecules. Fitted curves of the measured $\Delta f_n/n$ and ΔD_n data using the Voigt-based viscoelastic model are also shown

(**C**) Full thickness and mass plots indicated by solid black as well as dashed red curves, respectively. The mass was simply proportional with the thickness by the value of layer density ($\rho_A = 1000$ kg/m^3)

(**D**) Shear viscosity (black) and shear elastic modulus (red) curves calculated using the Voigt-based model

(**E**) Surface mass density measured by OWLS (solid red line) and QCM-I (solid black line). The thicker black line indicates the mass calculated by the Voigt-based model, the thinner black line refers to the third overtone mass calculated by the Sauerbrey equation. M_{water} indicates the mass of coupled water

(**F**) The illustration represents the final structure of the formed CMD-ut-Am layer together with coupled water molecules

Table 8.2 Summary of parallel OWLS and QCM-I results measured on CMD-ut layers: main physical properties and their comparison with literature values

	Measured	References
	CMD	PLL-*g*-D[c], D[d], CMC[e]
M_A^{OWLS} (ng/cm^2)	124 ± 23	–
$M_A^{QCM,S3}$ (ng/cm^2)	277 ± 28	–
$M_A^{QCM,V}$ (ng/cm^2)	1102 ± 487	–
φ_A (%)	89	57[c] (Nalam et al. 2013), 60–70[c] (Perrino 2009)
n_A	1.66 ± 0.22	–
d_A^{OWLS} (nm)	0.9 ± 0.5	–
$d_A^{QCM,S3}$ (nm)	2.8 ± 0.3	–
$d_A^{QCM,V}$ (nm)	11.0 ± 4.9	–
η_A (mPa·s)	1.43 ± 0.27	0.50 ± 0.26[d] (Kuhner and Sackmann 1996)
μ_A (MPa)	0.03 ± 0.01	0.1–0.2[e] (Liu et al. 2011)

The data are presented as averaged values ± standard deviations calculated from three repeated experiments. The data represent the remained and stably grafted layers reached at the end of washing section. The superscript indications refer to the following: [*], the reference data correspond to different polymers; [b], obtained on PLL-*g*-D (D as dextran); [c], on dextran (D); as well as [d], on carboxymethyl cellulose (CMC)

M_A^{OWLS}, surface mass density recorded by OWLS; $M_A^{QCM,S3}$, $M_A^{QCM,V}$, surface mass density recorded by QCM-I and evaluated using the Sauerbrey model or Voigt-based model, respectively; φ_A, hydration degree; n_A, d_A^{OWLS}, refractive index, as well as optical thickness of CMD adlayer obtained by OWLS; $d_A^{QCM,S3}$, $d_A^{QCM,V}$, thickness of CMD adlayer obtained by QCM-I and evaluated using the Sauerbrey equation or Voigt-based model, respectively; η_A, μ_A, shear viscosity and shear elastic modulus of CMD adlayer, calculated using the Voigt-based model. References are designated as shown

mass density ($M_A^{QCM,V}$), and viscoelastic data (η_A, μ_A) calculated by the fits are shown in Fig. 8.4. The resulting data representing the remained stable layer after washing are presented in Table 8.2. Herein, literature data about similar layers are also shown to provide a basis for comparison (Saftics et al. 2018).

8.3.5 Hydration Properties of Ultrathin CMD Layers: Combining QCM and OWLS Results

As a consequence of the lack of individual analytical techniques capable of measuring heavily hydrated objects at the nanometer scale, the nanostructure and dynamic behavior of ultrathin layers in aqueous environment is still poorly characterized. A main limitation of the optical techniques is that water molecules, coupled to the layer, cannot be distinguished from the aqueous background (bulk phase), and therefore they do not provide change in the refractive index signal.

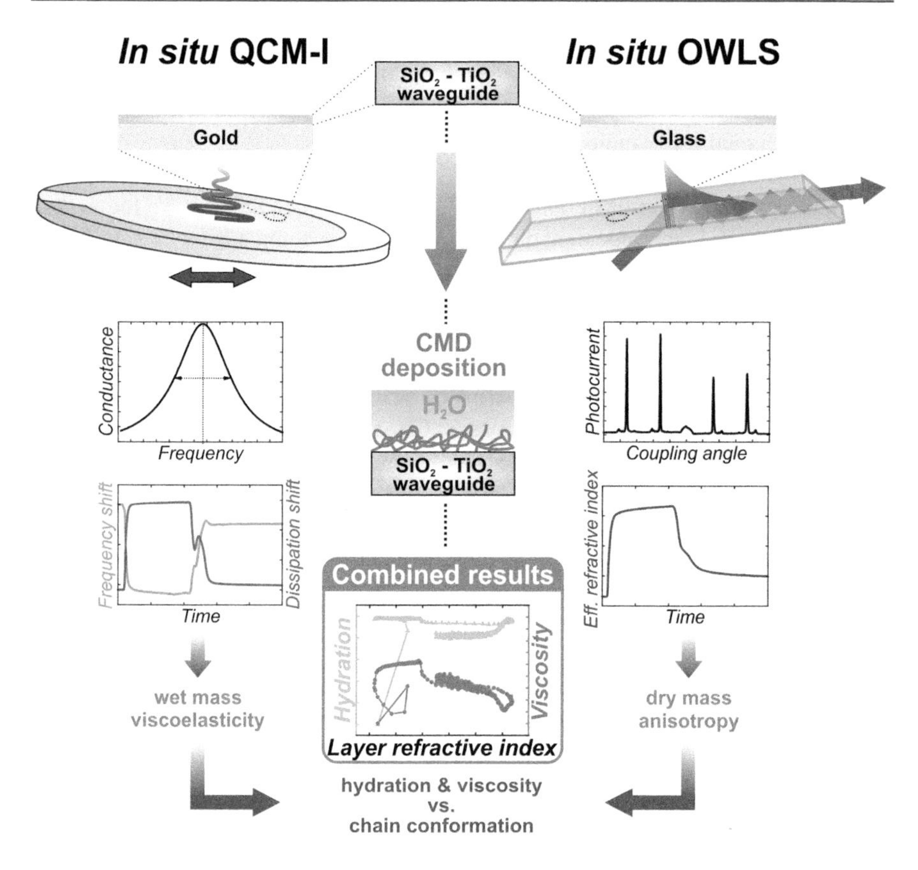

Fig. 8.5 Scheme about the methodology applied for exploring the formation as well as hydration and nanostructure of ultrathin CMD layers using parallel in situ OWLS and QCM-I measurements. We used chemically identical substrate surfaces in both measurement techniques. [Adapted with permission from (Saftics et al. 2018), copyright 2018 Springer Nature Publishing AG]

Consequently, only the "dry" mass of an analyte layer can be measured (Höök et al. 2002; Vörös 2004). In contrast to OWLS, QCM is sensitive to solvent molecules coupled to the layer, since it measures all the masses oscillating with the crystal. This combined "wet" mass cannot be uncoupled to the mass of dry adlayer and mass of bound solvent molecules. However, the combination of OWLS and QCM provides a special tool to determine the hydration degree of nanolayers (φ_A) by utilizing both the optical ("dry" OWLS mass, M_A^{OWLS}) and mechanical ("wet" QCM-I Voigt mass, M_A^{QCM}) mass data (Höök et al. 2001, 2002; Vörös 2004; Müller et al. 2005). A scheme of the combined evaluation methodology is shown in Fig. 8.5.

The dry and wet surface mass density values as a function of time measured during the formation of a CMD-ut-Am layer are plotted in Fig. 8.4E (specific mass data corresponding to the remained stable layer are shown in Table 8.2). The difference between the dry and wet mass, highlighted by M_{water}, refers to the mass of water molecules coupled to the CMD layer. The hydration degree (φ_A) of the stable layer was determined to be 89%. An assumed structure of the remained stable layer including coupled water molecules is also shown (**F**).

The combination of OWLS and QCM-I data also enabled us to characterize the layer nanostructure in terms of hydration and viscosity (see Fig. 8.5). We found that the conformational rearrangement during the washing accompanied with significant change in the layer hydration and viscosity. For more details see our related publication (Saftics et al. 2018).

8.3.6 Thickness and Composition of Spin-Coated and Crosslinked CMD Layers: Characterization by Ellipsometry and XPS

Spectroscopic ellipsometry is an optical technique that measures the polarization state of light reflected from the sample in order to measure the optical properties and thickness of the constituting layers. The wavelength spectra of the so-called ellipsometric angles (Ψ and Δ, characterizing the polarization state) are evaluated based on the fit of an adequate optical model, which is a function of the structure and physical parameters of the sample.

Ellipsometry could be effectively used to characterize the CMD-sc layers and determine their thickness and refractive index. It is important to emphasize that the thickness of these layers was over 10 nm. Below this value, in case of few nanometers thick CMD-ut coatings, CMD could not be separated from the substrate, and therefore, reliable layer parameters could not be determined. In order to control the thickness, the CMD-sc layers were fabricated with applying different spin-coating rotational speeds. Mapping mode spectroscopic ellipsometry measurements were performed to obtain thickness and refractive index maps about the examined samples. We developed an ellipsometric optical model that could be effectively used for the evaluation of the measured Ψ and Δ spectra with the main goal of determining the thickness of dry CMD-sc layers.

The calculations were performed in the 450–900 nm wavelength range. The refractive indices corresponding to the different layers were modeled using the Cauchy's equation. Fitted Ψ and Δ spectra of a measured CMD-sc sample (**A**) and evaluation results (**C**, **D**, and **E**) can be seen in Fig. 8.6, where details of the used optical model developed for the specific evaluations are also shown (**B**). The model presumes vertically inhomogeneous (gradient) refractive index in the SiO_2-TiO_2 waveguide layer which was divided into five slices with increasing refractive indices in depth. This presumption was based on the expectation that the density of the layer from the bottom to the top was decreased by the standard acidic cleaning applied on the OWLS sensor chips. The wavelength dependence of CMD refractive index characterized by the B parameter was determined in separate measurements. For these measurements, the CMD layers were spin-coated on gold substrates,

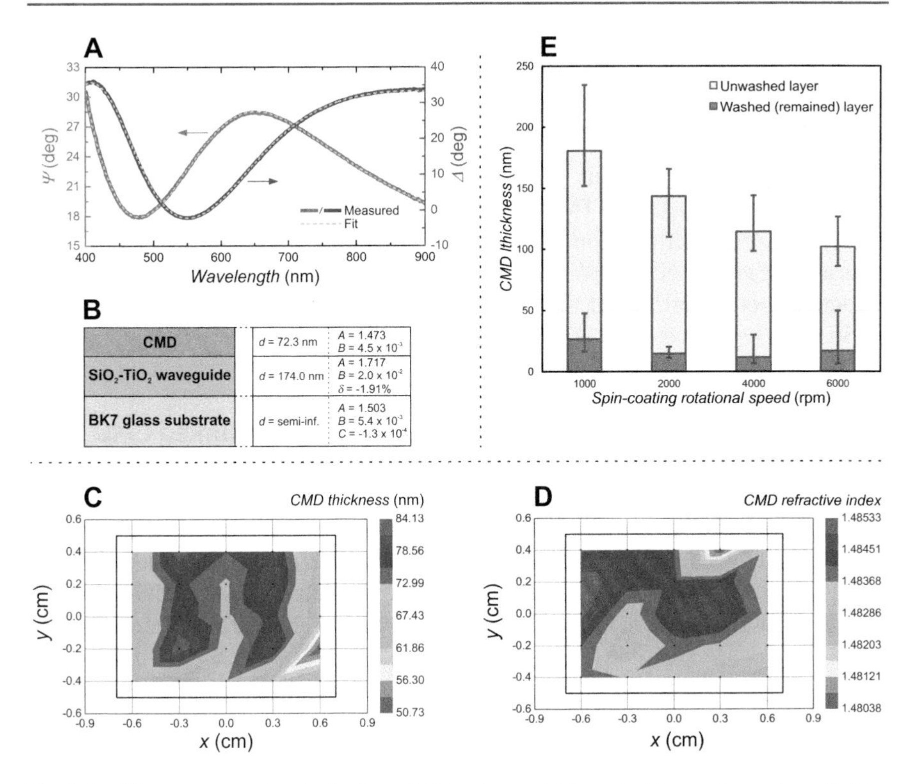

Fig. 8.6 Ellipsometry results obtained on CMD-sc layers. [Adapted with permission from (Saftics 2018) and from (Saftics et al. 2019), copyright 2019 American Chemical Society]
(**A**) Ψ and Δ spectra recorded on a CMD-sc layer that was prepared on an OWLS model substrate. The measured and fitted spectra are indicated by the solid and dashed lines, respectively (MSE = 1.9). The spectra correspond to the center of the measured sample (applied spin-coating rotational speed: 6000 rpm)
(**B**) Optical model and results obtained from the fit that is presented in graph **A** (the shown parameters are the following: d, thickness; A, B, C, parameters of Cauchy's equation; vertical inhomogeneity degree of the layer (δ))
(**C, D**) Lateral maps about thickness (d_{CMD}) and refractive index ($n_{\mathrm{CMD,\,632.8\,nm}}$) of the CMD layer. The refractive index was calculated using the Cauchy's equation at the wavelength of 632.8 nm
(**E**) Thickness of CMD-sc layers depending on the applied rotational speeds. The yellow-colored (top) columns refer to the thickness of CMD layers as prepared (unwashed). The orange (bottom) columns refer to the thickness of CMD layers remaining after the washing

where improved optical contrast and sensitivity could be achieved. As a result, $B_{\mathrm{CMD}} = 4.5 \times 10^{-3} \pm 7.4 \times 10^{-4}$ was found (average ± std., four analyzed samples) (Saftics et al. 2017).

Graphs **C** and **D** show the evaluated thickness and refractive index maps of an unwashed CMD-sc sample. The thickness values depending on the spin-coating rotational speeds both for unwashed and washed samples are presented in bar chart **E**. The rotational speed significantly affected the unwashed CMD layer thickness, and the effect followed the expected tendency. However, the effect of varying rotational speeds on the thickness of remained layers was not obvious. It was observed that the washing removed 80% of the CMD thickness resulting in 10–50 nm thickness for the remained layers (Saftics et al. 2019).

XPS was used to determine the elemental composition of CMD-sc layers with the main goal of phosphorous (P) detection in order to determine the crosslinking degree. Due to the fact that STMP crosslinker molecules were the only P-containing compounds, the determination of P could be used to detect the presence of crosslinks and measure their relative amount. As a result, the determined 0.2–0.4 ± 0.1% P content should correspond to an approx. 5% crosslinking degree (Saftics et al. 2019).

8.3.7 Wetting Properties of Spin-Coated and Crosslinked CMD Layers Characterized by Contact Angle Measurements

Measuring the CA of water droplets deposited on surfaces can be used to characterize the wettability and hydrophilic-hydrophobic nature (energetics) of a surface. Due to the fact that hydrogels can absorb a large amount of water, dynamic CA measurements are useful in demonstrating the hydrogel nature of the examined layer.

Figure 8.7 presents the results of CA measurements on CMD-sc surfaces. Compared to the aminosilane-coated surface (**A**), the CMD-sc layer (**B**) significantly affected the CA, resulting in CAs in the range of 10–30° (with respect to multiple samples). According to our observations, the drop shape remarkably varied in time after deposition on CMD-sc surface. The phenomenon was analyzed by dynamic CA measurements, and a typical outcome is shown in graph **C**. It was found that the CA continuously decreased until relaxation. The drop in CA can be attributed to a significant water uptake which characteristic is typical for hydrogels.

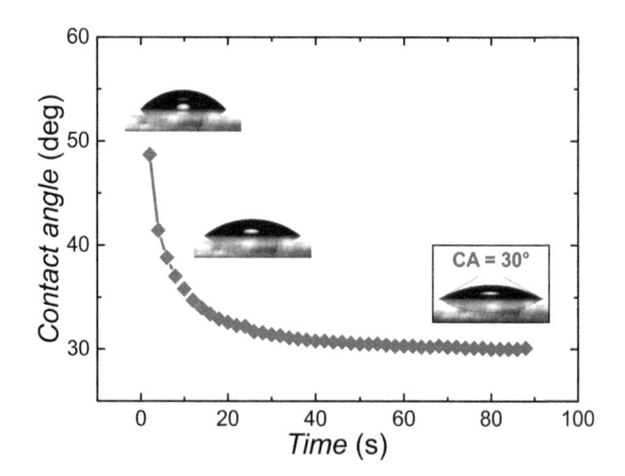

Fig. 8.7 Water CAs measured on a CMD-sc sample. [Adapted with permission from (Saftics et al. 2019), copyright 2019 American Chemical Society]
The graph shows the continuous change of CA in time, presenting the dynamic wetting behavior of the CMD-sc layer and its hydrogel nature.

It should be noted that such dynamic behavior was not detected for CMD-ut layers, which is an obvious verification of the difference (thickness, crosslinking) between the developed CMD-ut and CMD-sc layers (Saftics et al. 2019).

8.3.8 Protein- and Cell-Repellent Ability Characterized by OWLS and Phase Contrast Microscopy

A primary goal of the fabrication of CMD layers on SiO_2-TiO_2-type optical biosensor surfaces was the development of biofunctional sensor coatings with advanced antifouling abilities.

We applied the in situ OWLS method to characterize the protein-repellent (NSB-resistant) ability of CMD-ut and CMD-sc layers by measuring the amount (surface mass density) of adsorbed proteins. In these experiments, three different protein molecules with varying isoelectric point (pI) and molecular weight values were tested, including bovine serum albumin (BSA), fibrinogen (FGN), and lysozyme (LYZ). Typical adsorption curves measured on CMD-ut and CMD-sc surfaces are shown in Fig. 8.8A/a and A/b, respectively. Compared to the control measurements performed on bare SiO_2-TiO_2 surfaces, it is obvious that the CMD layers suppressed the adsorbed amount of each protein. The slightly higher deposited mass values of LYZ can be attributed to the attraction of positively charged LYZ (pI = 11.0) to the negatively charged CMD molecules ($pK_a \approx 3$) (Sidobre et al. 2002) in the used buffer environment (phosphate-buffered saline (PBS), pH 7.4) (Saftics et al. 2019).

The cell-repellent ability of CMD layers was tested by adhesion experiments using living HeLa cells, and the adhesion was observed by phase contrast microscopy. As shown in Fig. 8.8B, the presence of CMD provided considerable resistance against cell adhesion.

8.4 Summary

In this chapter, we have presented the surface analytical methodologies we applied to reveal the properties of our developed thin CMD layers. The proposed biosensor applications, for which the CMD layers were fabricated, demand to understand the layer structure and behavior in aqueous environment. Although the number of applicable techniques is very limited and there is no such method that could provide a full layer characterization, we demonstrated that using a collection of highly sensitive surface analytical techniques, several characteristics can be understood. It has been shown that the OWLS and QCM methods are capable to provide unique structural and mechanical data about heavily hydrated nanolayers. The presented measurement and data analysis methodologies are proposed to facilitate the analysis of biosensor coatings or other hydrated thin coatings used in various fields.

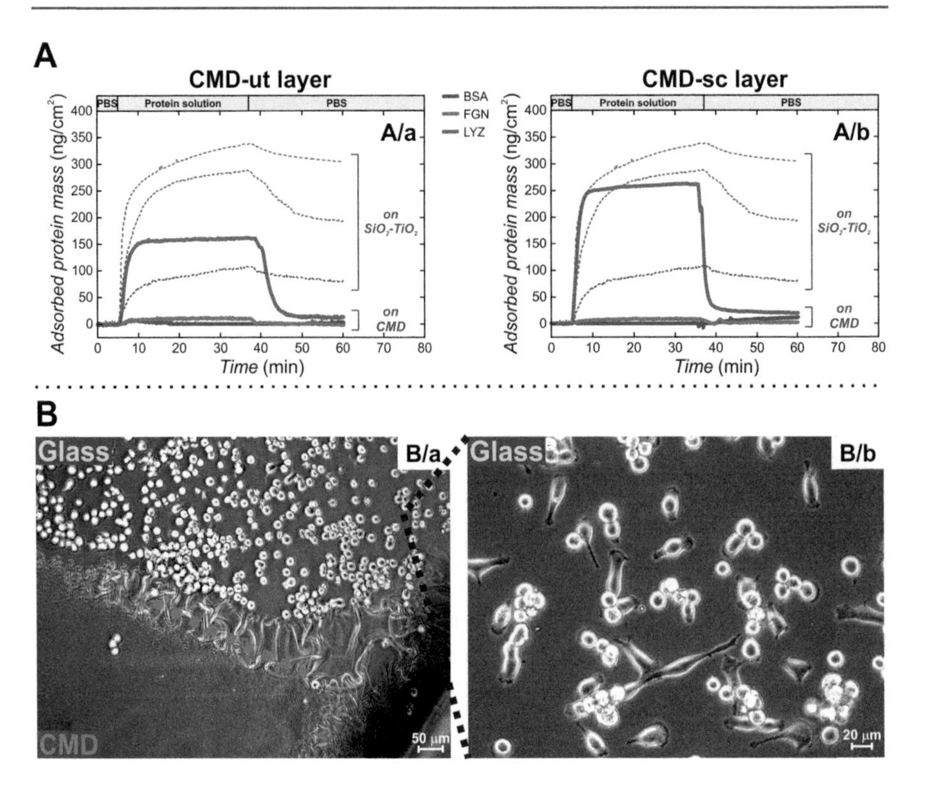

Fig. 8.8 Protein- and cell-repellent ability of CMD layers. [Adapted with permission from (Saftics 2018) and (Saftics et al. 2019), copyright 2019 American Chemical Society]
(**A**) Adsorbed mass of BSA, FGN, and LYZ proteins in situ monitored by OWLS. Graph **A/a** and **A/b** represent measurements performed on CMD-ut and CMD-sc layers, respectively. The different curve colors designate the specific proteins. While the solid curves were obtained on CMD-covered surfaces, the dashed ones indicate measurements carried out on unmodified SiO_2-TiO_2 substrate surfaces. The experimental phases are shown by the graph headers
(**B**) Image **B/a** corresponds to a sample partially coated with CMD-sc layer as indicated by the dashed line. While cells did not adhere on CMD surface, they could adhere on glass (see image **B/b** taken on glass surface). The images were captured after 4 (**B/a**) and 6 (**B/b**) hours of incubation in cell culture

Acknowledgments This study was funded by the Momentum ("Lendület") Program of the Hungarian Academy of Sciences, the National Research, Development and Innovation Office (NKFIH) [ERC_HU, KKP_19, KH and FK-128901 Programs] and János Bolyai Research Scholarship of the Hungarian Academy of Sciences are also gratefully acknowledged.

References

Akkoyun A, Bilitewski U (2002) Optimisation of glass surfaces for optical immunosensors. Biosens Bioelectron 17:655–664. https://doi.org/10.1016/S0956-5663(02)00029-5

Balakrishnan B, Banerjee R (2011) Biopolymer-based hydrogels for cartilage tissue engineering. Chem Rev 111:4453–4474. https://doi.org/10.1021/cr100123h

Chandra P (ed) (2016) Nanobiosensors for personalized and onsite biomedical diagnosis. Institution of Engineering and Technology, London

Chandra P, Tan YN, Singh SP (2017) Next generation point-of-care biomedical sensors technologies for cancer diagnosis. Springer, Singapore

Cutiongco MFA, Tan MH, Ng MYK et al (2014) Composite pullulan–dextran polysaccharide scaffold with interfacial polyelectrolyte complexation fibers: A platform with enhanced cell interaction and spatial distribution. Acta Biomater 10:4410–4418. https://doi.org/10.1016/j.actbio.2014.06.029

De Feijter J, Benjamins J, Veer F (1978) Ellipsometry as a tool to study the adsorption behavior of syntetic and biopolyers at the air water interface. Biopolymers 17:1759–1772. https://doi.org/10.1002/bip.1978.360170711

Dutta AK, Belfort G (2007) Adsorbed gels versus brushes: viscoelastic differences. Langmuir 23:3088–3094. https://doi.org/10.1021/la0624743

Elam JH, Nygren H, Stenberg M (1984) Covalent coupling of polysaccharides to silicon and silicon rubber surfaces. J Biomed Mater Res 18:953–959. https://doi.org/10.1002/jbm.820180809

Elender G, Kühner M, Sackmann E (1996) Functionalisation of Si/SiO2 and glass surfaces with ultrathin dextran films and deposition of lipid bilayers. Biosens Bioelectron 11:565–577. https://doi.org/10.1016/0956-5663(96)83292-1

Escorihuela J, González-Martínez MÁ, López-Paz JL et al (2015) Dual-polarization interferometry: a novel technique to light up the nanomolecular world. Chem Rev 115:265–294. https://doi.org/10.1021/cr5002063

Ferreira L, Rafael A, Lamghari M et al (2004) Biocompatibility of chemoenzymatically derived dextran-acrylate hydrogels. J Biomed Mater Res 68A:584–596. https://doi.org/10.1002/jbm.a.20102

Ferreira GNM, Da-Silva A-C, Tomé B (2009) Acoustic wave biosensors: physical models and biological applications of quartz crystal microbalance. Trends Biotechnol 27:689–697. https://doi.org/10.1016/j.tibtech.2009.09.003

Höök F, Kasemo B, Nylander T et al (2001) Variations in coupled water, viscoelastic properties, and film thickness of a Mefp-1 protein film during adsorption and cross-linking: a quartz crystal microbalance with dissipation monitoring, ellipsometry, and surface plasmon resonance study. Anal Chem 73:5796–5804. https://doi.org/10.1021/ac0106501

Höök F, Vörös J, Rodahl M et al (2002) A comparative study of protein adsorption on titanium oxide surfaces using in situ ellipsometry, optical waveguide lightmode spectroscopy, and quartz crystal microbalance/dissipation. Colloids Surf B Biointerfaces 24:155–170. https://doi.org/10.1016/S0927-7765(01)00236-3

Horvath R, Ramsden JJ (2007) Quasi-isotropic analysis of anisotropic thin films on optical waveguides. Langmuir 23:9330–9334. https://doi.org/10.1021/la701405n

Horvath R, McColl J, Yakubov GE, Ramsden JJ (2008) Structural hysteresis and hierarchy in adsorbed glycoproteins. J Chem Phys 129:071102. https://doi.org/10.1063/1.2968127

Horvath R, Gardner HC, Ramsden JJ (2015) Apparent self-accelerating alternating assembly of semiconductor nanoparticles and polymers. Appl Phys Lett 107:041604. https://doi.org/10.1063/1.4927403

Ismail IM, Gray ND, Owen JR (1996) A QCM analysis of water absorption in lithium polymer electrolytes. J Chem Soc Faraday Trans 92:4115. https://doi.org/10.1039/ft9969204115

Johannsmann D (2015) The quartz crystal microbalance in soft matter research. Springer, Cham

Kittle JD, Du X, Jiang F et al (2011) Equilibrium water contents of cellulose films determined via solvent exchange and quartz crystal microbalance with dissipation monitoring. Biomacromolecules 12:2881–2887. https://doi.org/10.1021/bm200352q

Knoll W (ed) (2013) Handbook of biofunctional surfaces, 1st edn. Pan Stanford Publishing, Boca Raton

Kovacs N, Patko D, Orgovan N et al (2013) Optical anisotropy of flagellin layers: in situ and label-free measurement of adsorbed protein orientation using OWLS. Anal Chem 85:5382–5389. https://doi.org/10.1021/ac3034322

Kuhner M, Sackmann E (1996) Ultrathin hydrated dextran films grafted on glass: preparation and characterization of structural, viscous, and elastic properties by quantitative microinterferometry. Langmuir 12:4866–4876. https://doi.org/10.1021/la960282+
Laos K, Parker R, Moffat J et al (2006) The adsorption of globular proteins, bovine serum albumin and β-lactoglobulin, on poly-l-lysine–furcellaran multilayers. Carbohydr Polym 65:235–242. https://doi.org/10.1016/j.carbpol.2006.01.010
Lee MH, Boettiger D, Composto RJ (2008) Biomimetic carbohydrate substrates of tunable properties using immobilized dextran hydrogels. Biomacromolecules 9:2315–2321. https://doi.org/10.1021/bm8002094
Lee T-H, Hirst DJ, Aguilar M-I (2015) New insights into the molecular mechanisms of biomembrane structural changes and interactions by optical biosensor technology. Biochim Biophys Acta Biomembr 1848:1868–1885. https://doi.org/10.1016/j.bbamem.2015.05.012
Lian M, Chen X, Lu Y, Yang W (2016) Self-assembled peptide hydrogel as a smart biointerface for enzyme-based electrochemical biosensing and cell monitoring. ACS Appl Mater Interfaces 8:25036–25042. https://doi.org/10.1021/acsami.6b05409
Linder V, Gates BD, Ryan D et al (2005) Water-soluble sacrificial layers for surface micromachining. Small 1:730–736. https://doi.org/10.1002/smll.200400159
Liu Z, Choi H, Gatenholm P, Esker AR (2011) Quartz crystal microbalance with dissipation monitoring and surface plasmon resonance studies of carboxymethyl cellulose adsorption onto regenerated cellulose surfaces. Langmuir 27:8718–8728. https://doi.org/10.1021/la200628a
Liu J, Qi C, Tao K et al (2016) Sericin/dextran injectable hydrogel as an optically trackable drug delivery system for malignant melanoma treatment. ACS Appl Mater Interfaces 8:6411–6422. https://doi.org/10.1021/acsami.6b00959
Liu M, Zeng X, Ma C et al (2017) Injectable hydrogels for cartilage and bone tissue engineering. Bone Res 5:17014. https://doi.org/10.1038/boneres.2017.14
Löfås S, Johnsson B (1990) A novel hydrogel matrix on gold surfaces in surface plasmon resonance sensors for fast and efficient covalent immobilization of ligands. J Chem Soc Chem Commun:1526–1528. https://doi.org/10.1039/C39900001526
Löfås S, Johnsson B, Tegendal K, Rönnberg I (1993) Dextran modified gold surfaces for surface plasmon resonance sensors: immunoreactivity of immobilized antibodies and antibody-surface interaction studies. Colloids Surf B Biointerfaces 1:83–89. https://doi.org/10.1016/0927-7765(93)80038-Z
Löfås S, Johnsson B, Edström Å et al (1995) Methods for site controlled coupling to carboxymethyldextran surfaces in surface plasmon resonance sensors. Biosens Bioelectron 10:813–822. https://doi.org/10.1016/0956-5663(95)99220-F
Marx KA (2003) Quartz crystal microbalance: a useful tool for studying thin polymer films and complex biomolecular systems at the solution−surface interface. Biomacromolecules 4:1099–1120. https://doi.org/10.1021/bm020116i
Massia SP, Stark J, Letbetter DS (2000) Surface-immobilized dextran limits cell adhesion and spreading. Biomaterials 21:2253–2261. https://doi.org/10.1016/S0142-9612(00)00151-4
Mateescu A, Wang Y, Dostalek J, Jonas U (2012) Thin hydrogel films for optical biosensor applications. Membranes (Basel) 2:40–69. https://doi.org/10.3390/membranes2010040
McArthur SL, McLean KM, Kingshott P et al (2000) Effect of polysaccharide structure on protein adsorption. Colloids Surf B Biointerfaces 17:37–48. https://doi.org/10.1016/S0927-7765(99)00086-7
Monchaux E, Vermette P (2007) Development of dextran-derivative arrays to identify physicochemical properties involved in biofouling from serum. Langmuir 23:3290–3297. https://doi.org/10.1021/la063012s
Müller MT, Yan X, Lee S et al (2005) Lubrication properties of a brushlike copolymer as a function of the amount of solvent absorbed within the brush. Macromolecules 38:5706–5713. https://doi.org/10.1021/ma0501545
Nalam PC, Daikhin L, Espinosa-Marzal RM et al (2013) Two-fluid model for the interpretation of quartz crystal microbalance response: tuning properties of polymer brushes with solvent mixtures. J Phys Chem C 117:4533–4543. https://doi.org/10.1021/jp310811a

Orgovan N, Patko D, Hos C et al (2014) Sample handling in surface sensitive chemical and biological sensing: a practical review of basic fluidics and analyte transport. Adv Colloid Interface Sci 211:1–16. https://doi.org/10.1016/j.cis.2014.03.011
Peppas NA, Van Blarcom DS (2016) Hydrogel-based biosensors and sensing devices for drug delivery. J Control Release 240:142–150. https://doi.org/10.1016/j.jconrel.2015.11.022
Perrino C (2009) Poly(L-lysine)-g-dextran (PLL-g-dex): brush-forming, biomimetic carbohydrate chains that inhibit fouling and promote lubricity
Piehler J, Brecht A, Hehl K, Gauglitz G (1999) Protein interactions in covalently attached dextran layers. Colloids Surf B Biointerfaces 13:325–336. https://doi.org/10.1016/S0927-7765(99)00046-6
Rivière JC, Myhra S (eds) (1998) Handbook of surface and interface analysis, 1st edn. Marcel Dekker, New York
Saftics A (2018) Development of dextran-based hydrogel layers for biosensor applications. Budapest University of Technology and Economics
Saftics A, Kurunczi S, Szekrényes Z et al (2016) Fabrication and characterization of ultrathin dextran layers: Time dependent nanostructure in aqueous environments revealed by OWLS. Colloids Surf B Biointerfaces 146:861–870. https://doi.org/10.1016/j.colsurfb.2016.06.057
Saftics A, Kurunczi S, Türk B et al (2017) Spin coated carboxymethyl dextran layers on TiO2-SiO2 optical waveguide surfaces. Rev Roum Chim 62:775–781
Saftics A, Prósz GA, Türk B et al (2018) In situ viscoelastic properties and chain conformations of heavily hydrated carboxymethyl dextran layers: a comparative study using OWLS and QCM-I chips coated with waveguide material. Sci Rep 8:11840. https://doi.org/10.1038/s41598-018-30201-6
Saftics A, Türk B, Sulyok A et al (2019) Biomimetic dextran-based hydrogel layers for cell micropatterning over large areas Using the FluidFM BOT technology. Langmuir 35:2412–2421. https://doi.org/10.1021/acs.langmuir.8b03249
Sauerbrey G (1959) Verwendung von Schwingquarzen zur Wagungdiinner Schichten und zur Mikrowagung. Zeitschrift fur Phys 155:206–222. https://doi.org/10.1007/BF01337937
Secret E, Kelly SJ, Crannell KE, Andrew JS (2014) Enzyme-responsive hydrogel microparticles for pulmonary drug delivery. ACS Appl Mater Interfaces 6:10313–10321. https://doi.org/10.1021/am501754s
Sidobre S, Puzo G, Rivière M (2002) Lipid-restricted recognition of mycobacterial lipoglycans by human pulmonary surfactant protein A: a surface-plasmon-resonance study. Biochem J 365:89–97. https://doi.org/10.1042/bj20011659
Sivakumaran D, Maitland D, Hoare T (2011) Injectable microgel-hydrogel composites for prolonged small-molecule drug delivery. Biomacromolecules 12:4112–4120. https://doi.org/10.1021/bm201170h
Stengel G, Höök F, Knoll W (2005) Viscoelastic modeling of template-directed DNA synthesis. Anal Chem 77:3709–3714. https://doi.org/10.1021/ac048302x
Tanaka M, Hayashi T, Morita S (2013) The roles of water molecules at the biointerface of medical polymers. Polym J 45:701–710. https://doi.org/10.1038/pj.2012.229
Tavakoli J, Tang Y (2017) Hydrogel based sensors for biomedical applications: an updated review. Polymers (Basel) 9:364. https://doi.org/10.3390/polym9080364
Vogt BD, Lin EK, Wu WI, White CC (2004) Effect of film thickness on the validity of the sauerbrey equation for hydrated polyelectrolyte films. J Phys Chem B 108:12685–12690. https://doi.org/10.1021/jp0481005
Voinova MV, Rodahl M, Jonson M, Kasemo B (1999) Viscoelastic acoustic response of layered polymer films at fluid-solid interfaces: continuum mechanics approach. Phys Scr 59:391–396. https://doi.org/10.1238/Physica.Regular.059a00391
Vörös J (2004) the density and refractive index of adsorbing protein layers. Biophys J 87:553–561. https://doi.org/10.1529/biophysj.103.030072
Xu C, Lee W, Dai G, Hong Y (2018) Highly elastic biodegradable single-network hydrogel for cell printing. ACS Appl Mater Interfaces acsami 8b01294. https://doi.org/10.1021/acsami.8b01294

Zhang J, Horváth C (2003) Capillary electrophoresis of proteins in dextran-coated columns. Electrophoresis 24:115–120. https://doi.org/10.1002/elps.200390002

Zhang YS, Khademhosseini A (2017) Advances in engineering hydrogels. Science (80–): 356:eaaf3627. https://doi.org/10.1126/science.aaf3627

Zhang R, Tang M, Bowyer A et al (2005) A novel pH- and ionic-strength-sensitive carboxy methyl dextran hydrogel. Biomaterials 26:4677–4683. https://doi.org/10.1016/j.biomaterials.2004.11.048

9 Electrochemical Nanoengineered Sensors in Infectious Disease Diagnosis

Suryasnata Tripathy, Patta Supraja, and Shiv Govind Singh

Abstract

This chapter reports a short review on electrochemical nanoengineered biosensors in infectious disease diagnosis. Early and timely diagnosis of infectious diseases has tremendous medical and social significance which advocates the development of new diagnostic tools. In this chapter, we discussed various electrochemical sensors for detection and diagnosis of tropical or subtropical fevers particularly dengue fever and malaria parasite. We also addressed the several important aspects of biosensors, namely, selectivity, sensitivity, and interference, and also the effect of engineering the nanomaterials (0D, 1D, 2D) on these aspects. In detail, we discussed the various techniques to immobilize the biomolecules on working electrode (glassy carbon, gold electrode, flexible substrates). Further, we discussed the several miniaturized sensing platforms with integrated microfluidic channels which can ensure for development of sensors for point-of-care applications.

Keywords

Electrochemical sensors · Nanomaterials · Biosensors · Infectious diseases · Tropical fevers

Authors Suryasnata Tripathy and Patta Supraja have equally contributed to this chapter.

S. Tripathy · P. Supraja · S. G. Singh (✉)
Department of Electrical Engineering, Indian Institute of Technology Hyderabad, Hyderabad, India
e-mail: sgsingh@iith.ac.in

P. Chandra, R. Prakash (eds.), *Nanobiomaterial Engineering*,
https://doi.org/10.1007/978-981-32-9840-8_9

9.1 Introduction

Infectious disease diagnosis is of tremendous medical and social significance, especially in the developing countries, where the current status of the healthcare sector is a matter of great concern. In particular, the number of people suffering from infectious diseases, such as the tropical fevers, has been alarming over the past decade, which advocates the development of new diagnostic tools for early and timely disease diagnosis. Toward this, a considerable amount of research has been devoted in the past, and a great many biosensing platforms for the detection of infectious diseases have been proposed. Among them, the electrochemical biosensors have been at the forefront, on account of their inherent advantages. Such sensors, covering both labelled and label-free transduction schemes, are amenable to miniaturization and hence are suitable candidates for developing point-of-care diagnostic tools for infectious disease detection. Further, a handshake between the electrochemical biosensors and nanotechnology has enhanced their performance in many folds. In the past, nano-sized multifunctional materials have been roped in to be used as the sensing/transducing elements in electrochemical sensors and have resulted in great sensitivity, with fairly low limits of detection. Also, nanoengineered sensing materials have been specifically designed for desired performance, be it improved biomolecule immobilization, selectivity, signal enhancement, or colorimetric transduction. Particularly, the 1D nanomaterials such as nanotubes (Minot et al. 2007; Wang 2005; Sotiropoulou and Chaniotakis 2003), nanoparticles (Sanvicens et al. 2009; Yanez-Sedeno and Pingarron 2005; Bhatnagar et al. 2018; Supraja et al. 2019a, b; Mandal et al. 2018), and nanofibers (Prakash et al. 2017; Wanekaya et al. 2006; Tripathy et al. 2018a, b; Vadnala et al. 2018; Guo et al. 2009; Matta et al. 2016; Supraja et al. 2019b) have been extensively used to decorate working electrodes in electrochemical sensors for enhancing the overall binding efficiency and the surface heterogeneity. This, in turn, results in very low limits of detection, even extending up to single molecules. In Fig. 9.1, the necessity of

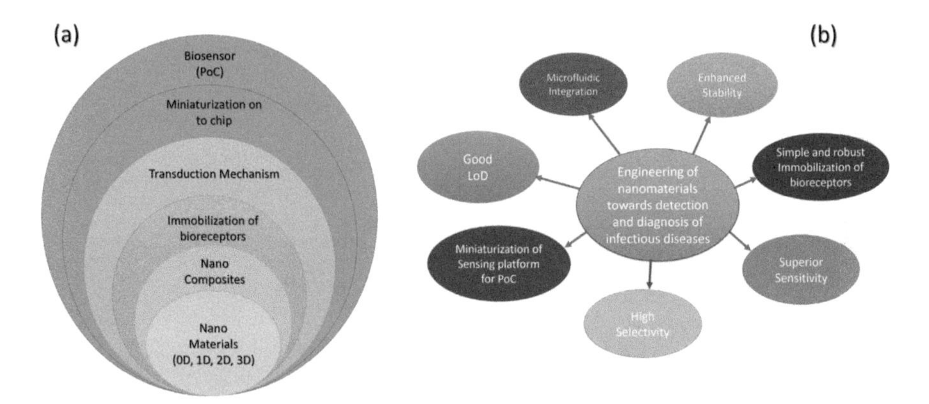

Fig. 9.1 Schematic representation of steps involved in preparation of biosensors (**a**) and advantages of nanoengineered materials (**b**)

engineering the nanomaterials to enhance their specified properties and the basic steps involved in preparing miniaturized biosensors for point-of-care applications has been shown.

In this chapter, we briefly discuss some of the nanoengineered electrochemical biosensing platforms associated with infectious disease detection. Specifically, we limit our discussion to two major infectious diseases, namely, malaria and dengue fever, on account of their global implications. Both these diseases account for a large number of rapid diagnosis tests being conducted in tropical counties annually, which is a testament of their ubiquitous nature. This calls for the development of biosensing platforms capable of early and accurate diagnosis of these infections, in an economic way, with minimum usage of the patient's blood. Thankfully, several such platforms have been reported in the literature. Such biosensors have employed a myriad of transduction principles, including optical, electrochemical, chemiresistive, and magnetic schemes. Among them, electrochemical biosensors have been a dominant choice, with a great fraction of the reported literature attributing to them. As reported, researchers over the years have explored several detection methods, namely, sandwich (with capture and signal antibodies)/single step (with only capture antibody), enzymatic/nonenzymatic, and labelled/label-free immunoassays, both on chip and off chip, for ensuring the detection of the target biomarkers. In these endeavors, nanomaterials have played a major role in ensuring signal amplification, surface area enhancement, sensitivity improvement, and lowering the limiting detection. Further, toward ensuring point-of-care biosensing, at low cost, disposable biosensors have been designed for malaria diagnosis, using screen-printed substrates. A majority of research have also been devoted toward ensuring lower detection limit in case of dengue diagnosis, as early detection of the same is critical to control the mortality rate. Herein, we intend to provide a brief discussion on such biosensing platforms. Section 9.2 of this chapter is dedicated toward discussing biosensing platforms associated with the diagnosis of dengue fever, whereas Sect. 9.3 is devoted to malarial parasite detection.

9.2 Detection of Dengue

Dengue fever is caused due to the female *Aedes* mosquitoes, which act as the carriers of the parasitic dengue virus (DENV). In the past decade, the dengue viral infection has remained as one of the most important health problems around the world, particularly the tropical and subtropical regions (Kyle and Harris 2008). In view of its severe effects on human health, it is highly desirable to ensure rapid and early detection of dengue virus in human blood. In this regard, conventionally, molecular biological-based clinical laboratory techniques such as monoclonal antibody-capture enzyme-linked immunosorbent assay (MAC-ELISA) (Datta and Wattal 2010), polymerase chain reaction (PCR) (Raengsakulrach et al. 2002), and reverse transcription polymerase chain reaction (RT-PCR) (Houng et al. 2001) are used for the detection of dengue virus. These methods have made limited impact due to several issues, namely, high time consumption, need of skilled operation, complex sample preparation, high cost, low sensitivity, and probability of false detection (Huy et al. 2011). On account of the lack of sophisticated individual methods for the detection of dengue virus, generally, two or more of the conventional methods/tests are performed to confirm the viral infection in human serum samples. This, in turn, affects the cost of detection and diagnosis. In this context, biosensors provide a suitable, and effective, alternative. During the early stages of the evolution of biosensors, researchers had used several optical transduction principles such as fluorescence, electrochemiluminescence, and surface plasmon resonance to improve the sensitivity (Xia et al. 2010; Rodriguez-Mozaz et al. 2004; Fan et al. 2015; Mauriz et al. 2006; Mahato et al. 2018). However, these optical methods seemed to be not suitable for miniaturization and, in turn, for point-of-care applications. This led the researchers to look beyond optical schemes for on-chip biosensing of analytes. Among several alternate principles, electrochemical transduction has been most widely used, on account of advantages previously described (Kryscio and Peppas 2012). Initially, label-free electrochemical immunosensors were restricted due to their low sensitivity. Keeping this in mind, in the past few years, researchers have tried to increase the sensitivity of electrochemical sensors by engineering the materials at nanoregime. As compared to the bulk materials, spatially downscaled materials result in significant change in the inherent properties, by increasing the effective surface area. Specifically, by engineering the materials at nano range, one can improve their surface to volume ratio. These materials, when used as electrochemical working electrodes, affect the overall heterogeneity of the sensing surfaces and thereby their ability to immobilize bioreceptors. As a result, the overall sensitivity of an immunosensor can be increased (Shinde et al. 2012). In such cases, the electroactive surface area of the nanomaterials can be calculated by using the Randles-Sevcik equation, given as:

$$I_P = 0.4463\, nFAC \left(nFvD/RT \right)^{0.5}$$

where:

Ip = peak current in cyclic voltammogram
n = number of electrons involved in redox reaction
F = Faraday constant
A = effective surface area of electrode
C = concentration of analyte
v = scan rate
D = diffusion coefficient
R = universal gas constant
T = temperature at which experiment has conducted

In addition to this, nanoengineered electrochemical biosensors also help in addressing another critical issue called specificity. For example, the detection of NS1 biomarker (a biomarker for dengue) in real-time human serum samples is extremely difficult due to the lack of specificity. Generally, human serum is a complex matrix containing several molecules and proteins, which can interfere with NS1 and cause false detection. However, specificity of the NS1 biosensors can be improved by using zwitterionic antifouling moieties. Nadiya et al. have reported that the best antifouling agents to ensure the specificity against both positive- and negative-charged proteins are the combination of 4-sulfophenyl, 4-trimethylammoniophenyl, and 1,4-phenylenediamine (SP:TMAP:PPD) in the molar ratio of 0.5:1.5:0.37 (Darwish et al. 2016). Initially, they electrodeposited the SP, TMAP, and PPD onto an ITO electrode and subsequently converted the terminal amine of the 1,4-phenylenediamine to diazonium group toward the immobilization of Au nanoparticles onto PPD via covalent bond. Modification of Au nanoparticles with 1,4-phenylenediamine results in amine ($-NH_2$) groups, which can be activated by *N*-(3-dimethylaminopropyl)-*N′*-ethylcarbodiimide hydrochloride (EDC) and N-hydroxysulfosuccinimide (NHS). Anti-NS1 antibody can be immobilized onto the functionally activated Au nanoparticles via the formation of peptide bond (–CO–NH). Mohato et al. group has reported the strategies for engineering the surface of gold nanoparticles for immobilization of biomolecules which inherently enhances the robustness and stability of biosensor (Mahato et al. 2019). Efficiency of as-prepared bioelectrode (antibody immobilized electrode) can be tested against different concentrations of NS1 protein with any of the popular electrochemical techniques, like differential pulse voltammetry (DPV), cyclic voltammetry (CV), and electrochemical impedance spectroscopy (EIS). As reported by the authors, by integrating nanomaterials and antifouling moieties onto the sensing electrode surface, one can improve the sensitivity and the specificity of biosensor. However, one major drawback of this method is that the antifouling molecules occupy large surface area on the substrate, which results in low limit of detection (LoD). Toward enhancing the LoD, in addition to improving the sensitivity, researchers have used the carbon nanotubes (CNTs). CNTs play an important role in the performance of sensors due to their prodigious physical, chemical, and electrical properties. Especially in electrochemical biosensors, the electrochemical reactivity of electroactive reagents can be enhanced with the fast electron transfer offered by CNTs. Also, as compared to nanoparticles, the electroactive surface area of CNTs is reasonably high. Further, one can easily functionalize CNTs with reactive groups, which inherently aids the immobilization of antibodies onto the electrode. In this context, Ana et al. (Dias et al. 2013) have reported the electrochemical detection of NS1 on screen-printed electrodes, wherein, as the active surface material, they have used carbon ink with dispersed carboxylated carbon nanotubes (CNT-COOH). In this work, ethylenediamine (EDA) has been used as biolinker (Dias et al. 2013). EDA ($C_2H_4(NH_2)_2$) is a linear molecule, which contains two amino groups at the two ends of its carbon chain. One amino end of EDA can form covalent bonds with the carboxylic group of the CNT, while the other end can favor the immobilization of antibody through the formation of strong covalent bond with the anti-NS1 antibody. Functionalized CNT and EDA film together results in stable and oriented sensor surface for immobilization of antibodies. For better understanding, the schematic representation of preparation of bioelectrode has been shown in Fig. 9.2. As shown in the

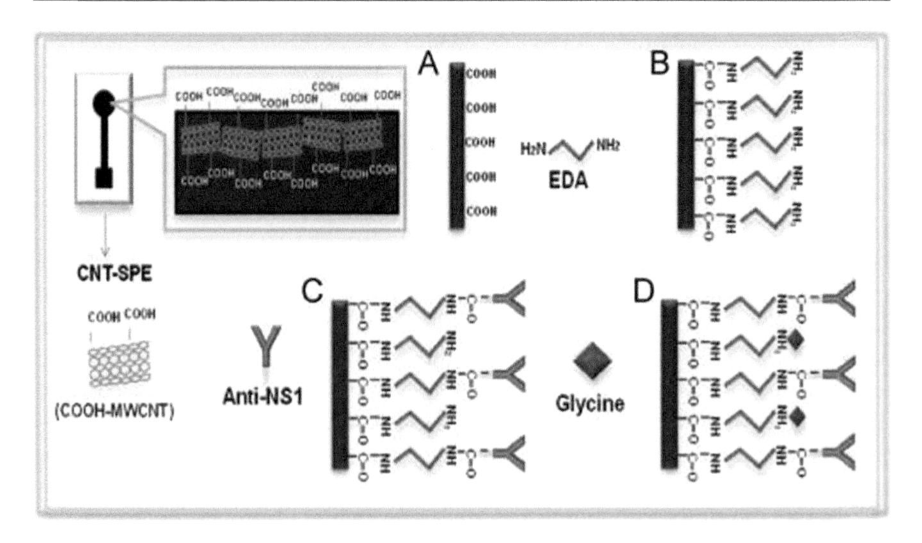

Fig. 9.2 Schematic representation of preparation of bioelectrode with carbon nanotubes on screen-printed platform (Dias et al. 2013) (Reprinted with permission)

schematic representation, one can directly deposit the CNTs without incorporating in any carrier matrix (like carbon ink) on the sensor surface. However, this would decrease the reproducibility and the stability of sensor, because of their weak interactions with electrode surface. In view of this, to improve reproducibility and stability of sensor without compromising the sensitivity, researchers moved toward composite nanomaterials.

The composites are materials resulting from the combination of two or more constituents, which are having different properties. There are two phases in composites, namely, the major phase (matrix) and the minor phase (reinforcement). The major phase can be either metallic or polymeric or ceramic, whereas the minor phase can be one-dimensional (nanoparticles, nanotubes, nanofibers, nanoribbons), two-dimensional (nanosheets, nanoflakes, nanodisks, nanowalls, nanotubes, nanoprisms, nanospheres, etc.), or three-dimensional (ball-like nanodendritic structures, nanoflowers, etc.) nanomaterials. In general, electrical, optical, electrochemical, mechanical, thermal, and catalytic properties of the nanocomposites can be engineered by selecting a suitable combination of the major and the minor phase materials. Using this principle, researchers have developed nanocomposite-based biosensor platforms for dengue virus detection. For instance, Silva et al. have developed bioelectrodes for electrochemical detection of NS1 protein by using polyallylamine (PAH) and carboxylated carbon nanotube (CNT) composite (Silva et al. 2015). PAH is a linear polymer which inherently contains amine groups (-NH_2) on the backbone of each monomer. These amine groups account for a "bi-linking" property that helps in avoiding the random immobilization of desired antibodies onto the sensor surface. PAH forms self-assembled monolayer on the carboxylated CNTs via formation of covalent bonds between the carboxylated group of CNT and the amine group of the monomer. By using PAH as the polymer matrix, one can

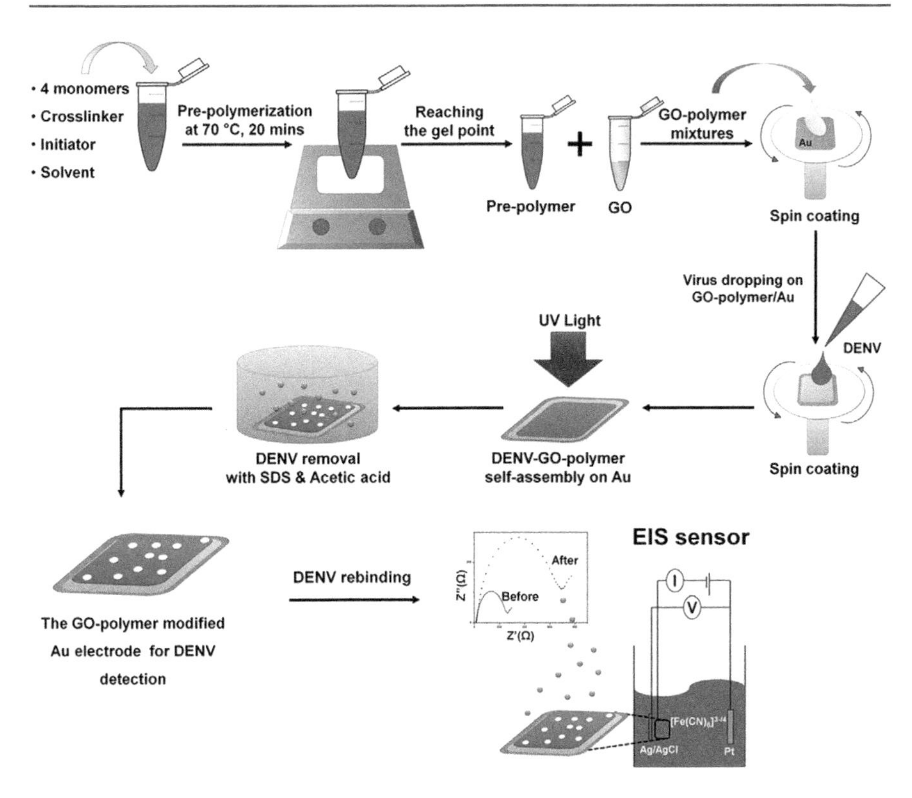

Fig. 9.3 Schematic representation of the preparation of GO-polymer on gold electrode for DENV detection (Navakul et al. 2017) (Reprinted with permission)

consequently increase the stability and reproducibility of the biosensor, without compromising its sensitivity.

Graphene oxide (GO) is a two-dimensional semiconducting material, accounting for high conductivity, high mechanical strength, and rapid functionalization ability with water-soluble polymers. It is the single monomolecular layer of graphite with various functional groups, which can easily functionalize (covalently or non-covalently) with the reactive moieties (such as virus, antibodies, enzymes, proteins, and polymers). On account of this, one can hope to develop biosensor for DENV with GO alone; however, the selectivity will be low because the surface of GO can readily absorb amino acids via hydrogen bonding, electrostatic, hydrophobic, and Π-Π stacking interactions. To address this issue, often, GO-reinforced polymer nanocomposites are used as sensing materials in electrochemical biosensing. In one such work, Navakul et al. developed a new sensing platform which can detect, classify, and quantify the DENV using graphene-reinforced polymer nanocomposite (Navakul et al. 2017). Figure 9.3 shows the schematic representation of the protocol for GO-reinforced polymer preparation.

Initially, the monomers (methacrylic acid, N-vinylpyrrolidone, acrylamide, and methyl methacrylate) were mixed (in the ratio 2.4:1.2:1:2.4) in dimethyl sulfoxide

(DMSO) solvent, along with a cross-linker and an initiator. Subsequently, the GO solution was added to the above polymer, prior to gelling of polymer mixture in a 2:3 molar ratio. The GO-reinforced polymeric mixture was then spin coated onto the gold electrode. Later, DENV was added onto the thus prepared electrode and exposed to UV for 3 h so as to allow polymerization. Post-polymerization, the electrodes were washed with sodium dodecyl sulfate and acetic acid (1:1) solution to remove polymer-bound virus. This creates necessary templates of DENV in the GO-polymer composite, which can later bind to DENV selectively. When DENV gets reabsorbed onto the template, the electrical conductivity of the nanocomposite decreases significantly. Based on change in the charge transfer resistance, one can detect DENV qualitatively as well as quantitatively. In such a detection scheme, the synthesis of polymer composites, which can respond selectively during the reabsorption of the virus onto the template, is a crucial step.

Till now, we have discussed the principles of platforms that can detect only single serotypes of DENV at a time. But in the areas of hyperendemic dengue, multiple serotypes are known to circulate simultaneously in blood, which increases the risk of multiple infections, organ failure, and finally death. Thus, development of platforms that can simultaneously detect multiple serotypes is highly essential. In this regard, Huang et al. have developed an n-type silicon nanowire array-based biosensing platform for simultaneous detection of all DENV serotypes (DENV1, DENV2, DENV3, DENV4) on a single chip (Huang et al. 2013). In their work, array of Si nanowires were patterned on SOI wafers using deep UV lithography, oxidation, and reactive-ion etching. Peptide nucleic acids (PNAs), with specificity to hybridize with the serotypes, were immobilized onto the Si nanowire through a two-step (creating amino groups, bi-linking of PNA to amino groups) silane chemistry. Here, the basic principle of detection is the change in the nanowire resistance upon hybridization of DENV serotypes to its complementary PNA. Generally, PNA probes are neutral in nature, and hybridization with their complementary serotypes results in a net negative charge on the n-type silicon nanowire. Due to this increasing of net negative charge, the depletion region in the n-type Si gets modified significantly, resulting in an increased resistance. However, binding of PNA probes to noncomplementary serotypes doesn't lead to any net change in charge on the nanowires, making the sensing approach highly selective. This sensing platform, though capable of simultaneously detecting multiple DENV serotypes, is disadvantageous on the grounds of cost and time. As an alternative approach, Rashid et al. have developed a low-cost disposable biosensing platform which can detect the dengue viral DNA using silicon nanowires (SiNWs) and gold nanoparticles (AuNPs) (Rashid et al. 2015). Here, the role of the SiNWs/AuNPs is to enhance the conductivity of screen-printed gold electrodes (SPGE), in addition to improving the surface area of SPGE for probe immobilization. As described by the authors, the SiNWs were initially mixed with APTES, and the mixture was drop cast onto hydroxyl functionalized SPGEs. Subsequently, the modified SPGE was dipped in dithiodipropionic acid (DTPA) and soaked immediately in AuNPs to form self-assembled monolayers through the formation of amide (-NH_2 group of APTES and -COOH group of DOTA) and thiol (covalent bond between S-S and Au) linkages. Finally, the thiolated probe DNA was immobilized onto DPTA-modified SPGE through covalent binding of thiol sulfur, for target DNA hybridization. Toward label-free detection of DENV-specific DNA targets, our research group has developed two distinct approaches, one being an electrochemical platform using electrospun semiconducting Mn_2O_3 nanofibers (Tripathy et al. 2017a) and the other being a chemiresistive platform derived from conducting polyaniline/polyethylene oxide (PANi/PEO) composite nanofibers (Tripathy et al. 2017b). Using both the platforms, we reported ultrasensitive detection of the DENV-specific consensus primer, using a single-stranded complementary probe anchored onto the nanomaterials. The Mn_2O_3 nanofiber-based electrochemical biosensor resulted in zeptomolar detection of the target DNA (LoD: 120 Zepto moles), using a standard glassy carbon working electrode. On the other hand, the conducting PANi/PEO nanofiber-based chemiresistive platform accounted for the target DNA detection in a wide concentration range of 10 pM–1 μM. Both these platforms are suitable for miniaturization, toward facilitating point-of-care diagnosis of DENV infection. Also, as a result of their high sensitivity and lower limits of detection, these platforms can ensure early detection of the viral infection.

9.3 Detection of Malaria

Malaria is one of the most important infectious diseases caused due to the protozoan of the genus *Plasmodium*, which is generally spread by the female anopheles mosquitoes (Cox 2010). According to a survey conducted by the world health organization (WHO), malaria is the fourth major endemic global health problem. Every year, more than 220 million people are infected due to malaria and nearly 500,000 deaths are reported (McBirney et al. 2018; Krampa et al. 2017). One of the major concerns in this regard is the lack of efficient analytical sensing platforms, which can detect and diagnose parasites sensitively, as well as selectively. It is therefore highly essential to develop platforms that can selectively detect malaria parasites with adequate sensitivity. Out of more than hundred species of *Plasmodium*, the *Plasmodium falciparum* parasite is known to cause severe form of malarial infection in human beings. Toward detecting the *Plasmodium falciparum* infection, the histidine-rich protein 2 (HRP2), the most abundant protein in the cell compartments of the parasite (Kifude et al. 2008), is considered to be a highly effective biomarker. In lieu of this, in this section, we discuss about nanoengineered sensing platforms targeting the detection of HRP2 in human blood. Hemben et al. have developed an immunosensor platform for the detection of *P. falciparum* histidine-rich protein 2 (pfHRP2) biomarker using Au nanoparticle (AuNPs)-modified screen-printed gold electrodes (SPGE) (Hemben et al. 2017). In this, the authors have drop casted the AuNPs onto the SPGE, followed by direct physical absorption of anti-pfHRP2 antibody. The unoccupied sites on the SPGE were blocked with bovine serum albumin (BSA). Efficiency of the as-prepared bioelectrodes was tested with different concentrations of HRP2 protein. Figure 9.4 shows the schematic representation of the sensor

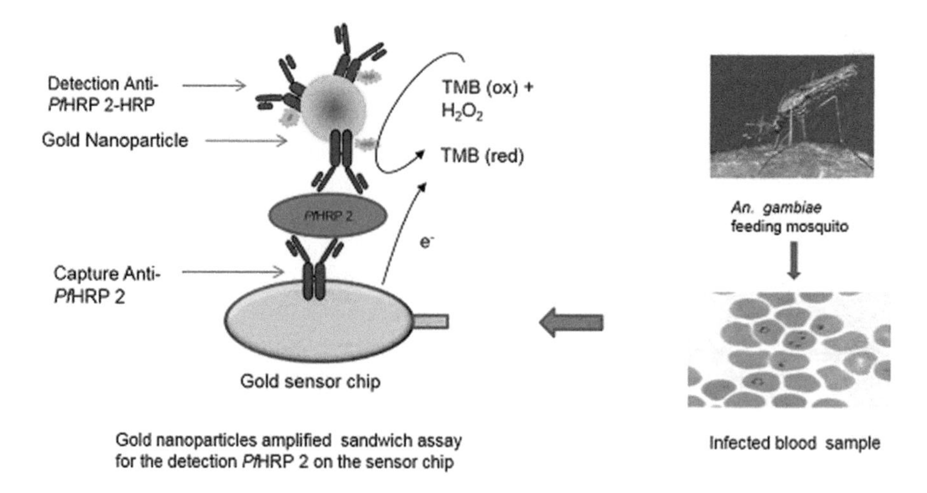

Fig. 9.4 Pictorial representation of preparation of sensing platform using Au nanoparticles for detecting malaria biomarkers (Hemben et al. 2017) (open access)

development protocol, as reported by the authors. As such, sensitivity offered by the said platform is reasonably good. However, the lack of stability is a major drawback.

The lack of stability arises due to the inconsistencies associated with the drop casting of AuNPs and the physical adsorption of antibody. In a separate study, Kumar et al. (Kumar et al. 2016) have addressed this issue using electrodeposited gold nanoparticles that can directly detect infected red blood cells qualitatively as well as quantitatively. Here, the electrodeposition of AuNPs using tetrachloroauric acid ensures strong adhesion of the nanoparticles to the SPGE, which inherently improves the stability of the biosensor. One of the major advantages of this platform is direct detection of the malaria-infected red blood cells in blood, without any specific sample preparation step (such as blood serum separation). Note the sensitivity of malaria sensing platforms can be further improved by increasing the effective surface area of the biosensor. Toward this, Sharma et al. have reported a disposable pfHRP2 immunosensor, derived from multi-walled carbon nanotubes (MWCNTs) and Au nanoparticles (Sharma et al. 2008). As reported, carboxylic functionalized MWCNTs were mixed with nafion (sulfonated tetrafluoroethylene polymer) and drop casted onto NaOH-treated SPEs. Due to its hydrophobic nature, nafion stabilizes the MWCNTs on the SPE surface. Next, AuNPs were electrodeposited onto the above prepared surface, using hydrochloroauric acid. Subsequently, desired antibody was immobilized onto the sensing surface via a sandwich immunoassay formation. Though, the disposable bioelectrode accounted for improved sensitivity and stability, it suffered from low limit of detection. In recent years, transition metal oxide-based nanofibers have garnered the attention of researchers for biosensing applications, due to their wide range of conductivity, enhanced sensitivity, and high biological activity. Among all metal oxides, zinc oxide (ZnO) has been extensively explored for its unique properties, such as free exciton binding energy (60 meV), wide bandgap (3.37 eV), non-toxicity, high diffusion coefficient, electron mobility, and isoelectric point (9.5). Note isoelectric point is the parameter which decides the degree of electrostatic attraction and binding of molecules. This property is critical in biosensors to facilitate the binding of the analytes (target antigens) electrostatically onto the metal oxide surfaces, which inherently improves the stability of biosensor. A major drawback of ZnO is its inherent low conductivity, a direct conclusion from its high bandgap. However, this can be enhanced by doping the metal oxide with secondary high conductive materials, such as Cu (Paul et al. 2016) and MWCNT (Paul et al. 2017). In light of this, Paul et al. have developed a biosensing platform using electrospun copper (Cu)-doped zinc oxide (ZnO) nanofibers, which can detect the malarial biomarker rapidly at trace levels (Paul et al. 2016). The Cu-doped ZnO nanofibers were synthesized using the simple, robust, and cost-effective method of electrospinning. For this, a homogeneous precursor consisting of zinc acetate and copper nitrate in a polyacrylonitrile (PAN)/DMF blend was electrospun using optimum parameters, so as to obtain a thick nanofiber mat. Subsequently, the same was subjected to a high-temperature calcination at 550 °C, to derive the Cu-doped ZnO nanofibers in a powdered form. Toward preparing the sensing electrodes, the as-prepared Cu-ZnO nanofibers were then drop casted onto glassy carbon electrodes (GCE). Later, immobilization of anti-HRP2 antibody was performed through covalent chemistry. For this, the nanofiber-modified GCEs were incubated in mercaptopropionic acid (MPA), for desired carboxylic functionalization of the surface via SAM formation. Following an NHS/EDC chemistry, the -COOH groups were subsequently activated, onto which the antibodies were covalently adsorbed. Using the thus prepared Cu-ZnO nanofiber-based biosensor, Paul et al. achieved a limiting detection of 6.8 ag/mL. The schematic representation of their reported protocol for preparation of the bioelectrode is shown below in Fig. 9.5.

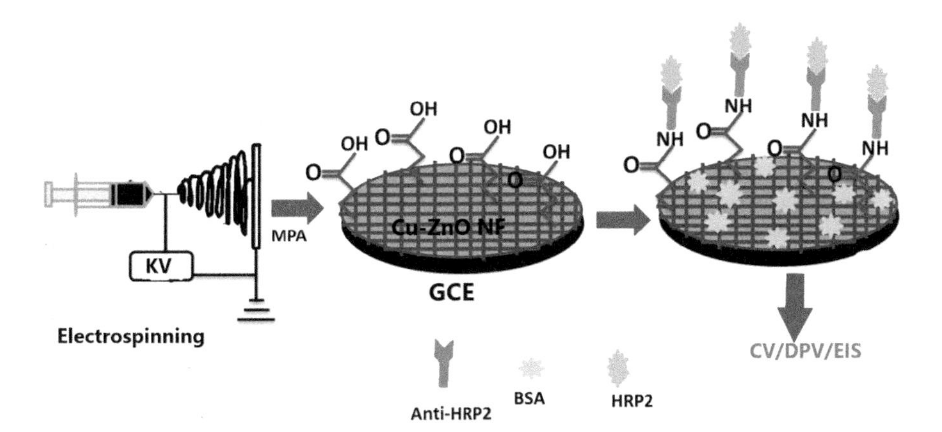

Fig. 9.5 Schematic representation of synthesis of Cu-ZnO nanofibers and preparation of bioelectrode using MPA and (EDC + NHS) chemistry. (Paul et al. 2016)

However, the above-discussed method suffers from the significantly higher time required toward preparing the bioelectrodes. To counter this, Paul et al. have further developed an approach for one-step functionalization of the nanomaterials, which ensures generation of carboxylic functional groups on the nanomaterial during its synthesis (Paul et al. 2017). In this work, they reported the synthesis of MWCNT-doped ZnO nanofibers, wherein, during the synthesis of the MWCNT-ZnO nanomaterial, high-temperature calcination (at 400 °C) in air converts the sp^2 hybridized carbon of MWCNT to -COOH groups. As a result, one can avoid the MPA treatment step, which essentially is a 15–18-h-long incubation step. This consequently reduces the overall time required to prepare the bioelectrodes, without compromising the sensitivity, selectivity, and stability. In recent times, several notable advances have been reported in engineered nanomaterial-based biosensing mechanisms. Further, researchers have targeted to miniaturize laboratory developed schemes in to handheld biochips without losing the efficiency, so as to realize point-of-care devices for infectious disease diagnosis (Chandra 2016; Chandra et al. 2017). In this endeavor, microfluidic channel integrated biosensing platforms have been extensively used. In microfluidics, the fundamental physics of systems changes rapidly with downscaling of the dimensions of the channel. This has given rise to a whole new field of research, which focuses on understanding the behavior of biofluids when the same is confined in micron-range channels. Particularly in biosensors, such devices have enabled the analysis of complex body fluids like the blood, via on-chip separation of plasma from whole blood (Tripathi et al. 2016). It is to be noted that the selection of materials for fabrication of microchannels is crucial in developing integrated biosensors. Generally, glass, silicon, and polymers (polydimethylsiloxane, polystyrene, etc.) are widely used substrate materials in relevant applications. Fabrication of microfluidic devices with aforementioned substrate materials requires clean room facilities (to perform process steps such as lithography and etching), which inherently increases the cost of devices. As an economic alternative, several fabless protocols have been reported in the literature for developing microfluidic integrated platforms. For instance, Paul et al. have developed a non-lithographic tune-transfer method to fabricate microfluidic channels on cost-effective and disposable plastic substrate (Brince Paul et al. 2017). Further, they integrated the microfluidic channel with a miniaturized electrochemical biosensor, to facilitate on-chip detection of HRP2 protein. Prior to the integration, the electrochemical biosensor with Au working, Ag/AgCl reference, and Pt counter electrodes was fabricated on a polyethylene terephthalate (PET) substrate through shadow masking and sputtering methods. The schematic representation of the reported non-lithographic fabrication process for the plastic-based biochip is shown in Fig. 9.6.

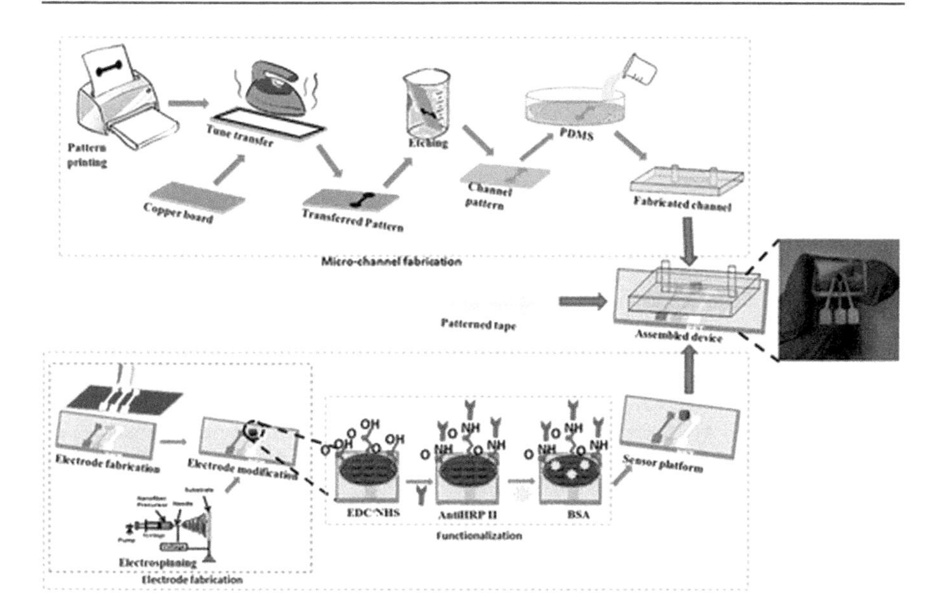

Fig. 9.6 Schematic representation of fabrication of plastic-based biochip for detection of infectious diseases (Brince Paul et al. 2017) (Reprinted with permission)

9.4 Conclusion

In this chapter, we have discussed several important literature concerning nanoengineered electrochemical biosensors for infectious disease diagnosis. Specifically, biosensors associated with dengue virus and malarial parasite detection have been explained in great detail, on account of the prevalence of these two infectious diseases in tropical and subtropical countries. In doing so, we have addressed several key aspects of the biosensors, namely, stability, selectivity, and sensitivity, and have reported the innovations made in the research world toward improving each of these aspects. Further, nanomaterial-based electrochemical biosensors, wherein the working electrode or the sensing surface consists of pristine or composite nanomaterials, have been discussed at length. Apart from this, we also discussed about several miniaturized biosensing platforms with integrated fluidic channels which have been reported in the literature targeting infectious disease detection. Truly, such platforms are a step toward ensuring point-of-care disease diagnosis in the future. However, as a prerequisite to this, prototyping such devices and evaluating their performance against real-time patient sample are essential. Further, to develop completely standalone diagnosis kits, an on-chip sample extraction unit should also be integrated with the sensing system. In addition, cost effectiveness of these systems is also an important aspect in deciding whether they can indeed make any significant social impact.

References

Bhatnagar I, Mahato K, Ealla KKR, Asthana A, Chandra P (2018) Chitosan stabilized gold nanoparticle mediated self-assembled gliP nanobiosensor for diagnosis of invasive Aspergillosis. Int J Biol Macromol 110:449–456

Brince Paul K, Panigrahi AK, Singh V, Singh SG (2017) Nonlithographic fabrication of plastic-based nanofibers integrated microfluidic biochip for sensitive detection of infectious biomarker. ACS Appl Mater Interfaces 9:39994–40005

Chandra P (2016) Nanobiosensors for personalized and onsite biomedical diagnosis. The Institution of Engineering and Technology

Chandra P, Tan YN, Singh SP (eds) (2017) Next generation point-of-care biomedical sensors technologies for cancer diagnosis. Springer, Singapore

Cox FEG (2010) History of the discovery of the malaria parasites and their vectors. Parasit Vectors 3(1):5

Darwish NT, Alrawi AH, Sekaran SD, Alias Y, Khor SM (2016) Electrochemical immunosensor based on antibody-nanoparticle hybrid for specific detection of the dengue virus NS1 biomarker. J Electrochem Soc 163(3):B19–B25

Datta S, Wattal C (2010) Dengue NS1 antigen detection: a useful tool in early diagnosis of dengue virus infection. Indian J Med Microbiol 28(2):107

Dias ACMS, Gomes-Filho SLR, Silva MMS, Dutra RF (2013) A sensor tip based on carbon nanotube-ink printed electrode for the dengue virus NS1 protein. Biosens Bioelectron 44:216–221

Fan Y, Tan X, Liu X, Ou X, Chen S, Wei S (2015) A novel non-enzymatic electrochemiluminescence sensor for the detection of glucose based on the competitive reaction between glucose and phenoxy dextran for concanavalin A binding sites. Electrochim Acta 180:471–478

Guo S, Dong S, Wang E (2009) Polyaniline/Pt hybrid nanofibers: high-efficiency nanoelectrocatalysts for electrochemical devices. Small 5(16):1869–1876

Hemben A, Ashley J, Tothill IE (2017) Development of an immunosensor for PfHRP2 as a biomarker for malaria detection. Biosensors 7(3):28

Houng H-SH, Chung-Ming Chen R, Vaughn DW, Kanesa-thasan N (2001) Development of a fluorogenic RT-PCR system for quantitative identification of dengue virus serotypes 1–4 using conserved and serotype-specific 3′ noncoding sequences. J Virol Methods 95(1–2):19–32

Huang MJ, Xie H, Wan Q, Zhang L, Ning Y, Zhang G-J (2013) Serotype-specific identification of dengue virus by silicon nanowire array biosensor. J Nanosci Nanotechnol 13(6):3810–3817

Huy TQ, Hanh NTH, Thuy NT, Van Chung P, Nga PT, Tuan MA (2011) A novel biosensor based on serum antibody immobilization for rapid detection of viral antigens. Talanta 86:271–277

Kifude CM, Rajasekariah HG, Sullivan DJ, Ann Stewart V, Angov E, Martin SK, Diggs CL, Waitumbi JN (2008) Enzyme-linked immunosorbent assay for detection of Plasmodium falciparum histidine-rich protein 2 in blood, plasma, and serum. Clin Vaccine Immunol 15(6):1012–1018

Krampa F, Aniweh Y, Awandare G, Kanyong P (2017) Recent progress in the development of diagnostic tests for malaria. Diagnostics 7(3):54

Kryscio DR, Peppas NA (2012) Critical review and perspective of macromolecularly imprinted polymers. Acta Biomater 8(2):461–473

Kumar B, Bhalla V, Singh Bhadoriya RP, Suri CR, Varshney GC (2016) Label-free electrochemical detection of malaria-infected red blood cells. RSC Adv 6(79):75862–75869

Kyle JL, Harris E (2008) Global spread and persistence of dengue. Annu Rev Microbiol 62:71–92

Mahato K, Maurya PK, Chandra P (2018) Fundamentals and commercial aspects of nanobiosensors in point-of-care clinical diagnostics. 3 Biotech 8(3):149

Mahato K, Nagpal S, Shah MA, Srivastava A, Maurya PK, Roy S, Jaiswal A, Singh R, Chandra P (2019) Gold nanoparticle surface engineering strategies and their applications in biomedicine and diagnostics. 3 Biotech 9(2):57

Mandal R, Baranwal A, Srivastava A, Chandra P (2018) Evolving trends in bio/chemical sensors fabrication incorporating bimetallic nanoparticles. Biosens Bioelectron 117:546–561

Matta DP, Tripathy S, Vanjari SRK, Sharma CS, Singh SG (2016) An ultrasensitive label free nano-biosensor platform for the detection of cardiac biomarkers. Biomed Microdevices 18(6):111

Mauriz E, Calle A, Abad A, Montoya A, Hildebrandt A, Barceló D, Lechuga LM (2006) Determination of carbaryl in natural water samples by a surface plasmon resonance flow-through immunosensor. Biosens Bioelectron 21:2129–2136

McBirney SE, Chen D, Scholtz A, Ameri H, Armani AM (2018) Rapid diagnostic for point-of-care malaria screening. ACS Sensors 3:1264–1270

Minot ED, Janssens AM, Heller I, Heering HA, Dekker C, Lemay SG (2007) Carbon nanotube biosensors: the critical role of the reference electrode. Appl Phys Lett 91(9):093507

Navakul K, Warakulwit C, Yenchitsomanus P-t, Panya A, Lieberzeit PA, Sangma C (2017) A novel method for dengue virus detection and antibody screening using a graphene-polymer based electrochemical biosensor. Nanomedicine 13(2):549–557

Paul KB, Panigrahi AK, Singh V, Singh SG (2017) A multi-walled carbon nanotube–zinc oxide nanofiber based flexible chemiresistive biosensor for malaria biomarker detection. Analyst 142:2128–2135. https://doi.org/10.1039/C7AN00243B

Paul KB, Kumar S, Tripathy S, Vanjari SRK, Singh V, Singh SG (2016) A highly sensitive self-assembled monolayer modified copper doped zinc oxide nanofiber interface for detection of Plasmodium falciparum histidine-rich protein-2: targeted towards rapid, early diagnosis of malaria. Biosens Bioelectron 80:39–46

Prakash MD, Singh SG, Sharma CS, Krishna VSR (2017) Electrochemical detection of cardiac biomarkers utilizing electrospun multiwalled carbon nanotubes embedded SU-8 Nanofibers. Electroanalysis 29(2):380–386

Raengsakulrach B, Nisalak A, Maneekarn N, Yenchitsomanus P-T, Limsomwong C, Jairungsri A, Thirawuth V et al (2002) Comparison of four reverse transcription-polymerase chain reaction procedures for the detection of dengue virus in clinical specimens. J Virol Methods 105(2):219–232

Rashid JIA, Yusof NA, Abdullah J, Hashim U, Hajian R (2015) A novel disposable biosensor based on SiNWs/AuNPs modified-screen printed electrode for dengue virus DNA oligomer detection. IEEE Sensors J 15(8):4420–4427

Rodriguez-Mozaz S, Reder S, Lopez de Alda M, Gauglitz G, Barceló D (2004) Simultaneous multi-analyte determination of estrone, isoproturon and atrazine in natural waters by the RIver ANAlyser (RIANA), an optical immunosensor. Biosens Bioelectron 19:633–640

Sanvicens N, Pastells C, Pascual N, Marco M-P (2009) Nanoparticle-based biosensors for detection of pathogenic bacteria. TrAC Trends Anal Chem 28(11):1243–1252

Sharma MK, Rao VK, Agarwal GS, Rai GP, Gopalan N, Prakash S, Sharma SK, Vijayaraghavan R (2008) Highly sensitive amperometric immunosensor for detection of Plasmodium falciparum histidine-rich protein 2 in serum of humans with malaria: comparison with a commercial kit. J Clin Microbiol 46(11):3759–3765

Shinde SB, Fernandes CB, Patravale VB (2012) Recent trends in in-vitro nanodiagnostics for detection of pathogens. J Control Release 159(2):164–180

Silva MMS, Dias ACMS, Silva BVM, Gomes Filho SLR, Kubota LT, Goulart MOF, Dutra RF (2015) Electrochemical detection of dengue virus NS1 protein with a poly (allylamine)/carbon nanotube layered immunoelectrode. J Chem Technol Biotechnol 90(1):194–200

Sotiropoulou S, Chaniotakis NA (2003) Carbon nanotube array-based biosensor. Anal Bioanal Chem 375(1):103–105

Supraja P, Sudarshan V, Tripathy S, Agrawal A, Singh SG (2019a) Label free electrochemical detection of cardiac biomarker troponin T using ZnSnO3 perovskite nanomaterials. Anal Methods 11:744–751

Supraja P, Tripathy S, Krishna Vanjari SR, Singh V, Singh SG (2019b) Label free, electrochemical detection of atrazine using electrospun Mn2O3 nanofibers: towards ultrasensitive small molecule detection. Sensors Actuators B Chem 285:317–325

Tripathi S, Kumar YVBV, Agrawal A, Prabhakar A, Joshi SS (2016) Microdevice for plasma separation from whole human blood using bio-physical and geometrical effects. Sci Rep 6:26749
Tripathy S, Krishna Vanjari SR, Singh V, Swaminathan S, Singh SG (2017a) Electrospun manganese (III) oxide nanofiber based electrochemical DNA-nanobiosensor for zeptomolar detection of dengue consensus primer. Biosens Bioelectron 90:378–387
Tripathy S, Naithani A, Vanjari SRK, Singh SG (2017b) Electrospun polyaniline nanofiber based chemiresistive nanobiosensor platform for DNA hybridization detection. In: SENSORS, 2017 IEEE. IEEE, pp 1–3
Tripathy S, Gangwar R, Supraja P, Rao AVSSN, Vanjari SRK, Singh SG (2018a) Graphene doped Mn_2O_3 nanofibers as a facile electroanalytical DNA point mutation detection platform for early diagnosis of breast/ovarian cancer. Electroanalysis 30:2110–2120
Tripathy S, Joseph J, Vanjari SRK, Rao AVSSN, Singh SG (2018b) Flexible ITO electrode with gold nanostructures for Femtomolar DNA hybridization detection. IEEE Sensors Lett 2(4):1–4
Vadnala S, Tripathy S, Paul N, Agrawal A, Singh SG (2018) Facile synthesis of electrospun nickel (II) oxide Nanofibers and its application for hydrogen peroxide sensing. ChemistrySelect 3(43):12263–12268
Wanekaya AK, Chen W, Myung NV, Mulchandani A (2006) Nanowire based electrochemical biosensors. Electroanalysis 18(6):533–550
Wang J (2005) Carbon nanotube based electrochemical biosensors: a review. Electroanalysis 17(1):7–14
Xia F, Zuo X, Yang R, Yi X, Di K, Vallée-Bélisle A, Gong X et al (2010) Colorimetric detection of DNA, small molecules, proteins, and ions using unmodified gold nanoparticles and conjugated polyelectrolytes. Proc Natl Acad Sci 107(24):10837–10841
Yanez-Sedeno P, Pingarron JM (2005) Gold nanoparticle-based electrochemical biosensors. Anal Bioanal Chem 382(4):884–886

Nanobiotechnology Advancements in Lateral Flow Immunodiagnostics

10

Vivek Borse and Rohit Srivastava

Abstract

Nanobiotechnology has emerged as an effective tool in the development of diagnostics, biomaterials, drug delivery systems, and numerous point-of-care (POC) applications in recent years. Proteins, antigens, antibodies, amino acids, DNA, etc. are conjugated with the nanomaterial to develop the potential biosensors. Lateral flow immunoassay (LFIA) is well-established technology that has been used in the development of the variety of POC techniques in the biomedical field for qualitative as well as quantitative diagnostics. Numerous LFIA test kits have been developed and scripted in the form of research publications, but most of them are not commercialized. In this chapter, the recent nanobiotechnological advancements in the development of LFIA with respect to the signal amplification and sensitivity enhancement are reviewed. Types of LFIA, advantages and disadvantages of advancements made in the detection of LFIA, smartphone-based detection techniques in LFIA, etc. are discussed with respect to their importance in the LFIA application. Various techniques used for the LFIA sensitivity enhancement are emphasized along with its effect on the application. The sensitivity of LFIA can be enhanced by implementing various mechanistic approaches such as the use of brighter sensor molecule, adjusting the position of test line on the LFIA strip, altering LFIA sample preparation procedure and physicochemical parameters, addition of additives to the LFIA reagents, using an additional component in LFIA strip assembly, post-assay test line color enhancement, removal of the interfering entities in the complex analyte, increase in

V. Borse (✉)
NanoBioSens Laboratory, Centre for Nanotechnology,
Indian Institute of Technology Guwahati, Guwahati, Assam, India
e-mail: vivek.borse@iitg.ac.in

R. Srivastava
NanoBios Laboratory, Department of Biosciences and Bioengineering,
Indian Institute of Technology Bombay, Mumbai, Maharashtra, India

P. Chandra, R. Prakash (eds.), *Nanobiomaterial Engineering*,
https://doi.org/10.1007/978-981-32-9840-8_10

number of the detection sites, increase in binding area of the detector-capture antibody pair, improved detection capabilities, highly sensitive imaging applications, etc. Selectivity and sensitivity enhancement has improved the detection capability, but continuous efforts are needed on the reproducibility, reliability, and simplicity of the LFIA techniques.

Keywords
Nanobiotechnology · Lateral flow immunoassay · Diagnostics · Biosensors

10.1 Introduction

Nanobiotechnology has shown remarkable progress in the recent years, especially in diagnostics, biomaterials, and drug delivery system. Nanobiotechnology can be defined as the application of nanotechnology in the developmental science of biotechnology, where the biological molecules are conjugated with the nanomaterial. The nanomaterial consists of the wide spectrum of materials including one of the dimensions in the order of nanometer. Nanomaterial exists either as a singular entity in solution form such as nanoparticles, nanospheres, nanobeads, nanostars, nanowires, nanofibers, nanopopcorns, nanohorns, nanoplates, nanoribbon, nanorods, nanotubes, nanocrystals, etc. or in association with other materials such as nanofoam, nanopore, nanocomposites, various surfaces modified using nanotechnology, etc. The primary purpose for fabricating nanomaterial is increasing the surface area of an object; penetrate, infiltrate, permeate, and invade an object to or through the particular system. In the medical field, the objective of the use of nanomaterial may be more specific such as reduce the dose of a drug to avoid the side effects or toxicity of therapeutic formulation. Nanomaterial seldom acts as the diagnostic agent but often as the therapeutic agent. Substantial research has been done, and a large number of publications have reported the development of nanoformulations of various drugs for therapeutic applications. Nanomaterial requires structural and functional modifications to be utilized as the diagnostic agent.

Nanodiagnostics is defined as the use of nanotechnology for clinical diagnostic purposes (Bellah et al. 2012). Proteins, antigens, antibodies, amino acids, DNA, etc. have been conjugated with the nanomaterial to develop the potential biosensors. Diagnosis based on the use of nanostructures has shown promising outcomes as they offer high sensitivity and specificity along with quick detection of the disease condition; thus, nanodiagnostics has enormous potential in developing numerous point-of-care (POC) applications.

Lateral flow immunoassay (LFIA) is very well-established technology that has been used in the development of the variety of POC techniques in the biomedical field. LFIA is an immunochromatographic assay where the analytes are detected on the paper surface. LFIA strip components include sample pad, where the sample is applied; conjugation pad, where the biosensors are immobilized; detection pad, which is nitrocellulose membrane and consists of the test and control line; and absorption pad, which pulls the fluid to the end of the strip. Additional parts can be added, or the structure of LFIA strip assembly can be modified as per requirement. Antibodies are immobilized on the paper surface, and fluid containing biosensors and analytes flow through the porous membrane, i.e., nitrocellulose membrane strip. The specific reaction occurs between antigen and antibody or enzyme and substrate or receptor and the ligand that lead to the appearance of response at the predesigned location on the strip. Presence of an analyte can be qualitatively determined by visual comparison, i.e., appearance of color at test and control line on the strip. The analysis can be made semiquantitative by using color comparison chart and quantitative using the specialized portable instrument.

LFIA technique has been used widely in the development of POC devices that have applications in qualitative as well as quantitative diagnostics. Various types of assays have been developed using lateral flow technique for detection of proteins, nucleic acid, viruses, bacteria, markers for infectious diseases, cancer, cardiac biomarkers, allergic components in food, water impurity, etc. The focus of this chapter is to review the recent nanobiotechnological advancements in the development of LFIA with respect to the signal amplification and sensitivity enhancement.

Advantages of LFIA technique are mentioned below (Sajid et al. 2015):

1. User-friendly: Easy to use technique, minimal steps required, and no need of a trained technician to interpret the results.
2. Cost: Cheaper technique as compared to the sophisticated devices.
3. Time: Gives results usually in minutes, which is a very important factor as far as the medical diagnostic field is concerned.
4. Sample requirement: Very small amount of sample (typically in μl) is required for the detection.
5. Multiplexing: Multiple analyte sensing can be achieved on the same strip.
6. Characteristics: LFIA is a robust technique and has sensitivity and specificity for the particular analyte.
7. Manufacturing: Easy to manufacture as the raw material, equipment, and processes are already well explored, developed, and tested.
8. Scale-up: Easy to scale up for manufacturing large number of assays in a single batch.
9. Stability: Manufactured test strips are stable for a long time, i.e., from 12 to 24 months, without any special storage condition.
10. Utility: The LFIA can be integrated with electronics detection devices or readers for the advanced quantitative sensing applications.
11. Regulatory: Timeline for LFIA development and regulatory approvals is shorter than other diagnostic techniques.
12. Market: Wide acceptance by the market as minimum training is required for users.

10.2 Types of LFIA

There are two basic types of the LFIA, standard and competitive LFIA as described in Fig. 10.1. In standard LFIA, the response in terms of color or fluorescence appears on the test line if the analyte is present in the sample, while in competitive-type LFIA, biosensors and analyte compete to bind on the test line, and the response is inversely proportional to the concentration of an analyte.

Most of the LFIA provides qualitative (yes/no-type) or semiquantitative results for analyte detection. There may be low sensitivity because of insufficient brightness appeared on the LFIA strip. Thus, various nanoparticle-based platforms have been developed for the improvement of the analytical performance of LFIA. Figure 10.2 explains the use of three types of nanoparticles that improve the analytical sensitivity of LFIA. Both qualitative and quantitative output can be obtained using these three different types of nanoparticles.

Several LFIA-based detection studies have been carried out since last decade. Most of these are based on the principle mentioned in the schematics in Fig. 10.2. Several review articles and books have been reporting the details of the methodological protocols for preparation of LFIA-based detection technique. Various technologies used for development and commercial aspects of the paper-based LFIA diagnostic kits with respect to personalized healthcare are reviewed in literature earlier (Mahato et al. 2017). Here in this chapter, we have focused on the recent nanobiotechnological advancements in LFIA development.

10.3 Advances in Detection Methods

LFIA process is completed with the result interpretation. Once the assay procedure is completed, the output is either qualitative, i.e., yes/no type, or semiquantitative, i.e., comparison with standard, or quantitative, i.e., concentration of the analyte is determined precisely in units. Since the evolution of the LFIA initiated with the

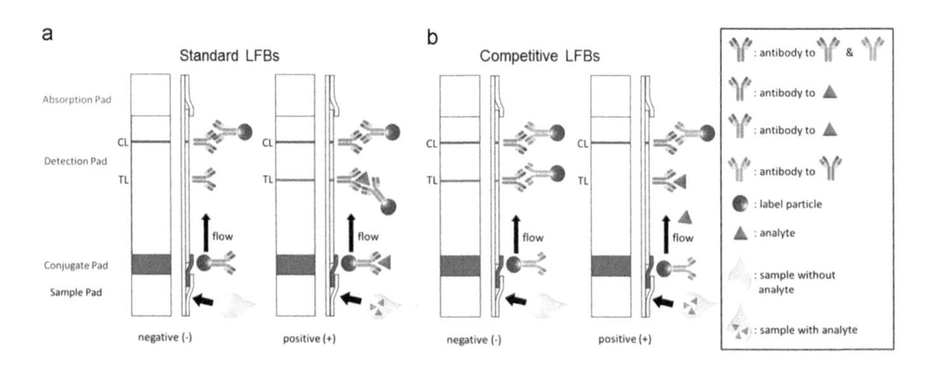

Fig. 10.1 Schematics for the different types of LFIA. Reprinted with permission from (Quesada-González and Merkoçi 2015), Copyright 2015 Elsevier

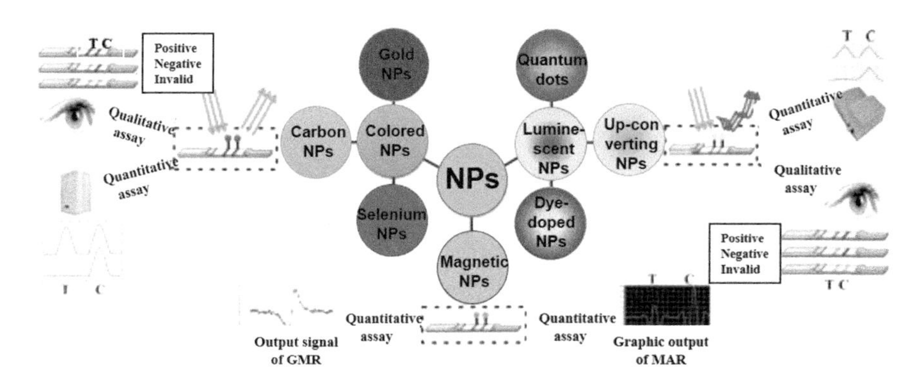

Fig. 10.2 Schematics for various LFIA detection strategies based on nanoparticles. Reprinted with permission from (Huang et al. 2016), Copyright 2016 Elsevier

pregnancy test kit, which is visual observation-based detection, the large number of LFIA is developed based on this method. Colorimetric detection is most common among the visual detection methods. Presence or absence of an analyte is revealed by the formation of color or change in color on the strip. The color on the strip is usually because of the colloidal gold nanoparticles that are conjugated with the antibody. In the case of semiquantitative detection, the result is interpreted as a range of concentration. Numerous studies have provided the standard comparison modules or charts to interpret the quantity of analyte present in the sample by comparing the intensity of the color formed on the strip. One of the most significant examples in this category is the marketed urine dipstick kit.

For quantitative analysis, sophisticated instruments are used; once the assay procedure is completed, the strip is inserted in the device, and after data processing, results are displayed in concentration units on the device panel or on the associated system. Our group is working on the synthesis, stability studies in various biochemical buffers, protein bioconjugation, and the utility of quantum dots; studies are reported in the literature (Borse et al. 2016a, b, 2017b, 2018). Recently, we have developed a quantitative LFIA for detection of orthopedic implant-associated infection biomarkers (Borse and Srivastava 2019). Quantum dots were used as a sensing platform, and antibodies were conjugated to the QDs. Schematics for the experimental protocol are shown in Fig. 10.3. We have used the standard protocols for LFIA development, and the analyte detection has been quantified using indigenously developed PorFloR™ (Borse et al. 2017a).

Recently, a portable fluorescence reader is designed by our group using silicon photomultiplier (SiPM) sensor (Makkar et al. 2018). This benchtop device is used to quantify the fluorescence intensity on the planar surfaces such as nitrocellulose membrane which is the most common platform for LFIA development. Bimetallic nanoparticles are also used in the fabrication of biochemical/chemical sensors that has applications in the LFIA development (Mandal et al. 2018). An amperometric immunosensor has been developed and evaluated for the detection of inducible nitric oxide synthase that is related to the toxicity of endocrine disrupters (Chandra

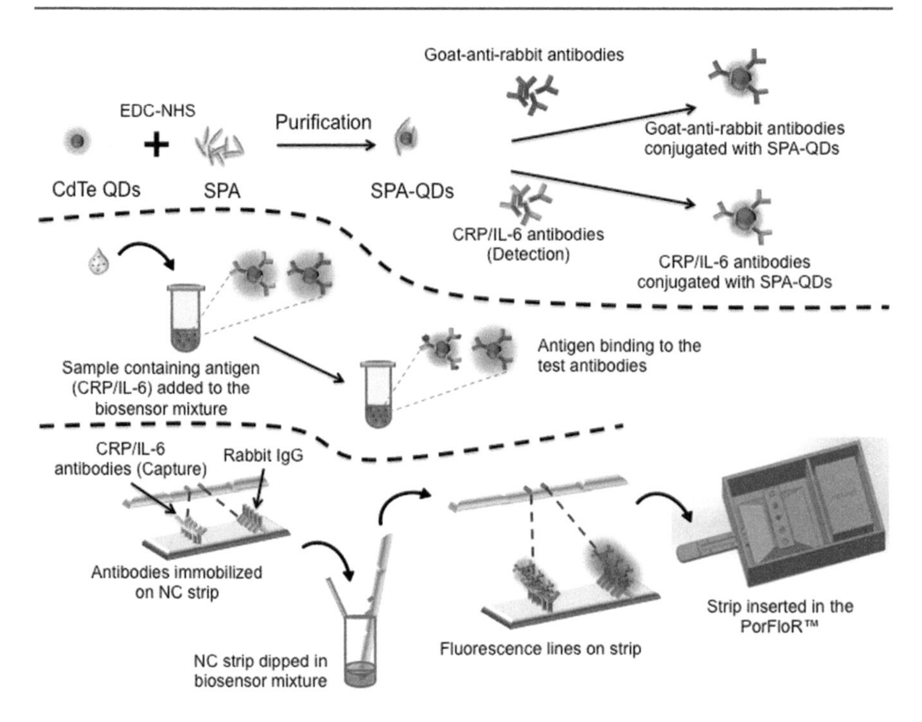

Fig. 10.3 Schematics for the LFIA development. Reprinted with permission from (Borse and Srivastava 2019), Copyright 2018 Elsevier

et al. 2012). A label-free impedimetric immunosensor has been developed for the detection of oral cancer biomarker (CD 59) in the early stage (Choudhary et al. 2016). Gold nanoparticle-based ultrasensitive immunosensor has been developed for the detection of brain-derived neurotrophic factor (Akhtar et al. 2018). Application of various electrochemical immunosensors in the clinical diagnostics is mentioned in the comprehensive literature with an essential discussion on the basic development of immunosensor to clinical testing (Mahato et al. 2018). Dual reading capability in LFIA has been employed for the detection of human thyroid-stimulating hormone. Hybrid nanocomposite particles having gold in core and europium in the shell were conjugated with the antibodies, and both colorimetric and fluorometric analysis are possible because of the different material properties (Preechakasedkit et al. 2018).

10.4 Smartphone-Based Detection Techniques in LFIA

The use of optical LFIA strip reader is widely reported; the devices are customized as per the requirement of signal detection. Nowadays researchers are working to develop the LFIA detection techniques that are compatible with mobile phones. The large population is using smart mobile phones and has an access to the Internet. Post-processing of LFIA, compact LFIA strips are integrated with the smartphone;

the strip is inserted into the holder attached to the mobile phone. The camera of the smartphone is used for the image processing purpose; a specialized application can process, analyze, and interpret the data and share the information through the network to the third party (e.g., expert consultant). The number of applications has been developed and connected to the server so that the expert clinician can access the data and consult to the patient immediately. This has brought the revolution to the healthcare sector as users at home or in the remote locations get access to the medical facilities.

A study has reported LFIA integrated with the smartphone for simultaneous detection of brain natriuretic peptide and suppression of tumorigenicity 2 (antigenic markers for heart failure) (You et al. 2017). Schematics for the study are shown in Fig. 10.4. Dual color core-shell upconversion nanoparticles (UCNP), with green and blue fluorescence, were conjugated with antibodies and used as sensing molecule. After conventional LFIA procedure, the fluorescence signal on the strip is read by the smartphone reader. The smartphone-based portable reader that comprises NIR laser, portable power and optical system, and the application that analyzes the fluorescence signal was developed. The results were displayed on the phone screen and can be shared with the doctors; thus, the patient can be continuously monitored. Recently, the alkaline phosphatase (ALP) detection is performed on the paper

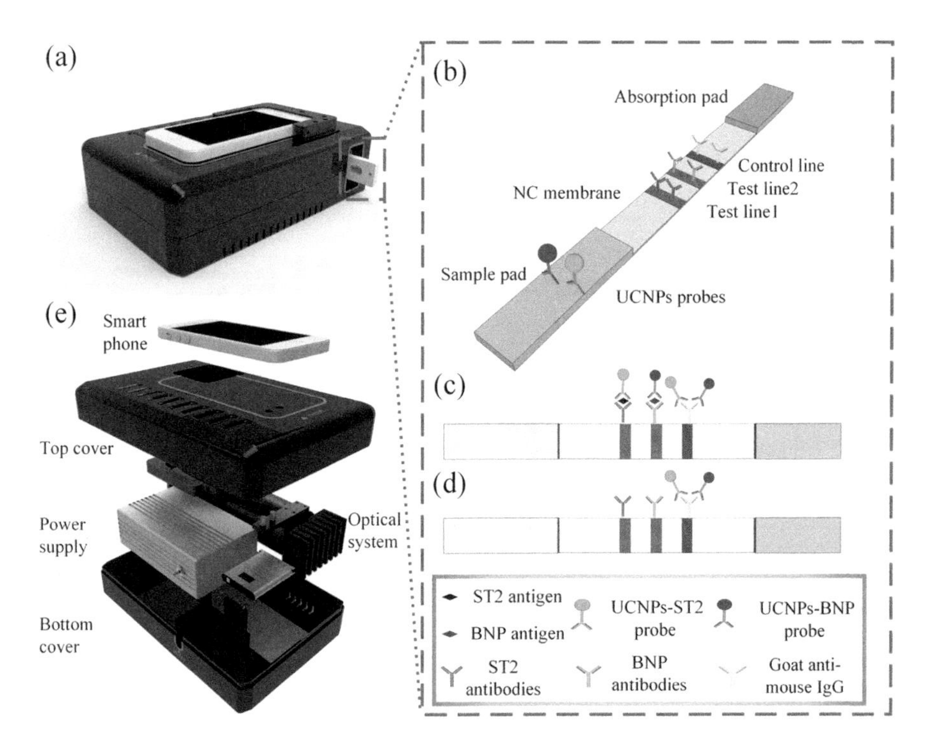

Fig. 10.4 Smartphone-based LFIA for heart failure monitoring. Reprinted with permission from (You et al. 2017), Copyright 2017 American Chemical Society

surface by miniaturized immunosensors. The color change chemistry is associated with the concentration of ALP in the milk sample, and the smartphone was integrated for signal image processing and quantitative analysis (Mahato and Chandra 2019). An ironPhone is a novel smartphone-based POC device that monitors iron deficiency by quantifying serum ferritin level (Srinivasan et al. 2018). ironPhone is gold nanoparticle-based disposable LFIA comprising of a smartphone and an app; the schematic is shown in Fig. 10.5. The smartphone is used for image processing purpose that receives the test strip, captures an image, and converts the results into concentration form, i.e., quantitation of ferritin level. Performance of ironPhone was calibrated and validated in comparison to the standard techniques. The combinatorial approach of using catalytic molecule for signal enhancement and smartphone to read the signal intensity is projecting in the LFIA development arena. Nanozymes (i.e., mesoporous Pd@Pt nanoparticles), which are excellent optical markers, were used as a signal amplifier in dual LFIA integrated with the smartphone for simultaneous detection of bacteria in food samples (e.g., *Salmonella enteritidis* and *Escherichia coli* O157:H7) (Cheng et al. 2017). The drawback of the reported method is requirement of the addition of dye solution at test line after processing of LFIA. Improved sensitivity of LFIA was attributed to the peroxidase-like catalytic activity of nanozymes for signal enhancement and eliminating the chance of cross-interference by parallel dual detection design.

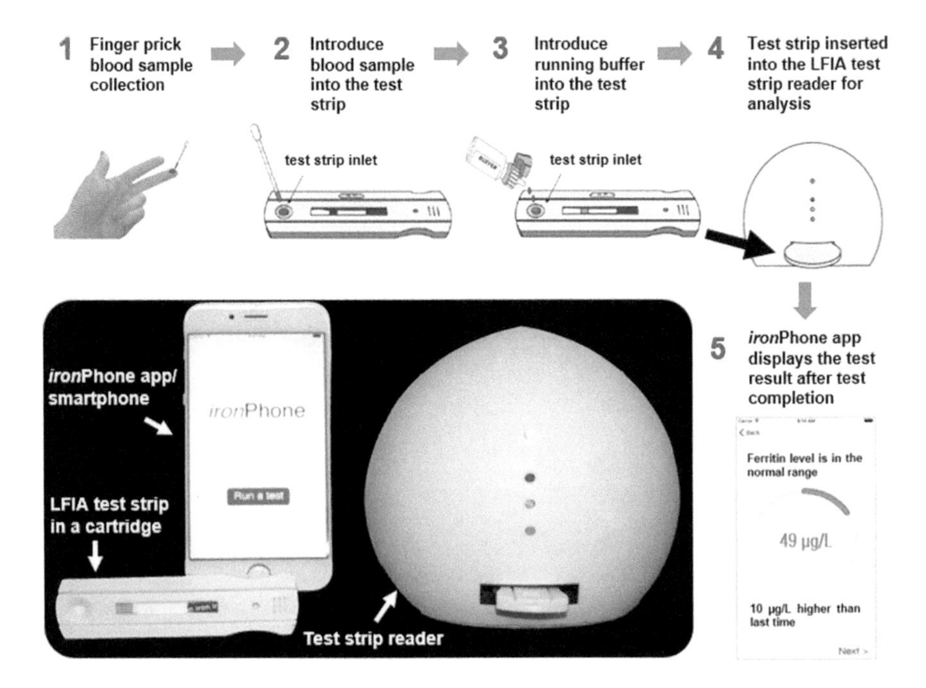

Fig. 10.5 Operation protocol and representation of smartphone based ironPhone. Reprinted with permission from (Srinivasan et al. 2018), Copyright 2017 Elsevier

Thus, the mobile phone-integrated LFIA techniques are emerging as a choice to improve the user-friendly modalities in diagnostics. Data processing and signal analysis are performed efficiently, and the results are interpreted within minutes, which can be shared with clinicians. This whole development is reducing the diagnosis time, and clinicians are able to take decisions for further treatment.

10.5 Sensitivity Enhancement in LFIA Technique

In the recent years, research has been focused on the improvement of LFIA sensitivity. The sensitivity of LFIA can be enhanced by implementing various mechanistic approaches mentioned below:

- Utilizing more bright sensor molecule
- Adjusting the position of test line on the LFIA strip
- Altering LFIA sample preparation procedure and physicochemical parameters
- Addition of additives to the LFIA reagents
- Using additional component in LFIA strip assembly
- Post-assay test line color enhancement
- Removal of the interfering entities in the complex analyte
- Increase in number of the detection sites
- Increase in binding area of the detector-capture antibody pair
- Improved detection capabilities
- Highly sensitive imaging applications

Type and morphology of the nanoparticle used for development of LFIA have a significant effect on the analytical outcome. Gold nanoparticles are being used in the LFIA diagnostics since long. Recently, the effect of the morphology of gold nanoparticles on the sensitivity of LFIA detection of procalcitonin (PCT), which is highly specific bacterial infection marker, has been reported (Serebrennikova et al. 2018). Spherical form of gold nanoparticle is more suitable to be utilized in the LFIA as antibodies can be conjugated uniformly on its surface. Increase in the size of gold nanospheres (GNS) increases the sensitivity of the immunoassay as the large number of the antibodies can be conjugated uniformly on the surface of nanosphere. But the cost of assay increases manyfold as the consumption of antibodies is increased, and also with the larger size of the gold nanospheres, color brightness and stability are reduced. Figure 10.6 explains the effect of the size of the label on the detection along with visual images of the LFIA strips. Instead of larger gold nanospheres, gold nanostars (GNSt) and nanopopcorns (GNPN) are found to be stable structures due to their three-dimensional configuration. Antibody consumption is lower for nanopopcorns as compared to nanospheres, and the color signal is higher, which results into the fivefold higher sensitivity for PCT detection.

In another study, the hollow nanogold microspheres were compared with gold nanoparticles for the detection of neurotoxin brevetoxin B (neurotoxin produced by algae) using LFIA (Zhang et al. 2014). The limit of detection was found to be

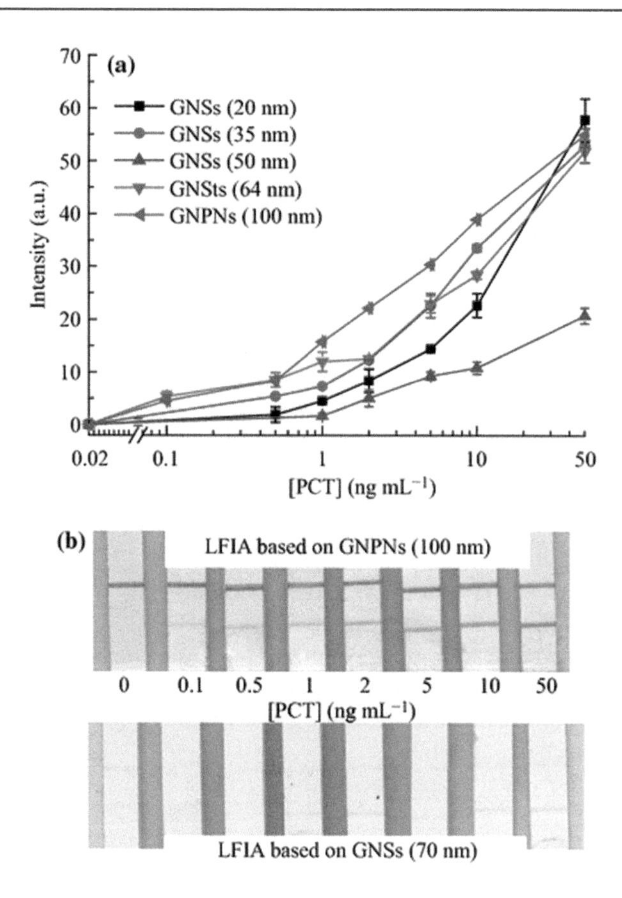

Fig. 10.6 (**a**) Calibration curves of LFIA for PCT detection, (**b**) color intensity comparison in the analytical zone. Reprinted with permission from (Serebrennikova et al. 2018), Copyright 2017 Author(s)

20-fold higher as compared to the colloidal gold nanoparticle. The LFIA has been developed for detection of influenza A (model antigen virus) where signal amplification is attempted using dual-layered and double-targeted nanogold (Wiriyachaiporn et al. 2015). Schematics of the LFIA are shown in Fig. 10.7. The detection capability was reported to be eight times higher than the conventional LFIA process even when the sample analyte is a complex biological matrix. Higher sensitivity of assay is attributed to the more number of target binding site, which was possible due to the two variants of antibody-conjugated gold nanoparticles; one variant binds to the antibody immobilized on the strip surface (along with the antigenic biomarker), while another antibody, conjugated antibody, binds on the primary conjugate, thus amplifying the signal without additional steps.

Molecules having advantages over regularly used gold nanoparticles have been explored for LFIA. Apart from gold nanoparticles, nowadays, novel, more bright, and quantitative signaling molecules are explored for application in LFIA. The use of these molecules not only improves the sensitivity (detection limit) of an assay but

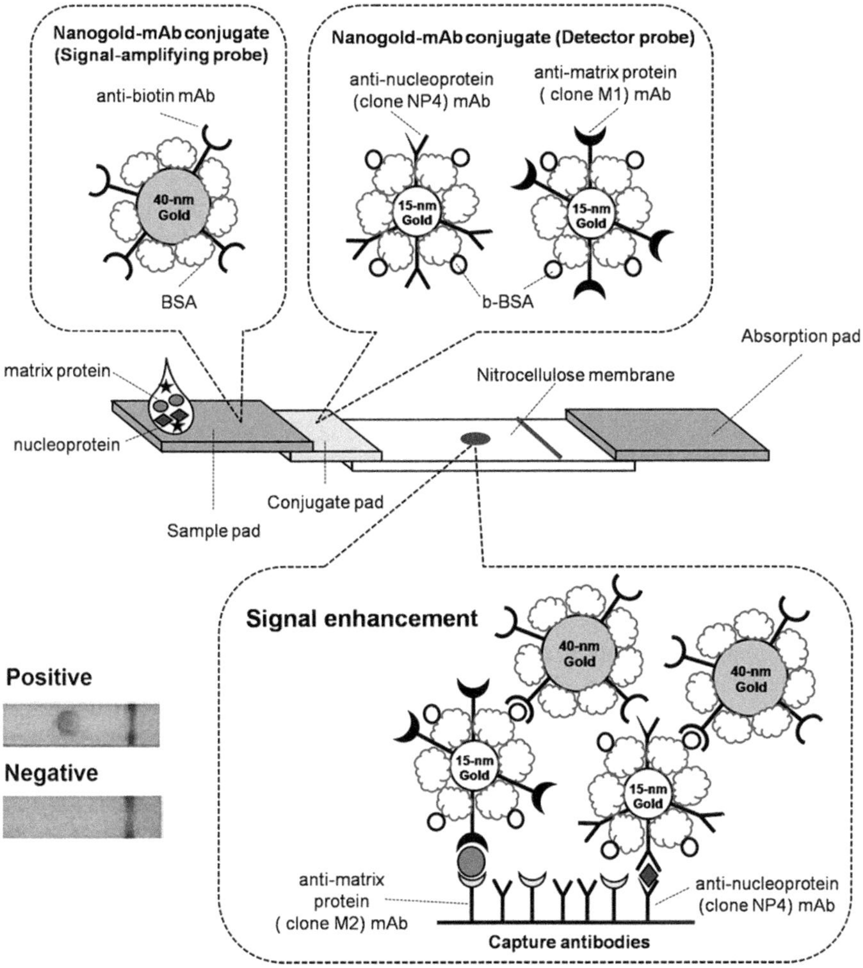

Fig. 10.7 Schematics of the dual-layered and double-targeted nanogold-based LFIA. Reprinted with permission from (Wiriyachaiporn et al. 2015), Copyright 2014 Springer-Verlag Wien

also offers easy quantitation capability. Colored nanoparticles such as colloidal carbon, colloidal selenium, etc.; fluorescent nanoparticles such as quantum dots, carbon dots, lanthanide-based upconversion nanoparticles, dye-doped nanoparticles, etc.; and magnetic nanoparticles are some of the examples. For example, recently, highly luminescent sodium yttrium fluoride upconversion nanoparticles (NaYF4: 30%Yb, 2%Er @NaLuF4 core-shell UCNPs) were used as sensing molecule to design the LFIA for myoglobin (cardiac marker) detection (Ji et al. 2019). The organic fluorescent molecules and quantum dots show stokes emission process for fluorescence, wherein lanthanide-doped UCNPs absorb light of longer wavelength and emit shorter wavelength, which offer advantages over organic fluorophores and QDs to use in LFIA such as the absence of autofluorescence, high quantum yield, very sharp emission spectra, and absence of photobleaching and photoblinking phenomenon. Fluorescent carbon nanoparticles were used in LFIA detection of DNA (Takalkar et al. 2017). And it is reported that the fluorescence-based LFIA had the limit of detection 4–6 orders lower as compared to the gold nanoparticle-based LFIA.

LFIA has been developed for detection of C-reactive protein (infection biomarker) using thermally imprinted microcone array (Aoyama et al. 2019). Schematics of the study are shown in Fig. 10.8. The study was performed to stabilize the number of immobilized antibody by increasing the binding area on the surface. This is achieved by fabricating polycarbonate microcones with nanoscale roughness. Because of the unique shape of the microcones, capture antibodies were immobilized stably on the strip surface as compared to the plane 2D surface.

Post-assay enhancement of test line can be achieved by the addition of an extra component to the assay that binds to the test line after the antigen-antibody complex binding reaction is completed. Such component enhances the already formed color on the test line, and the visual and quantitative detection signals are amplified. A study has reported sensitivity enhancement of LFIA for troponin I (cardiac biomarker) detection by using silver as color-improving agent (Kim et al. 2018). Schematics of the study are shown in Fig. 10.9. Core-shell hybrid nanofibers containing silver-reducing agent were deposited just before the test line. Regular immune-complex formation reaction of antibody-conjugated gold nanoparticle takes place, and the color change is induced. Subsequently, the silver ions are released from the polymer complex band; the silver ions are reduced to the metallic silver by gold nanoparticles that are present on the test band, thus darkening the color of the test line. Authors have claimed ten times increase in the visual detection limit of LFIA. In another LFIA study, the color formed on the strip after an assay was enhanced using copper deposition which improved the limit of detection of rabbit IgG (Tian et al. 2019). The red color formed on the test and control line is due to

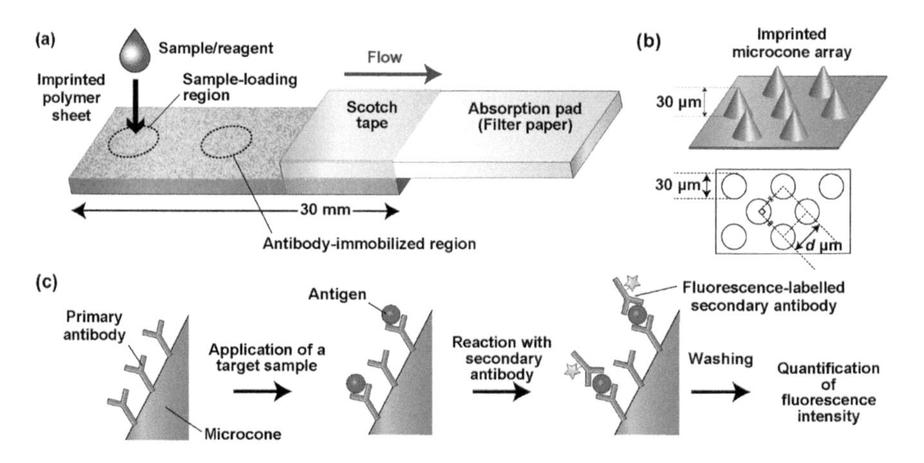

Fig. 10.8 (**a**) Schematics of the LFIA, (**b**) microcone arrays, (**c**) immunoassay protocol. Reprinted with permission from (Aoyama et al. 2019), Copyright 2019 The Royal Society of Chemistry

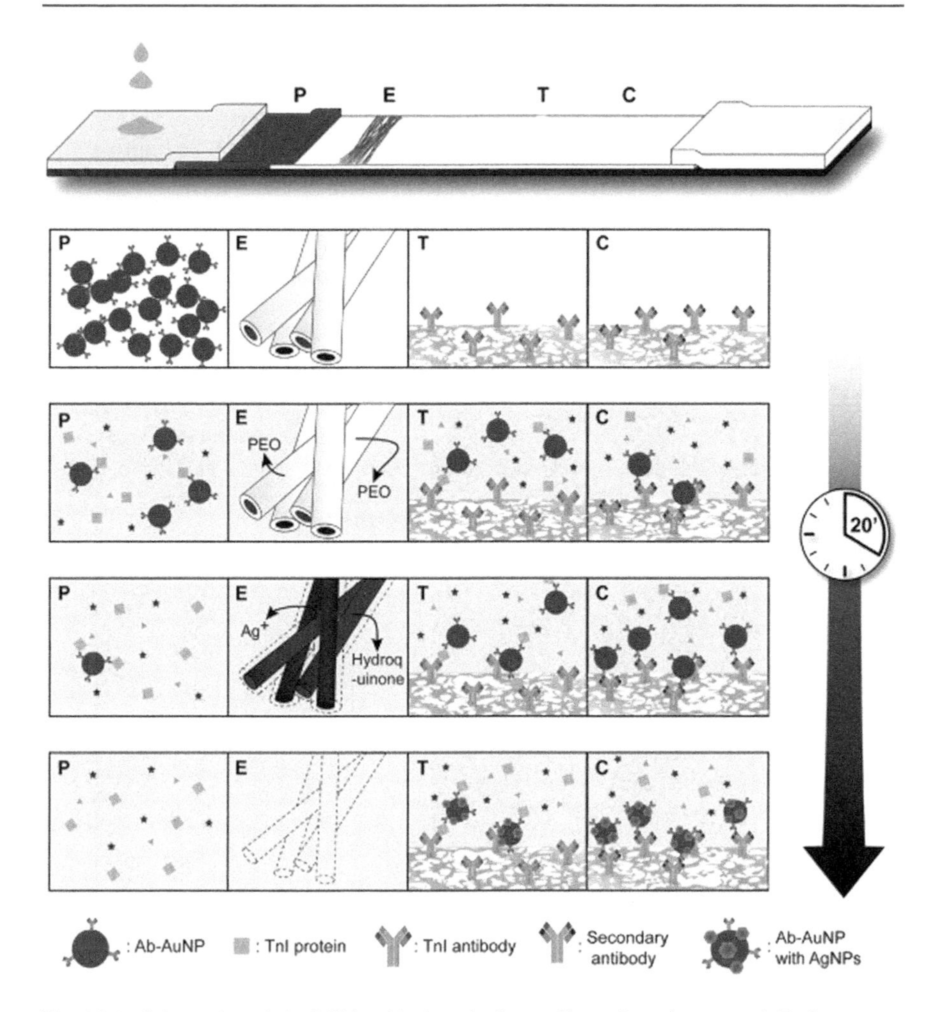

Fig. 10.9 Schematics of the LFIA with deposited nanofibers. P-conjugate pad, E-electrospun nanofibers, T-test line, and C-control line. Reprinted with permission from (Kim et al. 2018), Copyright 2018 Elsevier

the presence of the gold nanoparticle biosensor. The visual semiquantitative analysis is very difficult for the faded color appearance on the strip. The Cu^{+2} in solution reduces in presence of ascorbic acid to Cu^{+} which further reduced to copper in association with gold nanoparticle present on the strip. Thus, the signal of the faded red-colored weak line of gold nanoparticle is enhanced with copper deposition. Peroxidase-mimicking porous platinum core-shell nanocatalysts were conjugated with antibodies and utilized in LFIA to detect p24 (earliest HIV biomarker) (Loynachan et al. 2018). Pt nanocatalysts and biotinylated nanobody fragments were mixed with the serum sample; in presence of target, nanocrystals were biotinylated. Biotin-streptavidin high-affinity binding reaction causes deposition of Pt nanocrystals at the test line. Binding of Pt nanocrystals catalyzes the oxidation of a chemical in presence of hydrogen peroxide producing the insoluble black color. A femtomolar concentration of p24 can be detected in serum with naked eye due to the catalytic amplification process.

Removal of interfering impurities may be a potential approach to improve the sensitivity of LFIA. A study reports the development of fluorescence-based LFIA for detection of foodborne pathogens (*E. coli*), wherein detection antibodies are conjugated with the fluorescent magnetic nanobeads (Huang et al. 2019). Schematic is represented in Fig. 10.10. Prior to sample application on the LFIA,

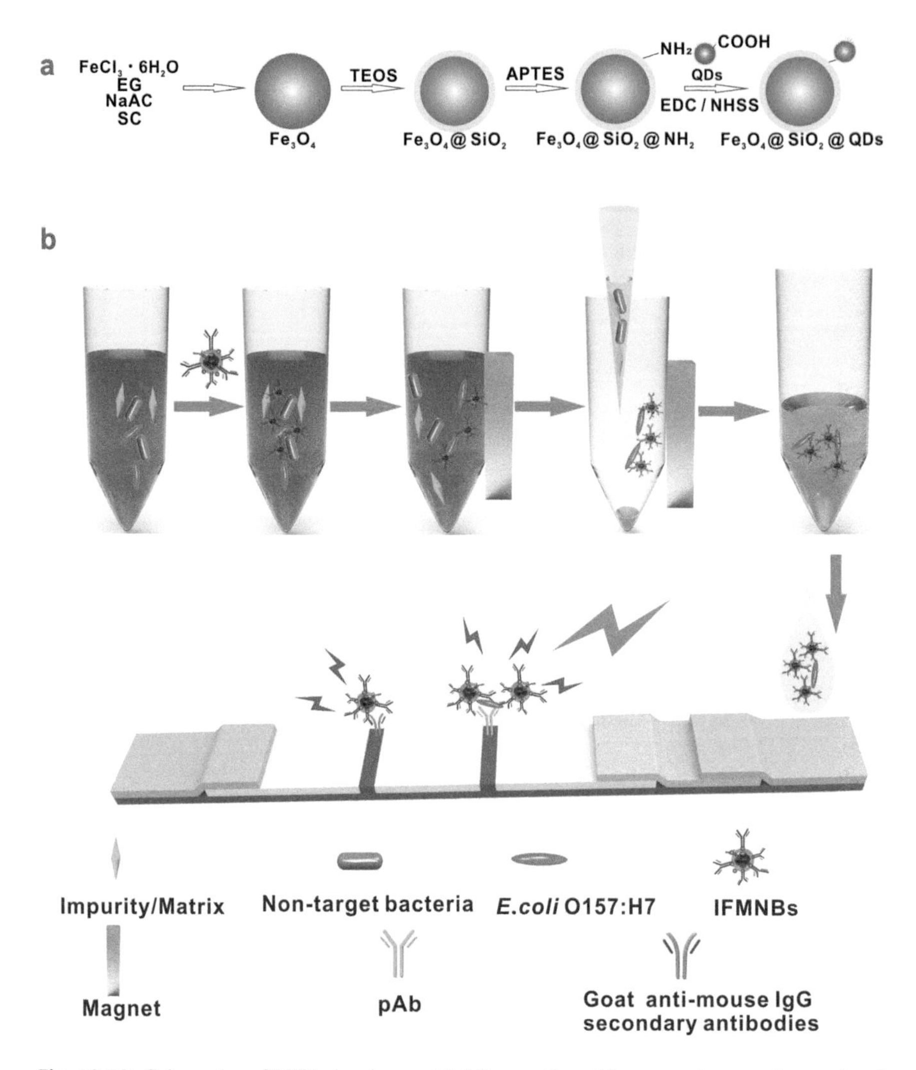

Fig. 10.10 Schematics of LFIA development (**a**) Preparation of fluorescent magnetic nanobeads (FMNBs), (**b**) immunomagnetic separation with fluorescent LFIA. Reprinted with permission from (Huang et al. 2019), Copyright 2019 Elsevier

immunomagnetic separation technique removes the unconjugated magnetic nanobeads and matrix impurities, so that the strong signal can be obtained on the strip. The concentration of an analyte presented for the immunoassay was increased, and the fluorescent property of magnetic nanobeads facilitated the detection process. Application of fluorescent magnetic nanobeads provided the dual advantage, as a detection agent for LFIA and for immunomagnetic separation. In another study, magnetic nanostructure (Fe_3O_4/Au core-shell nanostructures) conjugated with antibodies are used in LFIA, an external magnet slows down the conjugate movement in the detection zone on the LFIA membrane (prolonging the reaction time) (Ren et al. 2016). This method improved the limit of detection of microbial cells to an approximately single cell with the naked eye.

The sensitivity of LFIA could be improved by the improvement of detection capabilities by integrating mechanism that amplifies test results. Numerous LFIA strip readers are designed and developed for the customized detection for signal appearing on the LFIA strip. The instrument detects colorimetric, chemical, electric, or fluorescence signal and processes it to provide output in terms of concentration for the analyte. Recently, one such device has been developed by our research group, i.e., PorFloR™ (Borse et al. 2017a). It is a portable fluorescence reader that detects the color intensity of fluorescence emitted by the fluorescent molecules. Once the fluorescence-based LFIA is completed, the strip is inserted in the device, the fluorescence intensity at test and control line is measured, and concentration of the analyte is determined. It is very simple, easy-to-use, point-of-care device used to quantify the fluorescence intensity. In another report, thermophotonic lock-in imaging (TPLI) system has been incorporated in LFIA for detection of human chorionic gonadotropin (pregnancy hormone – hCG) (Ojaghi et al. 2018). Schematic for the LFIA is shown in Fig. 10.11. TPLI method is based on the detection of thermal infrared radiation that uses thermal waves as an indicator to study the subsurface inhomogeneity. According to the study, most of the optical immunoassay strip readers involve processing of the signal in terms of color intensity reflected from the strip surface. But at the same time, there is partial loss of signal due to entrapment of gold nanoparticles in the deeper part of the LFIA membrane, which affects the sensitivity of the detection. TPLI does not require optical transparency as it operates on the principle of thermal wave diffusion. The results of TPLI integrated LFIA for analysis of hCG have shown tenfold higher sensitivity in terms of detection limit.

Adjusting the position of test line on the LFIA strip is also an important parameter that affects the overall sensitivity of LFIA results. Commonly used LFIA protocols are structured in a manner that the test line appears prior to the control line when a sample is put for analysis. Thus, the analyte and biosensors react with test line first and then move to the control line. A study has been carried out to evaluate and quantify the flow velocity at various test line locations (Asiaei et al. 2018). As shown in Fig. 10.12, the test lines were printed at four different locations on the LFIA strip (2.5 mm, 7.5 mm, 12.5 mm, and 17.5 mm from the edge of the sample pad). Flow velocity was calculated by an analytical and numerical method and verified experimentally. Various concentrations of hCG were determined for calculation of sensitivity and limit of detection. Best results were observed for the test line printed at 12.5 mm location in terms of the color intensity. The exponential decrease

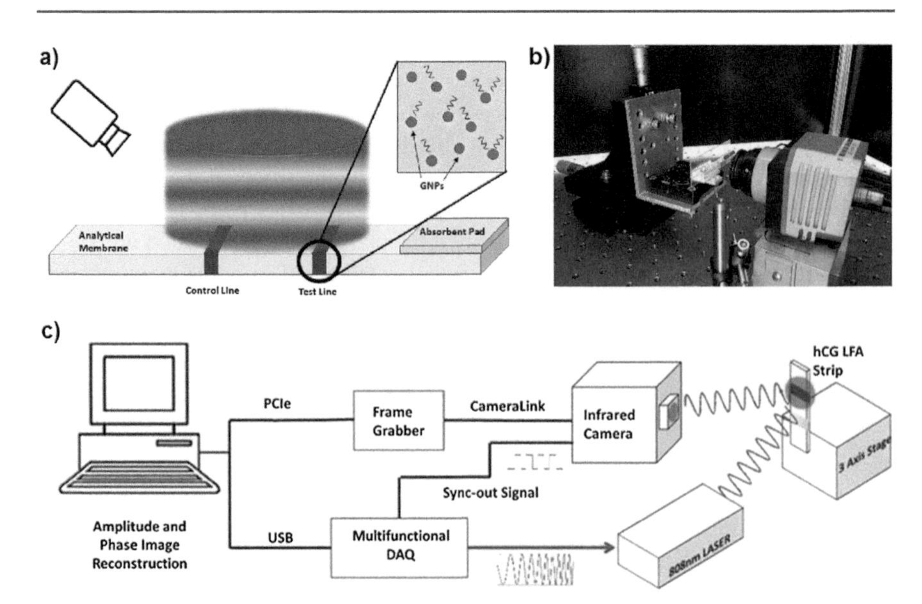

Fig. 10.11 Schematics of (**a**) assay working principle, (**b**) and (**c**) TPLI system components. Reprinted with permission from (Ojaghi et al. 2018), Copyright 2018 Elsevier

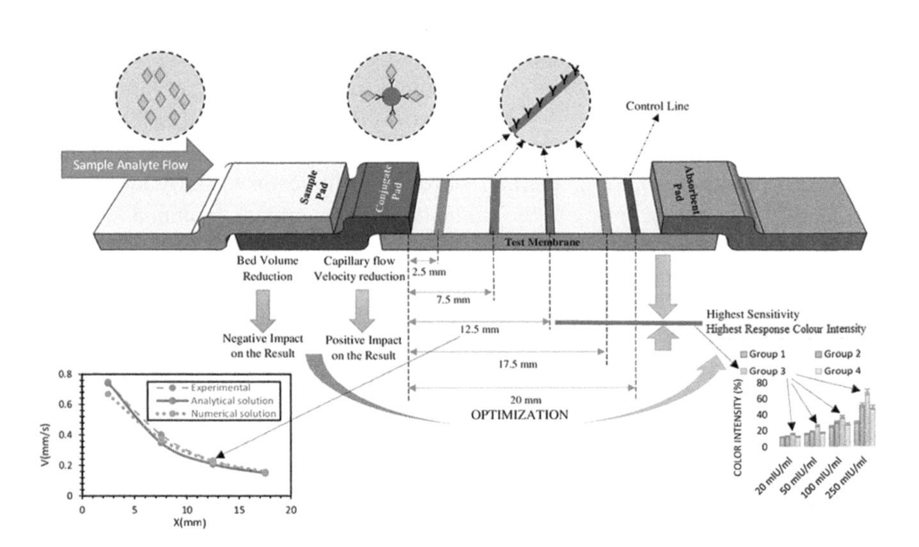

Fig. 10.12 Schematics of LFIA with varying test line position and results obtained. Reprinted with permission from (Asiaei et al. 2018), Copyright 2018 Elsevier

was observed in the flow velocity as the fluid moves through the strip membrane and it reaches a plateau at the end of the pad. Color intensity on test line increased with the decrease in flow velocity and found highest at 12.5 mm location. Increase in color intensity corresponds to the highest Ag-Ab immune-complex formation.

Thus, with optimal flow velocity (which is a function of the location of a test line on the LFIA strip), maximum gold-conjugated antibodies bind effectively at the test line producing the highest color signal. Proper assembly of LFIA components and optimal placement of the test line on the LFIA strip improve the limit of detection. This simple methodical adjustment can enhance LFIA signal without the use of strip reader, secondary signal amplification mechanism, or any other complex procedures and costs.

In another study "stacking pad," an additional membrane, is inserted between the conjugation pad and test pad to determine protein A and CRP using gold nanoparticle-based LFIA (Tsai et al. 2018). This additional membrane allows accumulation of antigen and antibodies that provides more time for the formation of immune complex and also increases interaction probability, thus enhancing the sensitivity of detection of markers. Remarkably, this concept was adapted from the use and function of stacking gel in polyacrylamide gel electrophoresis. Various materials such as polyester, cellulose, and glass fiber were evaluated that are suitable for preparation of stacking pad; cellulose was found to be superior. Results obtained by stacking pad modified LFIA illustrated two and five times more sensitivity for CRP and protein A, respectively, as compared to conventional LFIA.

Alteration in the sample processing methodology could also have an effect in the LFIA sensitivity. Changing the condition for binding antigen and gold nanoparticle-conjugated antibody has enhanced the LFIA sensitivity. As compared to conventional LFIA, where the sample containing antigen is applied on the dry LFIA strip, preincubation of a sample with gold conjugates for 30 s reduced the limit of detection of potato virus Y by 60-fold (Razo et al. 2018). This sensitivity improvement is attributed to the aggregate formation between polyvalent structures during the incubation time. Combination of additives can be used as an additive with the capture antibody to enhance the LFIA sensitivity (Upadhyay and Nara 2018). Additives such as anionic detergent (sodium dodecyl sulfate), nonionic detergent (Triton X-100), and hydrating agents such as methanol and glycerol were added in varying concentrations to the capture antibody mixture. It is reported that addition of additives, i.e., SDS (2.5%) and methanol (5%), to the capture antibody improved the lower limit of detection twofold as compared to LFIA processed without additives. *Staphylococcus enterotoxin* A was detected using gold nanoparticle-based LFIA, and the limit of detection was found to be 0.5 μg/mL. The additive such as detergent decreases non-specific binding and facilitates the fluid flow through the LFIA strip, while methanol increases the wettability of membrane, and glycerol increases density of capture antibody that helps in immobilization of the antibody on the strip surface. Optimization of physicochemical parameters such as pH of nanoparticle conjugate, size of the gold nanoparticle, and amount of antibody immobilized at the test line also improved the sensitivity of LFIA. Brain natriuretic peptide (a heart failure biomarker) was detected with the highest sensitivity using gold nanoparticle-based LFIA optimizing the physicochemical parameters (Gong et al. 2017). Antibody and gold nanoparticles interact through the hydrophobic and ionic bonds; thus, optimization of reaction condition is an important parameter for the success of LFIA. pH of gold nanoparticle solution and conjugating antibody was adjusted to the value 8, and approximately 20 µg antibody was added to gold nanoparticle solution. The sensitivity of LFIA increases with increasing concentration of capture antibody at the test line. But, the optimal concentration of capture antibody immobilized at test line was found to be 1.5 mg/mL. Higher concentration of capture antibody (more than 1.5 mg/mL) at the test line may result in the false negative results. The optimum diameter of gold nanoparticle was found to be is 35 ± 3 nm. The large size of the gold nanoparticle is preferred for better signal detection, but increase in size of particle more than 40 nm reduces colloidal stability of nanoparticles. Conjugate pad was used to improve the structure of LFIA strip. Almost 15-fold signal improvement was observed using these optimal conditions as compared to the unmodified LFIA.

Recent reports of nanobiotechnology-based LFIA along with the modifications and used techniques are mentioned in Table 10.1.

10.6 Summary and Future Perspectives

The pregnancy test kit has been the most successful and well-known LFIA used since decades. Use of nanomaterial in the development of LFIA permitted utility of varying detection approaches. Nanobiotechnology has improved the selectivity and sensitivity of the LFIA for detection of a wide range of markers in the medical diagnosis, veterinary, food, agriculture, environment, etc. sector. Selectivity and sensitivity enhancement has improved the detection capability, but continuous efforts are needed on the reproducibility, reliability, and simplicity of the LFIA techniques. Numerous optical sensing devices are developed indigenously or purchased from market to read the color or fluorescence signal on the LFIA strip. Parameters such as ease of use and cost shall be considered while integrating any advancement to the LFIA method. Numerous LFIA test kits have been developed and scripted in the form of research publications, but most of them are not commercialized. Scaling up and diverse regulatory regulations for clinical use may be major obstacles in the commercialization. Although these kits are categorized as point-of-care devices with advantages such as low cost, easy to use, easy to interpret the results within minutes, etc., it is not accepted by users as expected. The reason behind it may be ineffective marketing of the LFIA kit. There are many LFIA-based detection kits available in the market, but very few are sold, and users are hesitant to use them. The primary reason may be complicated test procedure and reliability of the test. Also, clinicians are not certain of using such LFIA because of false positive or false negative results. There is a strong need to popularize the LFIA kits so that it can reach the needy population at large. Clinicians are the prime stakeholders that can promote and recommend utility of the LFIA-based kits. Thus, it will take time for the clinicians to practice the routine use of LFIA kits with confidence for diagnostic purpose. Undoubtedly, featuring its advantages and advances, the future is very bright for the LFIA-based diagnosis.

Table 10.1 Recent reports of nanobiotechnology-based LFIA

Sr. No.	Substrate	Technique	Target analyte	Limit of detection	Assay time	References
1.	Fluorescent microspheres	LFIA	4(5)-methylimidazole in caramel color used in foods and beverages	0.18 mg/L	<15 min	Wu et al. (2016)
2.	Gold nanoparticles	LFIA	Parathyroid hormonelike hormone in cell cultures	1.42 ng/mL	20 min	Chamorro-Garcia et al. (2016)
3.	Magnetic Fe_3O_4/Au core-shell nanoprobes	Magnetic focus enhanced LFIA	Pathogenic microorganisms *Escherichia coli* O157:H7 and *Salmonella typhimurium*	Near single cell	<30 min	Ren et al. (2016)
4.	Fluorescent carbon nanoparticle	Fluorescence-based LFIA	DNA in buffer	0.4 fM	20 min	Takalkar et al. (2017)
5.	Mesoporous core-shell palladium@platinum nanoparticles	Smartphone-based LFIA	1. *Salmonella enteritidis* 2. *E. coli* O157:H7 spiked in food	1. ~20 cfu/mL 2. ~34 cfu/mL	NA	Cheng et al. (2017)
6.	Gold nanoparticles	LFIA with optimal physicochemical parameters	Brain natriuretic peptide spiked in human serum	0.1 ng/mL	10–15 min	Gong et al. (2017)
7.	Dual-color core-shell upconversion nanoparticles	LFIA integrated with smartphone-based portable reader and an analysis app	1. Brain natriuretic peptide 2. Suppression of tumorigenicity 2 in serum	1. 5 pg/mL 2. 1 ng/mL	NA	You et al. (2017)
8.	Gold nanoparticles	Microfluidic paper-based analytical device immunoassay	D-dimer in buffer/simulated plasma	15 ng D-dimer/mL	10–12 min	Ruivo et al. (2017)
9.	Raman reporter-labeled gold nanoparticles	Surface-enhanced Raman scattering (SERS)	Thyroid-stimulating hormone in clinical fluids	0.025 μIU/mL	10 min	Choi et al. (2017)
10.	Fluorescent lanthanide nanoparticle	Quantitative LFIA	Anti-pertussis toxin (PT) IgG in serum	157 IU/mL	30 min	Salminen et al. (2018)
11.	Gold nanostars	SERS-LFIA	Bisphenol A in aqueous media	0.073 ppb	NA	Lin and Stanciu (2018)

(continued)

Table 10.1 (continued)

Sr. No.	Substrate	Technique	Target analyte	Limit of detection	Assay time	References
12.	Nile-red doped nanoparticles	Quantitative sandwich immunoassay	C-reactive protein standard antigen in plasma	0.091 mg/L	3 min	Cai et al. (2018)
13.	Gold nanoparticles	LFIA with sample preincubation method	Potato virus Y in plant leaf extracts	5.4 ng/mL	NA	Razo et al. (2018)
14.	Gold nanoparticles	LFIA	*Staphylococcus enterotoxin* A in milk	0.5 μg/mL	NA	Upadhyay and Nara (2018)
15.	Multicolored upconversion nanoparticles	Lateral flow aptamer assay integrated smartphone-based portable device	1. Mercury ions 2. Ochratoxin A 3. Salmonella In tap water	1. 5 ppb 2. 3 ng/mL 3. 85 CFU/mL	30 min	Jin et al. (2018)
16.	Hybrid nanocomposite particles (AuNPs@ SiO_2–Eu^{3+})	Colorimetric and fluorometric dual read out LFIA	Human thyroid-stimulating hormone spiked in buffer and serum	5 μIU/mL for colorimetric and 0.1 μIU/mL for fluorometric	<30 min	Preechakasedkit et al. (2018)
17.	Porous platinum core-shell nanocatalysts	Catalytic amplification-based LFIA	p24 (HIV biomarker) spiked in serum	0.8 pg/mL	<20 min	Loynachan et al. (2018)
18.	Gold nanoparticles	LFIA with extra "stacking pad"	1. Protein A spiked in buffer 2. C-reactive protein In human serum and synovial fluid	1. 1 ng/mL 2. 15.5 ng/mL	20 min	Tsai et al. (2018)
19.	Gold nanoparticles	Gold nanoparticle-induced silver deposition on LFIA	Troponin I spiked in human serum	1 ng/mL	20 min	Kim et al. (2018)

20.	Gold nanoparticles	ironPhone	Ferritin spiked in buffer and serum	18 µg/L	NA	Srinivasan et al. (2018)
21.	Gold nanoparticles	LFIA integrated with thermophotonic lock-in imaging system	Human chorionic gonadotropin in urine	0.2 mIU	NA	Ojaghi et al. (2018)
22.	Quantum dots	Fluorescence-based LFIA	1. CRP 2. IL-6 In serum	1. 0.3 µg/mL 2. 0.9 pg/mL	30 min	Borse and Srivastava (2019)
23.	Gold nanopopcorns and nanostars	LFIA	Procalcitonin detection in serum	0.1 ng/mL	10–15 min	Serebrennikova et al. (2018)
24.	Fluorescent magnetic nanobeads	Fluorescent FLFIA based on immunomagnetic separation	*E. coli* O157:H7	2.39×10^2 CFU/mL	60 min	Huang et al. (2019)
25.	Core-shell upconversion nanoparticles	Luminescent LFIA	Myoglobin in blood plasma	0.21 ng/mL	10 min	Ji et al. (2019)
26.	Polycarbonate microcones array sheet	Thermal nano-imprinting techniques	C-reactive protein in human serum	~0.1 µg/mL	<15 min	Aoyama et al. (2019)
27.	Gold nanoparticles	Gold nanoparticle-induced copper deposition on LFIA	Rabbit IgG in human serum	1 pg/mL	30 min	Tian et al. (2019)
28.	Selenium nanoparticles	LFIA	Clenbuterol in pig urine	3 ng/mL	NA	Wang et al. (2019)

Acknowledgment Dr. Vivek Borse would like to thank the Department of Science and Technology, Government of India, for the INSPIRE Faculty Fellowship Award (IFA18-ENG266, DST/INSPIRE/04/2018/000991).

References

Akhtar MH, Hussain KK, Gurudatt NG, Chandra P, Shim Y-B (2018) Ultrasensitive dual probe immunosensor for the monitoring of nicotine induced-brain derived neurotrophic factor released from cancer cells. Biosens Bioelectron 116:108–115

Aoyama S, Akiyama Y, Monden K, Yamada M, Seki M (2019) Thermally imprinted microcone structure-assisted lateral-flow immunoassay platforms for detecting disease marker proteins. Analyst. https://doi.org/10.1039/C8AN01903G

Asiaei S, Bidgoli MR, ZadehKafi A, Saderi N, Siavashi M (2018) Sensitivity and colour intensity enhancement in lateral flow immunoassay tests by adjustment of test line position. Clin Chim Acta 487:210–215

Bellah MM, Christensen SM, Iqbal SM (2012) Nanostructures for medical diagnostics. J Nanomater 2012:1–21

Borse V, Srivastava R (2019) Fluorescence lateral flow immunoassay based point-of-care nanodiagnostics for orthopedic implant-associated infection. Sensors Actuators B Chem 280:24–33

Borse V, Jain P, Sadawana M, Srivastava R (2016a) 'Turn-on' fluorescence assay for inorganic phosphate sensing. Sensors Actuators B Chem 225:340–347

Borse V, Sadawana M, Srivastava R (2016b) CdTe quantum dots: aqueous phase synthesis, stability studies and protein conjugation for development of biosensors. In: Andrews DL, Nunzi J-M, Ostendorf A (eds) SPIE Photonics Europe. International Society for Optics and Photonics, p 988423

Borse V, Patil AS, Srivastava R (2017a) Development and testing of portable fluorescence reader (PorFloR™). In: 2017 ninth International Conference on Communication Systems and Networks, COMSNETS 2017

Borse V, Thakur M, Sengupta S, Srivastava R (2017b) N-doped multi-fluorescent carbon dots for 'turn off-on' silver-biothiol dual sensing and mammalian cell imaging application. Sensors Actuators, B Chem 248:481–492

Borse V, Kashikar A, Srivastava R (2018) Fluorescence stability of mercaptopropionic acid capped cadmium telluride quantum dots in various biochemical buffers. J Nanosci Nanotechnol 18:2582–2591

Cai Y, Kang K, Liu Y, Wang Y, He X (2018) Development of a lateral flow immunoassay of C-reactive protein detection based on red fluorescent nanoparticles. Anal Biochem 556:129–135

Chamorro-Garcia A, de la Escosura-Muñiz A, Espinoza-Castañeda M, Rodriguez-Hernandez CJ, de Torres C, Merkoçi A (2016) Detection of parathyroid hormone-like hormone in cancer cell cultures by gold nanoparticle-based lateral flow immunoassays. Nanomed Nanotechnol Biol Med 12:53–61

Chandra P, Koh WCA, Noh H-B, Shim Y-B (2012) In vitro monitoring of i-NOS concentrations with an immunosensor: the inhibitory effect of endocrine disruptors on i-NOS release. Biosens Bioelectron 32:278–282

Cheng N, Song Y, Zeinhom MMA, Chang Y-C, Sheng L, Li H, Du D, Li L, Zhu M-J, Luo Y, Xu W, Lin Y (2017) Nanozyme-mediated dual immunoassay integrated with smartphone for use in simultaneous detection of pathogens. ACS Appl Mater Interfaces 9:40671–40680

Choi S, Hwang J, Lee S, Lim DW, Joo H, Choo J (2017) Quantitative analysis of thyroid-stimulating hormone (TSH) using SERS-based lateral flow immunoassay. Sensors Actuators B Chem 240:358–364

Choudhary M, Yadav P, Singh A, Kaur S, Ramirez-Vick J, Chandra P, Arora K, Singh SP (2016) CD 59 Targeted ultrasensitive electrochemical immunosensor for fast and noninvasive diagnosis of oral cancer. Electroanalysis 28:2565–2574

Gong Y, Hu J, Choi JR, You M, Zheng Y, Xu B, Wen T, Xu F (2017) Improved LFIAs for highly sensitive detection of BNP at point-of-care. Int J Nanomedicine 12:4455–4466

Huang X, Aguilar ZP, Xu H, Lai W, Xiong Y (2016) Membrane-based lateral flow immunochromatographic strip with nanoparticles as reporters for detection: A review. Biosens Bioelectron 75:166–180

Huang Z, Peng J, Han J, Zhang G, Huang Y, Duan M, Liu D, Xiong Y, Xia S, Lai W (2019) A novel method based on fluorescent magnetic nanobeads for rapid detection of Escherichia coli O157:H7. Food Chem 276:333–341

Ji T, Xu X, Wang X, Zhou Q, Ding W, Chen B, Guo X, Hao Y, Chen G (2019) Point of care upconversion nanoparticles-based lateral flow assay quantifying myoglobin in clinical human blood samples. Sensors Actuators B Chem 282:309–316

Jin B, Yang Y, He R, Park Y II, Lee A, Bai D, Li F, Lu TJ, Xu F, Lin M (2018) Lateral flow aptamer assay integrated smartphone-based portable device for simultaneous detection of multiple targets using upconversion nanoparticles. Sensors Actuators B Chem 276:48–56

Kim W, Lee S, Jeon S (2018) Enhanced sensitivity of lateral flow immunoassays by using water-soluble nanofibers and silver-enhancement reactions. Sensors Actuators B Chem 273:1323–1327

Lin L-K, Stanciu LA (2018) Bisphenol a detection using gold nanostars in a SERS improved lateral flow immunochromatographic assay. Sensors Actuators B Chem 276:222–229

Loynachan CN, Thomas MR, Gray ER, Richards DA, Kim J, Miller BS, Brookes JC, Agarwal S, Chudasama V, McKendry RA, Stevens MM (2018) Platinum nanocatalyst amplification: redefining the gold standard for lateral flow immunoassays with ultrabroad dynamic range. ACS Nano 12:279–288

Mahato K, Chandra P (2019) Paper-based miniaturized immunosensor for naked eye ALP detection based on digital image colorimetry integrated with smartphone. Biosens Bioelectron 128:9–16

Mahato K, Srivastava A, Chandra P (2017) Paper based diagnostics for personalized health care: Emerging technologies and commercial aspects. Biosens Bioelectron 96:246–259

Mahato K, Kumar S, Srivastava A, Maurya PK, Singh R, Chandra P (2018) Electrochemical immunosensors: fundamentals and applications in clinical diagnostics. Handb Immunoass Technol:359–414

Makkar RL, Syeda Aliya S, Borse V, Srivastava R (2018) Design and development of portable fluorescence reader using silicon photo multiplier (SiPM) sensor. Opt Sens Detect V 106800B:12

Mandal R, Baranwal A, Srivastava A, Chandra P (2018) Evolving trends in bio/chemical sensor fabrication incorporating bimetallic nanoparticles. Biosens Bioelectron 117:546–561

Ojaghi A, Pallapa M, Tabatabaei N, Rezai P (2018) High-sensitivity interpretation of lateral flow immunoassays using thermophotonic lock-in imaging. Sensors Actuators A Phys 273:189–196

Preechakasedkit P, Osada K, Katayama Y, Ruecha N, Suzuki K, Chailapakul O, Citterio D (2018) Gold nanoparticle core–europium(iii) chelate fluorophore-doped silica shell hybrid nanocomposites for the lateral flow immunoassay of human thyroid stimulating hormone with a dual signal readout. Analyst 143:564–570

Quesada-González D, Merkoçi A (2015) Nanoparticle-based lateral flow biosensors. Biosens Bioelectron 73:47–63

Razo S, Panferov V, Safenkova I, Varitsev Y, Zherdev A, Pakina E, Dzantiev B, Razo SC, Panferov VG, Safenkova IV, Varitsev YA, Zherdev AV, Pakina EN, Dzantiev BB (2018) How to improve sensitivity of sandwich lateral flow immunoassay for corpuscular antigens on the example of potato Virus Y? Sensors 18:3975

Ren W, Cho I-H, Zhou Z, Irudayaraj J (2016) Ultrasensitive detection of microbial cells using magnetic focus enhanced lateral flow sensors. Chem Commun 52:4930–4933

Ruivo S, Azevedo AM, Prazeres DMF (2017) Colorimetric detection of D-dimer in a paper-based immunodetection device. Anal Biochem 538:5–12

Sajid M, Kawde A-N, Daud M (2015) Designs, formats and applications of lateral flow assay: a literature review. J Saudi Chem Soc 19:689–705

Salminen T, Knuutila A, Barkoff A-M, Mertsola J, He Q (2018) A rapid lateral flow immunoassay for serological diagnosis of pertussis. Vaccine 36:1429–1434

Serebrennikova K, Samsonova J, Osipov A (2018) Hierarchical nanogold labels to improve the sensitivity of lateral flow immunoassay. Nano-Micro Lett 10:24

Srinivasan B, O'Dell D, Finkelstein JL, Lee S, Erickson D, Mehta S (2018) ironPhone: mobile device-coupled point-of-care diagnostics for assessment of iron status by quantification of serum ferritin. Biosens Bioelectron 99:115–121

Takalkar S, Baryeh K, Liu G (2017) Fluorescent carbon nanoparticle-based lateral flow biosensor for ultrasensitive detection of DNA. Biosens Bioelectron 98:147–154

Tian M, Lei L, Xie W, Yang Q, Li CM, Liu Y (2019) Copper deposition-induced efficient signal amplification for ultrasensitive lateral flow immunoassay. Sensors Actuators B Chem 282:96–103

Tsai T-T, Huang T-H, Chen C-A, Ho NY-J, Chou Y-J, Chen C-F (2018) Development a stacking pad design for enhancing the sensitivity of lateral flow immunoassay. Sci Rep 8:17319

Upadhyay N, Nara S (2018) Lateral flow assay for rapid detection of Staphylococcus aureus enterotoxin A in milk. Microchem J 137:435–442

Wang Z, Jing J, Ren Y, Guo Y, Tao N, Zhou Q, Zhang H, Ma Y, Wang Y (2019) Preparation and application of selenium nanoparticles in a lateral flow immunoassay for clenbuterol detection. Mater Lett 234:212–215

Wiriyachaiporn N, Maneeprakorn W, Apiwat C, Dharakul T (2015) Dual-layered and Double-targeted Nanogold based Lateral Flow Immunoassay for Influenza Virus. Microchim Acta 182:85–93

Wu X, Huang M, Yu S, Kong F (2016) Rapid and quantitative detection of 4(5)-methylimidazole in caramel colours: a novel fluorescent-based immunochromatographic assay. Food Chem 190:843–847

You M, Lin M, Gong Y, Wang S, Li A, Ji L, Zhao H, Ling K, Wen T, Huang Y, Gao D, Ma Q, Wang T, Ma A, Li X, Xu F (2017) Household fluorescent lateral flow strip platform for sensitive and quantitative prognosis of heart failure using dual-color upconversion nanoparticles. ACS Nano 11:6261–6270

Zhang K, Wu J, Li Y, Wu Y, Huang T, Tang D (2014) Hollow nanogold microsphere-signalized lateral flow immunodipstick for the sensitive determination of the neurotoxin brevetoxin B. Microchim Acta 181:1447–1454

Biological Acoustic Sensors for Microbial Cell Detection

11

O. I. Guliy, B. D. Zaitsev, A. A. Teplykh, and I. A. Borodina

Abstract

One of the most popular areas in microbiology is the development of fast and sensitive methods for the detection of bacteria based on electrophysical analysis. The paper demonstrated the capabilities of various electroacoustic biological sensors for detection and identification of microbial cells. These sensors are based on the following main elements: the piezoelectric resonator with a longitudinal electric field, the piezoelectric resonator with a lateral electric field, the acoustic delay line with inhomogeneous piezoactive acoustic waves, and the delay line using a slot acoustic mode. They can conduct cell detection and identification of bacteria using immobilized microorganisms or directly in cell suspension. The principle of operation of such sensors is based on the registration of the interaction of microbial cells with specific antibodies, bacteriophages, and mini-antibodies. The sensitivity range of microbial cell detection is 10^3–10^8 cells/ml with the suspension conductivity of 5–50 μS/cm. At that the analysis time varies from 5 min to several hours. The presented possibilities of electroacoustic biological sensors for the detection of bacteria are focused on the clinical use of *onsite* as a personalized diagnostic device. The possibility of rapid detection of microflora allows timely diagnosis of the disease and timely medical assistance. In general, acoustic biological sensors form a wide class of detection systems and are very promising for use in microbiology, medicine, and veterinary medicine for solving the problems of detection and identification of bacteria.

O. I. Guliy (✉)
Institute of Biochemistry and Physiology of Plants and Microorganisms, RAS, Saratov, Russia

Saratov State Vavilov Agrarian University, Saratov, Russia

B. D. Zaitsev · A. A. Teplykh · I. A. Borodina
Kotel'nikov Institute of Radio Engineering and Electronics, RAS, Saratov Branch, Saratov, Russia

P. Chandra, R. Prakash (eds.), *Nanobiomaterial Engineering*,
https://doi.org/10.1007/978-981-32-9840-8_11

Keywords
Detection of microbial cells · Cell suspension · Piezoelectric resonators · Piezoactive acoustic waves · Slot acoustic mode · Antibodies · Bacteriophages · Mini-antibodies

11.1 Introduction

Methods for the detection of microbial cells are widely used in such fields of biology and medicine as the diagnosis of unknown microorganisms, the assessment and monitoring of the physiology and functioning of cells, the study of cellular waste products such as enzymes and antibiotics, and the development of new drugs.

Classical approaches such as microbiological and biochemical tests, methods of genetic engineering, and immunological methods are used to detect and identify microorganisms. These methods for the determination of bacteria, based on the cultivation of microorganisms with subsequent microbiological and biochemical analysis, are characterized by laborious and time-consuming use of expensive equipment. Another limitation of immunological methods and the method of polymerase chain reaction (PCR) is associated with the impossibility of screening a huge amount of material for the preliminary determination of bacteria.

Spectroscopy, fluorescence, and bioluminescence methods are also effective for the determination of microorganisms (Van Emon et al. 1995; Basile et al. 1998; Yousef 2008). However, they also require the use of expensive equipment and the involvement of highly qualified specialists.

Increasing attention of specialists is directed to the development of new express methods of indication of bacteria. In the development of modern methods for the diagnosis of microorganisms, two main directions of development can be distinguished.

The first direction is the development of the complicated automated systems. However, this way means the use of expensive equipment and the involvement of highly qualified specialists.

The second direction is connected with the development of new approaches that allow the rapid analysis of microbial cells, for example, using biosensor devices (Von Lode 2005; Hu et al. 2014). With the advent of biosensors, the traditional approaches to methods for determining microorganisms are changing significantly. The development of biosensor microbial cell detection technologies will be extremely useful for the early detection of diseases and the timely provision of medical care (Chandra 2016). The interest in biosensor systems for the detection of bacteria in aqueous solutions is due to their simplicity, cost-effectiveness, and relatively high sensitivity (Griffiths and Hall 1993; Li et al. 2010).

Obviously, new sensory technologies for the detection of microorganisms must have high sensitivity, reliability, and be able to get results in a short period of time. Therefore, among the electrophysical methods of analysis, electroacoustical methods are of particular interest (Andle and Vetelino 1994; Don et al. 2016).

The distinctive features of electroacoustic methods of the determination of microbial cells are:

- The possibility of analysis in solutions with a high content of ions, since their presence distorts the analytical signal
- The possibility of multiple use and cleaning of the liquid container from the sample being measured
- The short time of analysis

The principle of the operation of acoustic biological sensors is based on the registration of the biospecific reactions in a liquid suspension contacting with the surface of the piezoelectric. The interaction of the bacterial cells and specific agent causes a change in the conductivity of the suspension. When it contacts with the surface of the piezoelectric resonator, these changes lead to both a shift in its resonant frequency and a change in the frequency dependences of the electrical impedance (Zaitsev et al. 2012). Upon contact of the suspension with the surface of the piezoelectric waveguide with a propagating acoustic wave, the indicated change in conductivity leads to a change in the insertion loss and phase of output signal of the sensor (Borodina et al. 2018). All these effects are very sensitive for the detection of bacterial cells in suspension. It is obvious that the development and application of electroacoustical technology for the determination of bacteria will be extremely useful for the early detection of diseases.

So the chapter will discuss the main types and principles of operation of biological electroacoustic sensors. The main focus is pointed to the description of electroacoustic sensors for the detection of bacteria, which are oriented on the clinical use of *onsite* as a personalized diagnostic device. On the basis of literature data and our own experimental studies, the capabilities of the electroacoustic analysis method will be shown and compared with other ones.

11.2 Sensors Based on Piezoelectric Resonators with a Longitudinal Electric Field

Sensors based on piezoelectric resonators with a longitudinal electric field have been widely used for several decades. Such a sensor represents a piezoelectric plate with two metal electrodes deposited on each side of it. An active film containing immobilized microorganisms (usually bacterial cells) is applied directly to the electrode on one side of the sensor. Upon contact with a suspension containing any specific reagents (bacteriophages or antibodies), a layer of protein mass builds up on the active side of the resonator, which leads to a shift in the resonant frequency and a decrease in Q-factor. The distinctive features of such sensors are their reusability and low cost. However, these sensors have two significant drawbacks. Firstly, they do not allow the analysis of cells directly in suspension without applying the active layer of immobilized microorganisms. This is due to the fact that upon contact with a suspension, the parameters of the resonator respond only to changes in the mechanical properties of the fluid and practically do not respond to changes in its electrical properties. Secondly, the analysis time which leads to the appearance of the noticeable protein load on the resonator are several hours. Nevertheless, such sensors have been actively used for several decades (Ermolaeva and Kalmykova 2006; Ermolaeva and Kalmykova 2012).

The studies (Olsen et al. 2006; Ripp et al. 2008) show the possibility of using the affine-selective filamentous bacteriophage as a probe for detecting *S. typhimurium* cells using an acoustic biosensor based on a quartz resonator with a longitudinal electric field. The introduction of bacterial cells on the surface of the quartz resonator leads to a change in the resonant frequency, which is recorded by the device. The biosensor allows the detection of cells with a minimum number of 10^2 cells/ml, and the result is recorded in time less than 180 s.

By immobilization of antiviral antibodies on the surface of a piezoelectric resonator with a longitudinal electric field, an immunosensor was developed for the selective determination of herpes viruses in human blood (Koenig and Graetzel 1994) and biosensor for the detection of viruses in natural and artificial reservoirs (rivers, sewers, and wastewater) without prior processing of the analyzed substrate (Bisoffi et al. 2008).

11.3 Sensors Based on Piezoelectric Resonators with a Lateral Electric Field

Over the past two decades, the piezoelectric resonators with a lateral electric field have attracted particular attention from researches developing the biological sensors (Pinkham et al. 2005; York et al. 2005; Wark et al. 2007; Handa et al. 2008; Vetelino 2010).

Unlike conventional resonators with a longitudinal electric field, these sensors have electrodes only on one side of the piezoelectric plate, and the acoustic wave propagates mainly in the space between the electrodes. In this case, the sensitive surface bordering the liquid or the biochemical sample under study remains free, and the studied object does not contact the electrodes. The metallization-free surface allows the sensor to respond not only to the mechanical properties of the sample under study (viscosity) but also to its electrical properties (conductivity, permittivity).

Based on the lateral electric field resonators, there are two types of acoustic biosensors for the detection and identification of bacterial cells – sensors using bioreceptor films and sensors directly contacting with the suspension.

11.3.1 The Sensors Using the Bioreceptor Films

Sensors of this type use a bioreceptor film, which is fixed on the free surface of a piezoelectric plate. Biological material, for example, a suspension containing the object under study, is in contact with the biosensor through the specified film. The biological interaction between the measured object and the bioreceptor leads to a corresponding change in the analytical signal (resonant frequency or electrical impedance).

Traditionally, such sensors use active layers with immobilized bacterial cells. Due to the contact with a suspension containing any specific reagents (bacteriophages or antibodies), a layer of protein mass builds up on the active side of the resonator, which leads to a shift in the resonant frequency.

The paper (York et al. 2005) describes a sensor based on such a resonator, using anti-rabbit IgG and *E. coli* as test objects. The NH_2 film was deposited on the free surface of the resonator using a special vacuum technique, and then antibodies sensitive to the above antigens were immobilized. Experiments have shown that the interaction of a sensitive coating with an antigen solution resulted in a change in the resonant frequency of the order of 20–30 ppm in 5–8 h.

The biological sensor based on a resonator with a lateral electric field for the study of pesticide (organophosphates), contained in vegetables and fruits, was described in Pinkham et al. (2005). The free surface of the resonator has been covered with polyepichlorohydrin film, which is selectively sensitive to phosmet ($C_{11}H_{12}No_2PS_2$). It was shown that the shift of the resonant frequency with an increase in the phosphate concentration exceeded the frequency shift in a conventional quartz resonator with a longitudinal electric field by almost 30%. The analysis time in both cases was ~ 40 min.

The possibility of developing a sensor based on a resonator with a lateral electric field for determining the content of saxitoxin in water is described in Wark et al. (2007). The free side of the resonator was covered with a special multilayer sensitive film. The shift of the resonant frequency varied in the range of 0–20 ppm with a change in the concentration from 0 to 200 μM. The analysis time was ~ 2–3 min.

The possibility of developing a biological sensor based on a resonator with a lateral electric field with a sensitive multilayer film was shown in Vetelino (2010). The interaction of a sensitive coating with a suspension of *E. coli* cells resulted in a change in the resonant frequency within 20 ppm in 6 h.

It should be noted that sensors using active coatings have a number of significant drawbacks: a long detection time (several hours) and the impossibility of multiple use of the same active layer. After the first experiment, it is necessary to remove the used active film and apply a new one. These facts significantly limit the possibility of sensors using film bioreceptors.

11.3.2 The Sensors Directly Contacting with Suspension

Sensors of this type are free from the above disadvantages. They record the change in the physical properties of a biological liquid or suspension without the use of active films. These changes lead to a corresponding change in the analytical signal (resonant frequency or electrical impedance). A biological sensor based on a resonator with a lateral electric field, which allows the detection and identification of bacterial cells without an active reagent directly in the liquid phase, is described in

Handa et al. (2008). This possibility was demonstrated by the detection of *S. typhimurium* cells using the specific bacteriophages.

It has been pointed above that acoustic sensors based on piezoelectric resonators with a lateral electric field have an advantage over other types of sensors. But the main difficulty in designing such resonators is the suppression of unwanted oscillations in order to ensure a high-quality factor for the selected resonant frequency. In this regard, the choice of the optimal shape of the electrodes and their exact orientation relative to the crystallographic axes of the plate and the edges of the crystal were traditionally used (Wark et al. 2007). However, the realization of this approach involves a number of technical difficulties.

Another approach to the problem of suppressing unwanted oscillations in a resonator with a lateral electric field by partially covering the electrodes with a damping layer allowed to develop a new biological sensor described in Zaitsev et al. (2012). The sensor is designed to detect bacterial cells directly in the liquid phase by registering specific interactions such as "bacterial cells–antibodies," "bacterial cells–bacteriophages," and "bacterial cells–mini-antibodies." Next, we focus on a brief description of the design of this sensor and on the features of the experiments.

Figure 11.1 shows the scheme of the biosensor used, containing a resonator with two rectangular electrodes on X-cut lithium niobate plate and a liquid container. Studies have shown that for the indicated orientation of the plate, a stable resonance is observed on a longitudinal wave propagating along the X axis and excited by lateral components of the electric field. The aluminum electrodes with a width of 5 mm and a length of 10 mm were deposited on a lithium niobate plate through a special mask in a vacuum. The frequency dependences of the real and imaginary parts of the electrical impedance of the resonator were measured using a precision LCR meter 4285A (Agilent Company, USA). The quality factor of such resonator turned out to be ~ 630 at a resonant frequency of ~ 6.6 MHz.

The idea of experiments with an electroacoustic sensor is based on the fact that the physical properties of a cell suspension after adding the bioselective agents (antibodies, bacteriophages, or mini-antibodies) change, which allow to detect and identify the microbial cells in the sample under study. The general scheme of the experiments is presented in Fig. 11.1.

11.3.3 Interaction "Cells–Antibodies"

Registration of the interaction of microbial cells with specific antibodies (Abs) is an important condition for their successful use for the detection of cells. The principle of interaction "antigen–antibody" is widely used in sensory systems to identify microorganisms (Gascoyne et al. 1997; Vaughan et al. 2001; Yousef 2008).

To study the interaction of antibodies with microbial cells by the developed acoustic sensor, polyclonal antibodies (pAbs) and microbial cells of *Azospirillum brasilense* Sp7 with different concentrations in suspension were used (Guliy et al. 2013). The frequency dependences of the real and imaginary parts of the electrical impedance were measured before and after the addition of specific Abs (Fig. 11.2a

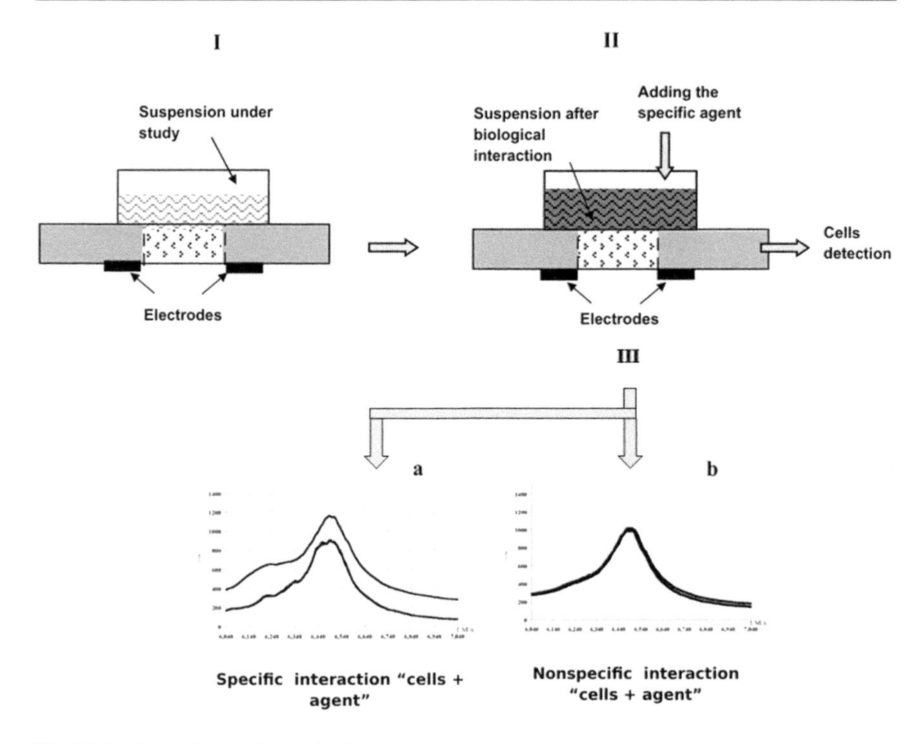

Fig. 11.1 General experimental scheme
The process of the measurement consists of the following steps:
I – a suspension cells is placed in the liquid container and the analytical signal of the sensor is measured
II – the specific/nonspecific (for studied cells) antibodies/or bacteriophages/or phage mini-antibodies are added in liquid container and the analytical signal is measured
III – the obtained results are analyzed and this allows to make conclusion about the presence of the cells under study in suspension:
III (**a**) if the specific agent interacts with the cells, the analytical signal being recorded for suspensions of cells with specific agent (*curve 1*) and without them (*curve 2*) significantly differs
III (**b**) if the specific agent do not interact with the cells, the analytical signal being recorded for suspensions of cells with specific agent (*curve 1*) and without them (*curve 2*) is practically the same

and b). Additionally, the transmission electron microscopy (TEM) identification of the interaction of *A. brasilense* Sp7 cells with the used antibodies labeled with colloidal gold was carried out. In the micrograph (Fig. 11.2e), one can see that the antibodies obtained interact with the *A. brasilense* Sp7, and the accumulation of the marker occurs over the entire cell surface. It has been found that an increase in the electrical conductivity of a cell suspension during their interaction with specific antibodies is due to the adsorption of antibodies on the cell surface (Guliy et al. 2013). Also it was shown that the interaction of *A. brasilense* Sp7 cells with specific antibodies in the presence of extraneous microflora (*E. coli* BL-Ril cells) leads to a change in the abovementioned recorded dependencies (Fig. 11.2c, d). The results of

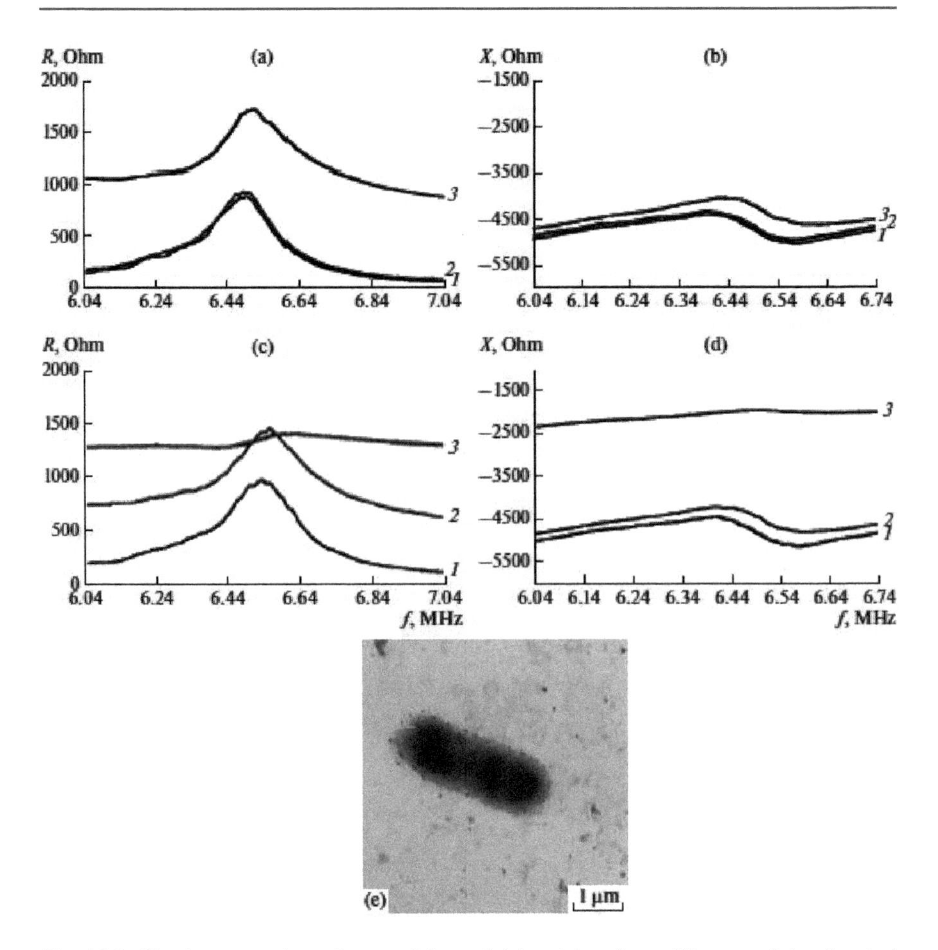

Fig. 11.2 The frequency dependences of the real (**a**) and imaginary (**b**) parts of the electrical impedance of the resonator (the cell contains 10^4 cells/mL) for the interaction between cell suspension of *A. brasilense* Sp7 and specific antibodies. The frequency dependences of the real (**c**) and imaginary (**d**) parts of the electrical impedance of the resonator for the interaction of a mixed cell suspension containing *A. brasilense* Sp7 and *E. coli* BL-Ril and Abs specific to *A. brasilense* Sp7: (1) distilled water; (2) cell suspension without Abs; (3) cell suspension with Abs; (**e**) a TEM image of *A. brasilense* Sp7 cells labeled with the conjugate of protein A and colloidal gold after reaction with the Abs

electroacoustic analysis are in good agreement with the results of electron microscopy.

Therefore, the use of an acoustic method for the detection of microbial cells by registering a specific interaction "bacterial cells–antibodies" has great prospects for analysis of microbial cells with a lower detection limit of ~ 10^4 cells/ml.

11.3.4 Interaction "Cells–Bacteriophages"

Bacteriophages have a certain selectivity of interaction with the cells surface, and they are very accurate indicators that determine the species and type of bacteria. Therefore, they are widely used in medical practice to identify bacteria secreted from the body of the patient and infected objects of the environment (Jung et al. 1999; Summers 2005).

In this regard, the methods of phage indication and phage identification of cells *Bacillus subtilis* in food raw materials and food were developed (Yuan et al. 2012). A method of identifying *Pseudomonas mallei* bacteria of glanders pathogens using *P. pseudomallei* bacteriophages is widely used (Manzeniuk et al. 1994). Method of microbial cell detection based on specific binding to bacteriophages was developed for pathogenic cells such as *Bacillus cereus* and *Clostridium perfringens* (Kretzer et al. 2007) and also other bacteria (Chatterjee et al. 2000; Low et al. 2005; Schmelcher 2008).

Despite the fact that the methods of phage typing of microbial cells are quite specific, simple in formulation, and generally accessible, they require a lot of time (from 48 h to 5 days). Therefore, the development of alternative methods for the detection of bacteria using bacteriophages with a shorter analysis time is very important.

The experiments investigating the possibility of an acoustic sensor for the detection of microbial cells *Escherichia coli* during their interaction with specific bacteriophages were conducted in Guliy et al. (2015). Also as a model system, the cells *A. brasilense* Sp7 for infection by bacteriophage FAb-Sp7 were used in Guliy et al. (2017). The scheme of the experiment was similar to the case of the interaction the microbial cells with specific antibodies (see Fig. 11.1).

As an example Fig. 11.3 shows the dependences of the real part of the electrical impedance for cells *A. brasilense* Sp7, which are infected with the specific bacteriophage FAb-Sp7. The number of cells in the liquid container was 10^6 (a) and 10^4 (b) cells/ml, respectively. As a result, it has been shown that the developed sensor allows distinguishing the cases of infection of bacterial cells by specific bacteriophages from the control experiments without such infection (Guliy et al. 2015, 2017).

To confirm the results of electroacoustic measurements, an electron microscopic study of the interaction of the bacteriophage FAb-Sp7 with microbial cells *A. brasilense* Sp7 was performed (Fig. 11.3c). The presented data shows that bacteriophages are uniformly adsorbed on the entire surface of the bacterial cell. To improve the perception of the obtained data, Fig. 11.3d shows electron microscopic image of the bacteriophage FAb-Sp7.

It was also shown that a significant change in the physical properties of the suspension, leading to a change in the analyzed frequency dependencies of the real and imaginary parts of the electrical impedance, occurred even in the presence of extraneous microflora. This means that the sensor detects the infection of bacterial cells by specific bacteriophages even in the presence of extraneous microflora.

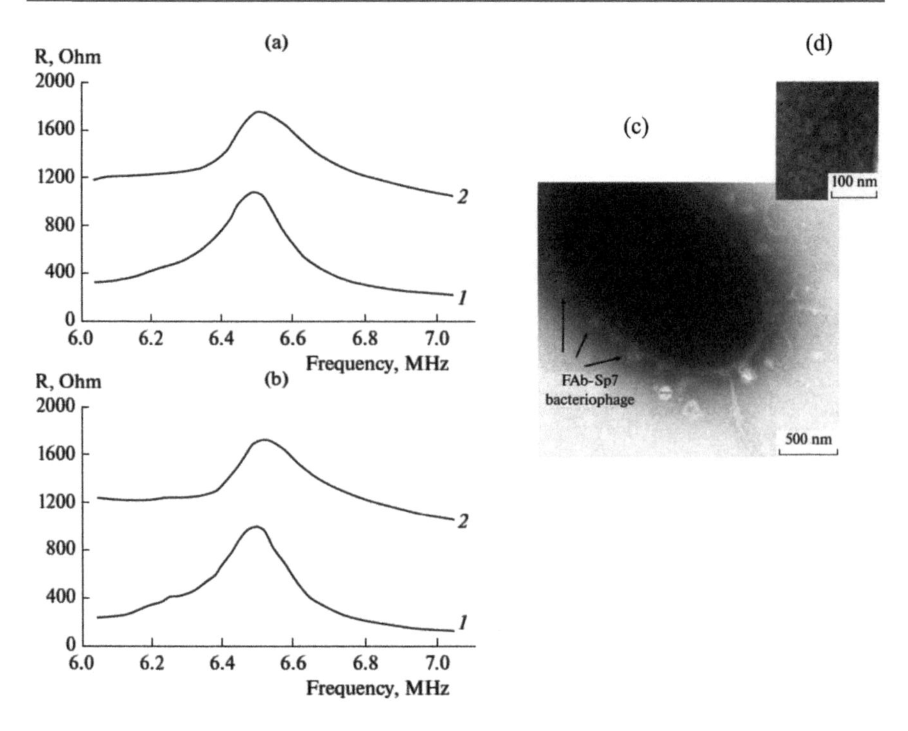

Fig. 11.3 The dependence of the real part of the electrical impedance of the resonator for bacterial suspension of *A. brasilense* Sp7 infected by FAb-Sp7 bacteriophage for the number of cells in the container: 10^6 cells/mL (**a**) and 10^4 cells/mL (**b**). Dependencies (1) and (2) refer to the suspension before and after addition of bacteriophages, respectively. (**c**) A TEM image of *A. brasilense* Sp7 cells at their interaction with the FAb-Sp7 bacteriophage; (**d**) the sidebar contains an image of the FAb-Sp7 bacteriophage

Thus, changes in the parameters of the cell suspensions under the action of bacteriophage occur only if the microbial cells under study are sensitive to bacteriophages. So the electroacoustic sensor allows distinguishing situations when infection of bacterial cells by specific bacteriophages occurs from the control experiments when such an infection is absent. It also allowed determining the spectrum of lytic activity of bacteriophages (Guliy et al. 2015). Based on numerous experiments and statistical data processing, an approximate criterion was developed for the presence of specific interaction of the bacteriophages and cells in the analyzed suspension. It consists in the following: the change in the module of the electrical impedance of the sensor should not be less than ~ 10% when a minimum number of bacteriophages are added to the cell suspension (five phage particles per cell) (Guliy et al. 2015, 2017). The results show the possibility of creating a rapid method for detecting microbial cells when they are infected with a specific bacteriophages and determining their bacteriophage resistance.

11.3.5 Interaction "Cells–Mini-antibodies"

Traditionally, biological components used to identify cells are polyclonal and monoclonal antibodies (Abs), which are widely used in the production of diagnostic test systems. However, Abs, obtained by phage display technology, also can be used to determine microbial cells (Smith and Petrenko 1997; Petrenko 2003, Petrenko and Sorokulova 2004; Williams et al. 2003; Paoli et al. 2004; Nanduri et al. 2007).

We have obtained phage-displayed mini-antibodies (mini-Abs) by phage display technology to whole cells of *A. brasilense* Sp245 and investigated their interaction with cells *A. brasilense* Sp245 with help of this sensor. It has been found that the frequency dependences of the real and imaginary parts of the impedance for the cell suspensions with mini-Abs and without them are very different. It should be noted that during the measurements no changes in the resonator impedance were observed after ~ 10 min, i.e., the impedance change time was fairly short. These measurements recorded the final state of the suspension, since repeated experiments led to the same results (Guliy et al. 2012).

These studies have shown the possibility of cell detection using specific mini-Abs, and the limit of detection is approximately 10^3 cells/ml. The degree of change in the characteristics of the resonator depends on the concentration of cells, which opens up the possibility of carrying out not only qualitative but also quantitative analysis of bacteria.

11.4 Sensors Based on the Use of Inhomogeneous Piezoactive Acoustic Waves

11.4.1 Sensors Using the Active Films with Immobilized Microorganisms

Sensors based on the surface acoustic waves are widely used to measure various physical quantities and chemical composition of objects contacting with the waveguides. In order to develop liquid and biological sensors, basically, Love waves with shear-horizontal polarization, characterized by low attenuation due to the absence of radiation losses, are used (Gaso et al. 2012). The principle of operation of sensors based on surface acoustic waves is that the measured parameter affects the wave propagation velocity, which in turn causes changes in the time interval between the input and output signals or in the phase of the output signal.

In acoustic biosensors, the recognizing reagent is usually a macromolecule immobilized inside a membrane chemically bound to a surface that contacts the solution of the analyte. A specific chemical reaction takes place between the reagent and the substance under study. This can be either direct interaction of the reagent with the analyte, as in the case of an antigen–antibody reaction, or catalytic interaction of the immobilized enzyme with the analyte leading to the formation of an easily identifiable product. Piezoelectric sensors register a mass change of the order of 10–12 ng. Such biosensors contain an enzyme immobilized on a surface of piezocrystal, an antibody, an antigen, or a DNA (RNA) strand. The presence of a detectable agent entering into an enzymatic immune response or forming a hybrid molecule with an immobilized bioobject is evidenced by the formation of the corresponding complex on the piezocrystal, which changes the wave velocity. With the help of a piezoelectric biosensor with an immobilized antigen, it can be detected antibodies in serum diluted 1000 times.

The first experiments using Love waves for biochemical research were described in Kovacs et al. (1992) and Gizeli et al. (1992). In 1997, acoustic devices based on Love waves were used to study the real-time antigen–antibody interaction in liquid media (Harding et al. 1997).

A two-channel biosensor on the Love wave was developed for simultaneous detection of *Legionella* and *E. coli* cells with the help of antibodies (Howe and Harding 2000). The detection limit for both types of bacteria was 10^6 cells/ml at 3 h analysis time.

An immunosensor on the Love wave for the detection of viruses or bacteria in liquids (in food and process water, beverages, etc.) in real time is also developed and described in Tamarin et al. (2003). The authors used monoclonal antibodies against bacteriophage M13 immobilized on the surface of the waveguide and showed the possibility of determining bacteriophage M13.

In has been shown that a sensor with a Love wave, containing a layer of zinc oxide, is promising for use as immunosensor (Kalantar-Zadeh et al. 2003). The authors successfully controlled the process of adsorption of rat immunoglobulin G.

A sensory platform based on immobilized cells and a sensor using a Love wave for detecting heavy metals in a liquid medium was also successfully used (Gammoudi et al. 2010, 2011). The acoustic delay line was included in the generator feedback circuit, and the generation frequency was recorded in real time. A polydimethylsiloxane chip with a liquid chamber was in contact with a waveguide with a propagating acoustic wave. Bacteria (*E. coli*) were fixed in the form of bioreceptors on a sensitive sensor surface coated with a multilayer polyelectrolyte using a simple and efficient electrostatic self-assembly procedure between layers. The response of the sensor in real time to various concentrations of cadmium and mercury ions was investigated. It turned out that the detection limit was 10–12 mol/l. The analytical signal of the sensor depended on changes in the viscoelastic properties of the fluid associated with changes in the metabolism of bacteria.

A similar principle with immobilized cells was also used to determine ions of heavy metals (Tekaya et al. 2012). In this case, the immobilization of microalgae *Arthrospira platensis* was used as a receptor.

An innovative method of detecting *E. coli* using a sensor with a Love wave and the antibodies immobilized on the surface of the waveguide with the threshold of detection of 10^6 bacteria/ml is described in Moll et al. (2007). The same group of researchers demonstrated a multifunctional immunosensor on the Love wave for the detection of bacteria, viruses, and proteins (Moll et al. 2008). They successfully detected bacteriophages and proteins up to 4 ng/mm^2 and *E. coli* bacteria up to 5×10^5 cells in a volume of 500 μl with good specificity and reproducibility. The authors stated that whole bacteria can be detected no later than 1 h.

In the work of Andrä et al. (2008), a sensor for determining the lipid specificity of human antimicrobial peptides is described. In this case, the membranes were attached to the sensor surface, imitating the cytoplasmic and outer bacterial membranes.

The resonators on the surface acoustic wave were also used as biological sensors. For example, it has been shown that the formation of the biofilm of bacteria due to the interaction with an antibiotic on the surface of a piezocrystal can be controlled with high sensitivity by measuring the resonant frequency of a resonator on a surface acoustic wave (Kim et al. 2016). However, the total time of biofilm growth and its removal was 48 h, during which a change in a temperature could lead to an uncontrolled change in the resonant frequency.

11.4.2 Sensors Based on the Use of Surface and Plate Piezoactive Acoustic Waves Without Immobilized Microorganisms

Another approach is the development of sensors for the analysis of bacteria without the immobilization of microorganisms.

In this regard, a sensor using the Love wave based on the aptamer, which allowed detecting small biomolecules, was developed (Schlensog et al. 2004). This biosensor has an advantage over immunosensors, since it does not require the production of antibodies against toxic substances.

A biosensor based on the Love wave was developed to detect pathogenic spores at the level of infections by inhalation (Branch and Brozik 2004). Monoclonal Abs with a high degree of selectivity were used to determine anthrax spores under water conditions. The authors claim that using the developed method, whole cells can also be detected.

An acoustic sensor for the microbial cell detection using the plate acoustic waves is described in Zaitsev et al. (2001). It represented a two-channel delay line made of a Y-X plate of lithium niobate 0.2 mm thick (Fig. 11.4). Two pairs of interdigital transducers (IDT) for excitation and reception of an acoustic wave with shear-horizontal polarization in each channel were applied by the photolithography method. One of the channels of the delay line was electrically shorted by depositing a layer of aluminum on the surface between the IDTs; the second channel remained electrically free. The liquid container was fixed on the surface of the plate between the IDTs. The volume of the suspension container was ~ 5 ml. The specified sensor was connected to 4-port meter of S parameters 5071C (Agilent Technologies, USA).

With the help of this sensor, an experimental study of a specific interaction of the types "bacterial cells–bacteriophages," "bacterial cells–antibodies," and "bacterial cells–mini-Abs" directly in suspension with different initial electrical conductivities was carried out. As an example Fig. 11.5 (a, b) presents the time dependencies of insertion loss of sensor (a) and phase of output signal (b) before and after adding the specific mini-Abs in the cell suspension for electrically open (1) and electrically shorted (2) channels, respectively. Figure 11.6 (a–f) shows the dependencies of the change in insertion loss ($\Delta\alpha$) (a, c, e) and phase of output signal ($\Delta\Phi$) (b, d, f) on

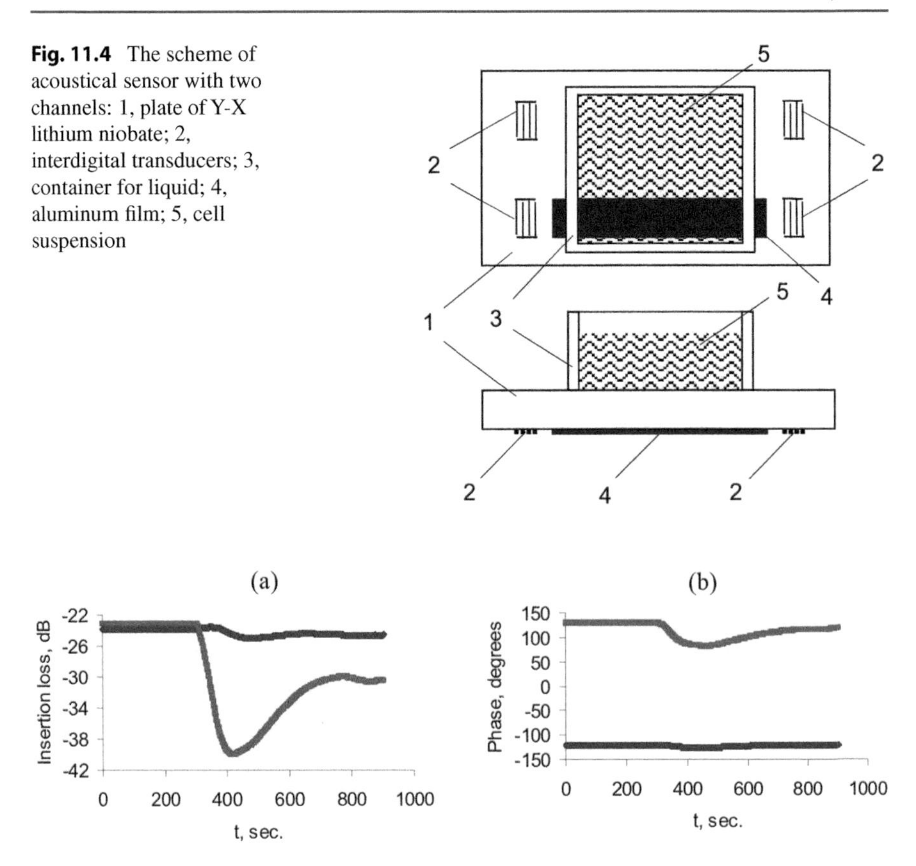

Fig. 11.4 The scheme of acoustical sensor with two channels: 1, plate of Y-X lithium niobate; 2, interdigital transducers; 3, container for liquid; 4, aluminum film; 5, cell suspension

Fig. 11.5 The time dependencies of insertion loss (**a**) and phase of output signal (**b**) before and after adding the specific mini-Abs in cell suspension for electrically open (1) and electrically shorted (2) channels, respectively. Mini-Abs were added at the point of time t = 300 s

the conductivity of the buffer solution (σ) at specific interaction "bacterial cells *A. brasilense* Sp245–mini-Abs" for the various cell concentrations for electrically short and open channels. One can see that these changes for electrically shorted channel are significantly less in comparison with electrically open channel. This confirms the earlier statement that a specific interaction increases the conductivity of the cell suspension. The mini-Abs interaction specificity was controlled by electron microscopic identification of the interaction of *A. brasilense* Sp245 with CG-labeled mini-Abs (Fig. 11.6g). One can see that the labeled mini-Abs interacted with *Azospirilla* and the CG label was distributed throughout the cell surface (Guliy et al. 2018).

The value of the minimum concentration of cells, for which their detection is possible, turned out to be 10^4 cells/ml for each type of the interaction. At that the initial conductivity of the buffer solution was changed in the range 2–50 μS/cm.

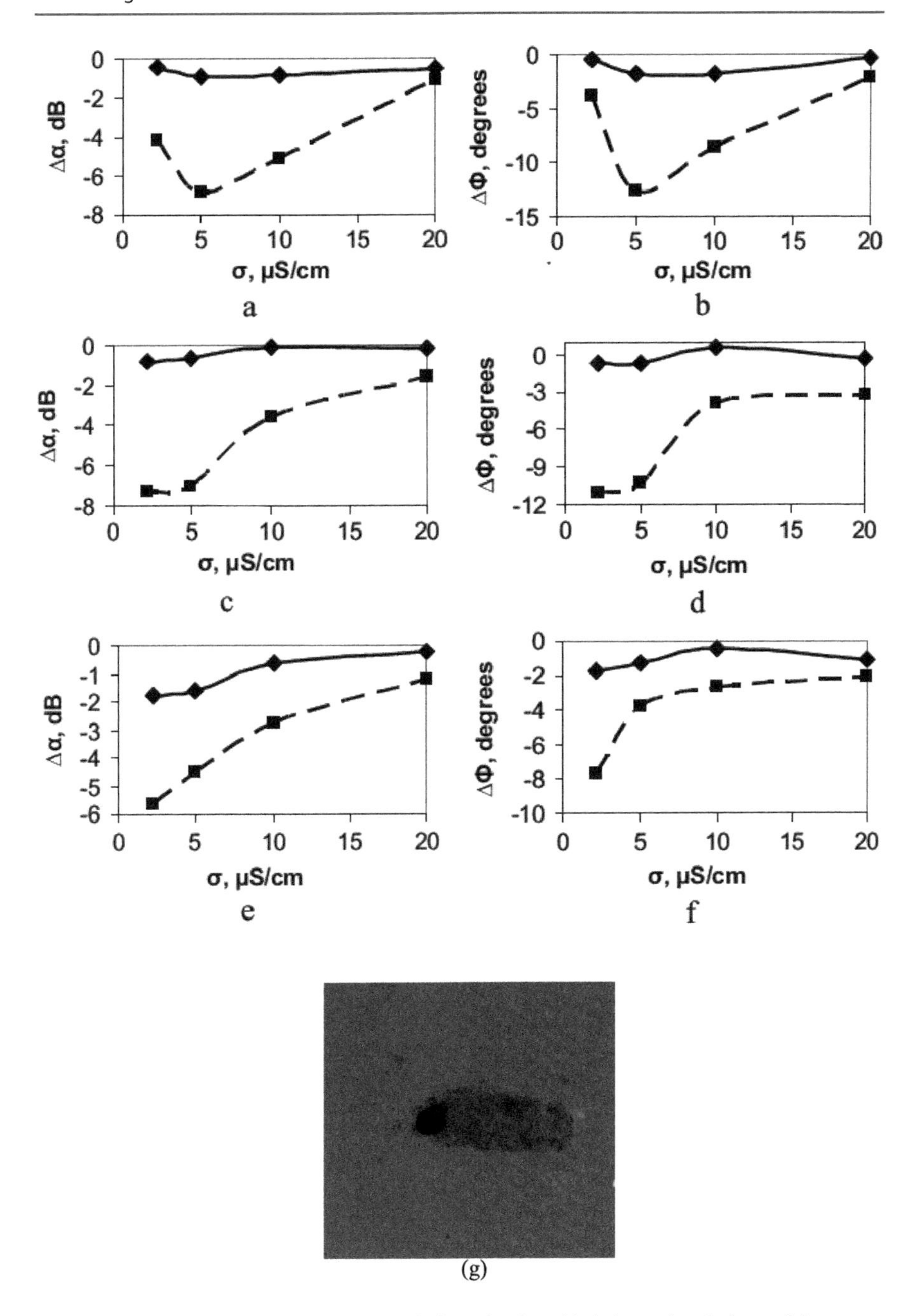

Fig. 11.6 The dependencies of the change in insertion loss (Δα) (**a**, **c**, **e**) and phase of the output signals (ΔΦ) (**b**, **d**, **f**) on the conductivity of the buffer solution (σ) at specific interaction "bacterial cells *A. brasilense* Sp245–mini-Abs" for cell concentrations of 10^4 cells/ml (**a**, **b**), 10^6 cells/ml (**c**, **d**), and 10^8 cells/ml (**e**, **f**). Dotted and solid lines correspond to electrically open and shorted channels, respectively

TEM of bacterial cells *A. brasilense* Sp245 marked by conjugant of antiphage antibodies after interaction with phage antibodies (×10,000) (g)

11.5 Noncontact Sensor Based on a Slot Acoustic Mode

As already noted, over the past few years, acoustic biological sensors with good speed and sensitivity for detecting bacterial cells directly in the liquid phase have been developed. But it should be noted that in most articles describing sensors, there is practically no data on the conditions for cleaning the sensors from the spent biological sample (Turner et al. 1987; Gaso et al. 2012; Chen and Cheng 2017).

However, the known methods of cell detection have a significant drawback associated with the need for thorough cleaning the surface of the piezocrystal after contact with the investigated suspension. Often this requires the use of special methods to remove residues of previous biological samples and control the quality of cleaning, which can lead to damage to the sensor elements and, as a result, distortion of the analytical signal. This problem could be solved using contactless analysis methods, in which the container with the investigated suspension is isolated from the sensor surface. To solve this problem, a contactless biological sensor was developed (Borodina et al. 2018) based on a slot mode in a single-channel delay line with a shear-horizontal acoustic wave of zero order. In accordance with Fig. 11.7, a delay line based on a Y-X piezoelectric plate of lithium niobate (1) with a thickness of 200 μm was used as the main element of the sensor for contactless examination of bacterial cells in conducting suspensions. Two interdigital transducers (2) were applied to the surface of the plate to excite and receive an acoustic wave with a shear-horizontal polarization of zero order at a center frequency of ~ 3.5 MHz. A liquid container (5) with a volume of 1.5 ml was placed above the acoustic delay line between the IDTs. The bottom of the cell was made of a plate of Z-X lithium niobate (4) (Borodina et al. 2013). A fixed gap between the surface of the delay line and the bottom of the liquid container was provided by means of strips of aluminum foil with a thickness of 8 μm. To study the specific interaction of the bacterial cells with the polyclonal antibodies, the sensor was connected to the meter of S-parameters E5071C ("Agilent," USA). This meter operated in the regime of measuring the

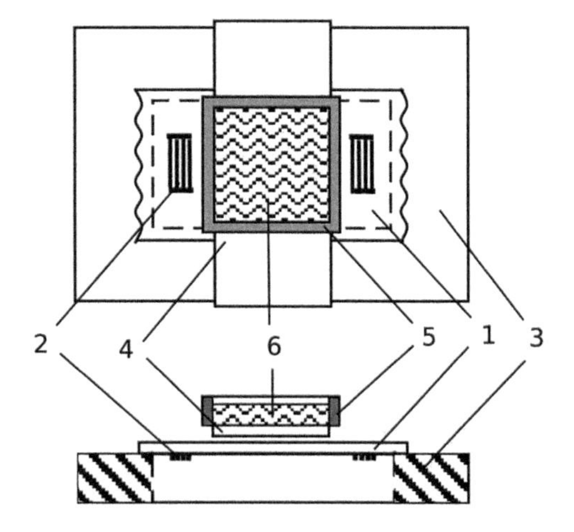

Fig. 11.7 The scheme of the contactless sensor: 1, piezoelectric plate of Y-X $LiNbO_3$; 2, IDTs; 3, the holder of the plexiglass; 4, piezoelectric plate of Z-X $LiNbO_3$; 5, liquid container; 6, suspension of the cells under study

frequency dependence of the insertion loss of the sensor. It has been found (Borodina et al. 2013) that such structure is characterized by the clearly pronounced resonant peaks on the frequency dependence of the insertion loss associated with the excitation of the slot mode.

As an example Fig. 11.8a shows the frequency dependence of the insertion loss of the sensor loaded with a buffer solution with a conductivity of 20 μS/cm (Borodina et al. 2018). Figure 11.8b presents the same dependencies of a sensor loaded by a buffer solution with the cells before (green line) and after (pink line) the addition of antibodies. One can see that addition of antibodies leads to significant change in values of depth of resonant peaks. The control experiments including nonspecific interaction "microbial cells–antibodies" showed the absence of changes in peaks parameters. This sensor showed the possibility of recording the interaction of microbial cells with specific antibodies in suspensions with a conductivity of 5–50 μS/cm and with a minimum cell concentration of 10^4–10^3 cells/ml. A

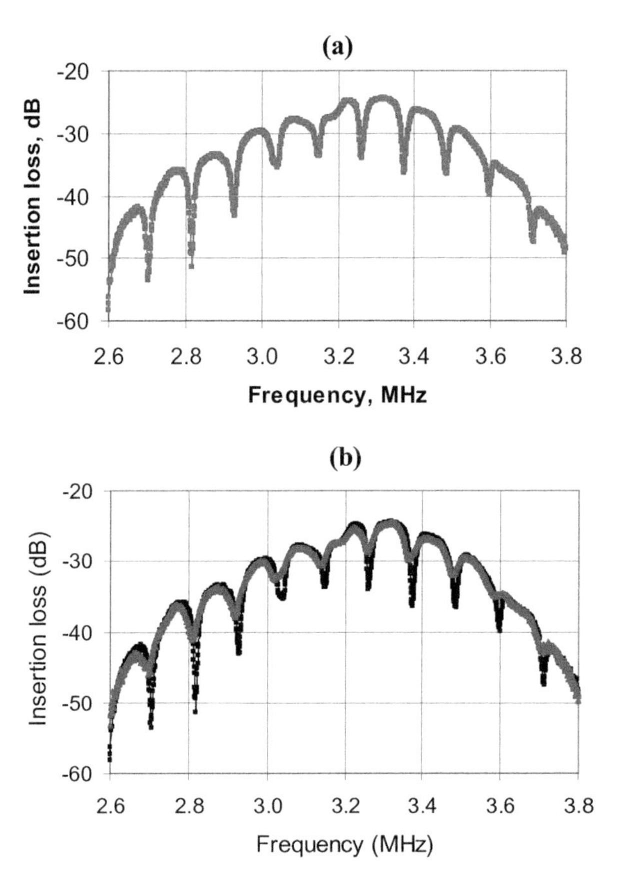

Fig. 11.8 The frequency dependence of the insertion loss of the sensor loaded by the buffer solution with the conductivity of 20 μS/cm without (**a**) and (**b**) with the cells *A. lipoferum* Sp59b before (black line) and after the addition of the specific Abs (pink line)

distinctive feature of the sensor is the lack of contact of the investigated suspension with a thin (200 μm) acoustic waveguide. An additional advantage of the sensor is the presence of a removable liquid container, which allows to reuse it and facilitates the process of cleaning the container from the sample.

11.6 Conclusion

The capabilities of electroacoustic biological sensors based on resonators with a longitudinal and lateral electric field, based on delay lines with a propagating piezo-active acoustic wave, and based on a delay line with a slot mode are demonstrated. These sensors can conduct cell detection and identification of bacteria using immobilized microorganisms or directly in cell suspension. The sensitivity range of microbial cell detection is 10^3–10^8 cells/ml and for suspension with the conductivity of 5–50 μS/cm. At that the analysis time varies from 5 min to several hours. Obtaining reliable information is carried out by registering and processing changes in the analytical signal of the sensor when specific agents are introduced into the cell suspension. The presented possibilities of electroacoustic biological sensors for the detection of bacteria are focused on the clinical use of *onsite* as a personalized diagnostic device. The possibility of rapid detection of microflora will allow timely diagnosis of the disease and timely medical assistance. An important feature of acoustic biological sensors is the detection of any biochemical and/or biophysical signal associated with a particular microorganism. Further studies in the development of acoustic biological sensors are aimed at adapting the method for detection of microbial cells of different taxonomic groups, optimizing the measurement procedure and developing criteria for specific binding to the selective agent in real conditions.

In general, acoustic biological sensors form a wide class of detection systems and are very promising for use in microbiology, medicine, and veterinary medicine for solving the problems of detection and identification of bacteria.

Acknowledgment This work was supported in part by the grant of the Russian Foundation of Basic Research (projects nos. 19-07-00304 and 19-07-00300). Authors are grateful to Semyonov A.P. and Karavaeva O.A. for assisting in conducting the experiment.

References

Andle JC, Vetelino JF (1994) Acoustic wave biosensors. Sensor Actuators A Phys 44(3):167–176

Andrä J, Böhling A, Gronewold TMA, Schlecht U, Perpeet M, Gutsmann T (2008) Surface acoustic wave biosensor as a tool to study the interactions of antimicrobial peptides with phospholipid and lipopolysaccharide model membranes. Langmuir 24(16):9148–9153

Basile F, Beverly MB, Hadfield TL, Voorhees KJ (1998) Pathogenic Bacteria: their detection and differentiation by rapid liquid profiling with pyrolysis mass spectrometry. Trends Anal Chem 17:95–109

Bisoffi M, Hjelle B, Brown DC, Branch DW, Edwards TL, Brozik SM, Bondu-Hawkins VS, Larson RS (2008) Detection of viral bioagents using a shear horizontal surface acoustic wave biosensor. Biosens Bioelectron 23(9):1397–1403

Borodina IA, Zaitsev BD, Kuznetsova IE, Teplykh AA (2013) Acoustic waves in a structure containing two piezoelectric plates separated by an air (vacuum) gap. IEEE Trans Ultrason Ferroelectr Freq Control 60(12):2677–2681

Borodina IA, Zaitsev BD, Burygin GL, Guliy OI (2018) Sensor based on the slot acoustic wave for the non-contact analysis of the bacterial cells in the conducting suspensions. Sensors Actuators B 268:217–222

Branch DW, Brozik SM (2004) Low-level detection of a Bacillus anthracis simulant using love-wave biosensors on 36 YX LiTaO3. Biosens Bioelectron 19:849–859

Chandra P (ed) (2016) Nanobiosensors for personalized and onsite biomedical diagnosis. IET London

Chatterjee S, Mitra M, Das Gupta SK (2000) A high yielding mutant of mycobacteriophage L1 and its application as a diagnostic tool. FEMS Microbiol Lett 188:47–53

Chen S, Cheng YF (2017) Biosensors for bacterial detection. Int J Biosen Bioelectron 2(6):197–199

Don E, Farafonova O, Pokhil S, Barykina D, Nikiforova M, Shulga D, Borshcheva A, Tarasov S, Ermolaeva T, Epstein O (2016) Use of piezoelectric immunosensors for detection of interferon–gamma interaction with specific antibodies in the presence of released–active forms of antibodies to interferon–gamma. Sensors 16(1):96. https://doi.org/10.3390/s16010096

Ermolaeva TN, Kalmykova EN (2006) Piezoelectric immunosensors: analytical potentials and outlooks. J Russ Chem Rev 75(5):397–409

Ermolaeva TN, Kalmykova EN (2012) Capabilities of piezoelectric immunosensors for detecting infections and for early clinical diagnostics. In: Chiu NHL, Christopoulos TK (eds) Advances in immunoassay technology. In TeO

Gammoudi I, Tarbague H, Othmane A, Moynet D, Rebière D, Kalfat R, Dejous C (2010) Love-wave bacteria-based sensor for the detection of heavy metal toxicity in liquid medium. Biosens Bioelectron 26(4):1723–1726. https://doi.org/10.1016/j.bios.2010.07.118

Gammoudi I, Tarbague H, Lachaud JL, Destor S, Othmane A, Moynet D, Kalfat R, Rebière D, Dejous C (2011) Love wave bacterial biosensors and microfluidic network for detection of heavy metal toxicity. Sensors Lett 9(2):816–818

Gascoyne P, Pethig R, Satayavivad J, Becker FF, Ruchirawat M (1997) Dielectrophoretic detection of changes in erythrocyte membranes following malarial infection. Biochim Biophys Acta 1323:240–252

Gaso Rocha MI, Jiménez Y, Francis A (2012) Laurent and Antonio Arnau Love wave biosensors: a review. Chapter 11. https://doi.org/10.5772/53077

Gizeli E, Goddard NJ, Lowe CR, Stevenson AC (1992) A love plate biosensor utilising a polymer layer. Sensors Actuators B Chem 6:131–137

Griffiths D, Hall G (1993) Biosensors – what real progress is being made? Trends Biotechnol 11:122–130

Guliy OI, Zaitsev BD, Kuznetsova IE, Shikhabudinov AM, Karavaeva OA, Dykman LA, Staroverov SA, Ignatov OV (2012) Phage mini-antibodies and their use for detection of microbial cells by using electro-acoustic sensor. Biophysics 57(3):336–342

Guliy OI, Zaitsev BD, Kuznetsova IE, Shikhabudinov AM, Matora LY, Makarikhina SS, Ignatov OV (2013) Investigation of specific interactions between microbial cells and polyclonal antibodies using a resonator with lateral electric field. Microbiology 82(2):215–223

Guliy OI, Zaitsev BD, Kuznetsova IE, Shikhabudinov AM, Dykman LA, Staroverov SA, Karavaeva OA, Pavliy SA, Ignatov OV (2015) Determination of the spectrum of lytic activity of bacteriophages by the method of acoustic analysis. Biophysics 60(4):592–597

Guliy OI, Zaitsev BD, Shikhabudinov AM, Borodina IA, Karavaeva OA, Larionova OS, Volkov AA, Teplykh AA (2017) A method of acoustic analysis for detection of bacteriophage-infected microbial cells. Biophysics 62(4):580–587

Guliy OI, Zaitsev BD, Borodina IA, Shikhabudinov AM, Teplykh AA, Staroverov SA, Fomin AS (2018) The biological acoustic sensor to record the interactions of the microbial cells with the phage antibodies in conducting suspensions. Talanta 178:569–576
Handa H, Gurczynski S, Jackson MP, Auner G, Mao G (2008) Recognition of *Salmonella typhimurium* by immobilized phage P22 monolayers. Surf Sci 602:1393–1400
Harding GL, Du J, Dencher PR, Barnett D, Howe E (1997) Love wave acoustic immunosensor operation in liquid. Sensors Actuators A 61(1–3):279–286
Howe E, Harding G (2000) A comparison of protocols for the optimisation of detection of bacteria using a surface acoustic wave (SAW) biosensor. Biosens Bioelectron 15(11–12):641–649
Hu J, Wang S, Wang L, Li F, Pingguan-Murphy B, Lu TJ, Xu F (2014) Advances in paperbased point-of-care diagnostics. Biosens Bioelectron 54:585–597
Jung S, Arndt KM, Müller KM, Plückthun A (1999) Selectively infective phage (SIP) technology: scope and limitations. J Immunol Methods 231:93–104
Kalantar-Zadeh K, Wlodarski W, Chen YY, Fry BN, Galatsis K (2003) Novel love mode surface acoustic wave based immunosensors. Sensors Actuators B 91:143–147
Kim YW, Meyer MT, Berkovich A, Subramanian S, Iliadis AA, Bentley WE, Ghodssi R (2016) A surface acoustic wave biofilm sensor integrated with a treatment method based on the bioelectric effect. Sensors Actuators A 238:140–149
Koenig B, Graetzel M (1994) A novel immunosensor for herpes virus. Anal Chem 66(3):341–348
Kovacs G, Lubking GW, Vellekoop MJ, Venema A (1992) Love waves for (bio)chemical sensing in liquids. In: Proceedings of the IEEE Ultrasonics Symposium, Tucson, USA
Kretzer JW, Lehmann R, Schmelcher M, Banz M, Kim K, Korn C, Loessner MJ (2007) Use of high-affinity cell wall-binding domains of bacteriophage endolysins for immobilization and separation of bacterial cells. Appl Environ Microbiol 73:1992–2000
Li S, Lib Y, Chen H, Horikawa S, Shen W, Simonian A, China BA (2010) Direct detection of *Salmonella typhimurium* on fresh produce using phage–based magneto elastic biosensors. Biosens Bioelectron 26(4):1313–1319
Low LY, Yang C, Perego M, Osterman A, Liddington RC (2005) Structure and lytic activity of a *Bacillus anthracis* prophage endolysin. J Biol Chem 280:35433–35439
Manzeniuk O, Volozhantsev NV, Svetoch EA (1994) Identification of the bacterium Pseudomonas mallei using Pseudomonas pseudomallei bacteriophages. Mikrobiologiia 63(3):537–544
Moll N, Pascal E, Dinh DH, Pillot JP, Bennetau B, Rebiere D, Moynet D, Mas Y, Mossalayi D, Pistre J, Dejous C (2007) A love wave immunosensor for whole *E. coli* bacteria detection using an innovative two-step immobilisation approach. Biosens Bioelectron 22(9–10):2145–2150
Moll N, Pascal E, Dinh DH, Lachaud J-L, Vellutini L, Pillot J-P, Rebière D, Moynet D, Pistré J, Mossalayi D, Mas Y, Bennetau B, Déjous C (2008) Multipurpose love acoustic wave immunosensor for bacteria, virus or proteins detection. ITBM-RBM 29:155–161
Nanduri V, Sorokulova IB, Samoylov AM, Simonian AL, Petrenko VA, Vodyanoy V (2007) Phage as a molecular recognition element in biosensors immobilized by physical adsorption. Biosens Bioelectron 22:986–992
Olsen EV, Sorokulova IB, Petrenko VA, Chen IH, Barbaree JM, Vodyanoy VJ (2006) Affinity-selected filamentous bacteriophage as a probe for acoustic wave biodetectors of *Salmonella typhimurium*. Biosens Bioelectron 21:1434–1442
Paoli GC, Chen CY, Brewster JD (2004) Single-chain Fv antibody with specificity for Listeria monocytogenes. J Immunol Methods 289:147–166
Petrenko VA (2003) Phage display for detection of biological threat agents. J Microbiol Methods 53:253–262
Petrenko VA, Sorokulova IB (2004) Detection of biological threats. A challenge for directed molecular evolution. J Microbiol Methods 58:147–168
Pinkham W, Wark M, Winters S (2005) A lateral field excited acoustic wave pesticide sensor. In: Proceedings of the IEEE Ultrasonics Symposium, pp 2279–2283
Ripp S, Jegier P, Johnson CM, Brigati JR, Sayler GS (2008) Bacteriophage-amplified bioluminescent sensing of *Escherichia coli* O157:H7. Anal Bioanal Chem 391(2):507–514

Schlensog MD, Thomas MA, Gronewold TM, Tewes M, Famulok M, Quandt E (2004) A love-wave biosensor using nucleic acids as ligands. Sensors Actuators B Chem 101:308–315

Schmelcher M (2008) Bacteriophage: powerful tools for the detection of bacterial pathogens. In: Principles of bacterial detection: biosensors, recognition receptors and microsystems. Springer, New York, pp 731–754

Smith GP, Petrenko VA (1997) Phage display. Chem Rev 97(2):391–410

Summers WS (2005) In: Kutter E, Sulakvelidze A (eds) Bacteriophage research: early history in bacteriophages: biology and applications. CRP Press, Boca Raton

Tamarin O, Comeau S, Déjous C, Moynet D, Rebière D, Bezian J, Pistré J (2003) Real time device for biosensing: design of a bacteriophage model using love acoustic wave. Biosens Bioelectron 18:755–763

Tekaya N, Tarbague H, Moroté F, Gammoudi I, Sakly N, Hat BO, Raimbault V, Rebière D, Ben OH, Jaffrezic-Renault N, Lagarde F, Dejous C, Cohen-Bouhacina T (2012) Optimization of spirulina biofilm for in-situ heavy metals detection with microfluidic-acoustic sensor. In: IMCS 2012 – The 14th International meeting on chemical sensors, pp 92–95. https://doi.org/10.5162/IMCS2012/1.2.5

Turner APF, Karube I, Wilson GS (1987) Biosensors: fundamentals and applications. Oxford University Press, Oxford

Van Emon JM, Gerlach CL, Johnson JC (1995) Environmental immunochemical methods. ACS, Washington, DC

Vaughan RD, O'Sullivan CK, Cuilbault GG (2001) Development of a quartz crystal microbalance (QCM) immunosensor for the detection of *Listeria monocytogenes*. Enzym Microb Technol 29:635–638

Vetelino JF (2010) A lateral field excited acoustic wave sensor platform. In: Proceedings of the IEEE Ultrasonics Symposium, San-Diego, pp 2269–2272

Von Lode P (2005) Point-of-care immunotesting: approaching the analytical performance of central laboratory methods. Clin Biochem 38(7):591–606

Wark M, Kalanyan B, Ellis L (2007) A lateral field exited acoustic wave sensor for the detection of saxitoxin in water. In: Proceedings of the IEEE Ultrasonics Symposium, pp 1217–1220

Williams DD, Benedek O Jr, Turnbough CL (2003) Species-specific peptide ligands for the detection of *Bacillus anthracis* spores. Appl Environ Microbiol 69:6288–6293

York C, French LA, Millard P, Vetelino JF (2005) A lateral field exited acoustic wave biosensor. In: Proceedings of the IEEE Ultrasonics Symposium, pp 44–48

Yousef E (2008) Detection of bacterial pathogens in different matrices: current practices and challenges. In: Principles of bacterial detection: biosensors, recognition receptors and microsystems. Springer, New York, pp 31–48

Yuan Y, Peng Q, Gao M (2012) Characteristics of a broad lytic spectrum endolysin from phage BtCS33 of *Bacillus thuringiensis*. BMC Microbiol 12:297–305

Zaitsev BD, Joshi SG, Kuznetsova IE, Borodina IA (2001) Acoustic waves in piezoelectric plates bordered with viscous and conductive liquids. Ultrasonics 39(1):45–50

Zaitsev BD, Kuznetsova IE, Shikhabudinov AM, Ignatov OV, Guliy OI (2012) Biological sensor based on the lateral electric field excited resonator. Trans Ultrason Ferroelectr Freq Control 59(5):963–969

Nanorobots for In Vivo Monitoring: The Future of Nano-Implantable Devices

12

Mamta Gandhi and Preeti Nigam Joshi

Abstract

Innovation is important for the healthcare system advancement, in order to continue delivering the high-quality care at an affordable cost to the society. It can be achieved by nonconventional thinking, tapping into creative minds, and extensive collaborative work to make better use of existing facilities and designing new technologies. Nanorobotics is such an innovation that can revolutionize the current face of medicine and biomedical sciences with their state-of-the-art technology. Improved outcomes of nanorobots-based treatments for diabetes, drug delivery for pancreatic and ovarian cancer, and laparoscopic treatment of skin cancer have already been reported. This book chapter will cover the recent advancements of this emerging field with their biomedical applications.

Keywords

Nanorobots · DNA origami · Nano - implantats · Biomedical magic bullets · Drug delivery

12.1 Introduction of Nanorobotics: Magic Bullets of Future

Innovation is important for the advancement of medicine and healthcare segment in order to continue delivering the high-quality care at an affordable cost to the society. It can be achieved by unconventional thinking, tapping into creative minds, and extensive collaborative work to make better use of existing facilities and designing new technologies. Nanorobotics is such an innovation that can revolutionize the current face of drug delivery systems and biomedical sciences with its

M. Gandhi · P. N. Joshi (✉)
NCL Venture Center, FastSense Diagnostics Pvt. Ltd, Pune, India
e-mail: preetijoshi@fastsensediagnostics.com

P. Chandra, R. Prakash (eds.), *Nanobiomaterial Engineering*,
https://doi.org/10.1007/978-981-32-9840-8_12

state-of-the-art technology. Nanorobots are futuristic concept, and by definition, nanorobots are electromechanical systems in size range of 10^{-9} nanometers and are functionally capable of carrying out humanly impossible tasks with precision at nanoscale dimensions. Typically miniaturized computers, onboard sensing devices, motors, power supplies, and manipulators are basic component of a nanorobot. The construction of nanorobots is a complex engineering process and still in its infancy, but with the emerging new technologies and major research going on in this area, soon nanorobots will be a significant reality with their real-world applications, and they will be introduced in every field of industry, especially in the medical industry. Improved outcomes of nanorobots based drug delivery for ovarian and pancreatic cancer have already been reported. This book chapter will cover the recent advancements in this emerging field with emphasis on their applications in cancer therapy. Although this concept is in its infancy but the technology of nanorobotics is on the right direction and healthcare and medical sectors will be the immediate beneficiary. The promising future applications of robotic science are attracting government interest in many countries (Reppesgaard 2002; Wieland III 2004; Cavalcanti et al. 2004). Recently the US National Science Foundation has launched a program on "Scientific Visualization" on how nanorobotics can be portrayed by supercomputers. In 2015 itself, nanosystems and embedded nanodevices had a market projection of 1 trillion US$, while global market of nanobots has CAGR of 21% and is expected to reach US$ 100 billion in 2023 from US$ 74 billion in 2016. IBM, PARC, Hewlett Packard, Bell Laboratories, and Intel Corp. are few companies involved in nanoelectronics fabrication. With technological advancement in molecular computing, processing of logic task is possible by nano-bio-computers (Adleman 1995; Hagiya 2000), which is a promising first step to enable future nanoprocessors with increased complexity. Nano-kinetic devices and miniaturized sensing devices are the stepping stone for nanorobotics operation and locomotion (Sun et al. 2001; Stracke et al. 2000).

Apparently, nanorobotics sounds a new word, but in the early 1900s, German Nobel laureate Paul Ehrlich laid the foundation of this field by virtually describing an in vivo missile kind of therapy module that would deliver the drug at specific site to cure the diseases. Ehrlich named this concept as "Magische Kugel," meaning "Magic Bullet," which would have numerous applications in treatment of diseases, including cancer. In those times, it was just hypothetical, but with the onset of nanotechnology advancements, we now have commercial magic bullets or nano-based multifunctional targeted drug delivery systems to perform the task described by Dr. Ehrlich.

Nanotechnology advancements are the major driving force behind the fast pace of nanorobotics. Although it's a highly interdisciplinary field, but better understanding of nanofabrication and emergence of nano electronics and NEMS have enabled scientists to develop different nanodevices. Currently this field is in its infancy, but several substantial steps have been taken by great researchers all over the world who are contributing to this ever challenging and exciting field. The ability to manipulate matter at the nanoscale is one core application for which nanorobots could be the technological solution (Ummat et al. 2005a, b).

Although in many references, different nanoparticles were also included while describing nanorobotics task in different fields involving defense, medical, and environmental applications, in this chapter we only focused on basics related with emergence of nanorobots and their possible medical applications as nanorobots are better substitutes of nanoparticles in performing cell repair, programmed drug release, and fighting infections in vivo. Apparently these nanodevices will build a better future with their enormous applications in medicine and biomedical engineering.

12.1.1 Origin of Nanorobots: From Fiction to Reality

The concept of nanorobots is a technological breakthrough that would evolve new therapy modules and many in vivo applications that are beyond imagination in current medical scenario. Science fiction movies and novels have already talked about such devices long back in 1966, where in a movie "The Fantastic Voyage," a submarine was shrunk up to the limit to be injected into bloodstream for repair tasks. Although it was pure imagination, with the immense technological advancement in material science, MEMS, microfabrication, and biomolecular science, researchers are working to materialize this important aspect of device fabrication from micro to nano level that will be a boon for medical sciences to send minuscule nanorobots inside the human body for repair tasks. The concept of bio-nanorobotics is now evolved as micro devices in surgery and medical treatments are frequently used and improving clinical procedures in recent years (Murphy et al. 2007; Cavalcanti et al. 2008a, b, c). Catheterization is one such successful biomedical device application for heart and intracranial surgery (Roue 2002; Ikeda et al. 2005; Fann et al. 2004).

Although technical advancements in this field have shown a promising picture of the future, the real-time applications are still far from reality. In recent years, many new ideas have come into existence. Nanorobotics is a multidisciplinary field and needs combined efforts of specialists from many fields of science and technology to work achieving the common target of nurturing this field to maturity. If we look a little back in time, Sir Richard Feynman in his famous lecture "There is plenty of room at the bottom" discussed about the possibility of creating tiny objects mimicking tasks of biological cells in vivo (Feynman 1996). This thought provoking speech of Dr. Feynman turned into reality two decades later, when Eric Drexler suggested that construction of nanodevices is possible from biological parts to inspect and repair the cells of a living human being. Today's molecular machine systems and molecular manufacturing foundation were laid by Drexler's book on *Nanosystems: Molecular Machinery, Manufacturing, and Computation* (Drexler 1992; Freitas Jr 2005a, b; Saxena et al. 2015).

Owing to its high precision, efficacy robustness, affordability, and rapid mode of action, molecular nanotechnology (MNT) or nanorobotics would enhance the limits of existing branches of science. With the advent of MNT, in future, direct in vivo surgery on individual human cells by deploying large numbers of microscopic medical nanorobots would be possible (Freitas Jr 2005a, b).

12.1.2 Nature: The Inspiration Behind Nanorobotics

Nanorobotics is an emerging interdisciplinary field based on nanotechnology and nanofabrication and governs the applications and control of nanodevices. This field is still in its infancy despite all hypes as any robotic device with all artificial components hasn't been generated yet. "Nature" is the prime inspiration of designing nanorobots because natural processes are immensely optimized in terms of energy and material usage that inspires human quest of designing artificial in vitro and in vivo nanodevices to perform tasks equivalent to natural biological processes.

Protein and DNAs are the natural molecular machines of nature to accomplish various synthetic mechanisms and are deployed to execute several cellular tasks, from moving payload to catalyzing reactions and as cellular information carrier. With the scientific advancement, today we have a far better understanding of working principles and mechanisms that is the prime governing factor to use the natural machines (proteins and DNA), or creating artificial ones, using nature's modules. An optimum assembly of bio or biosimilar component would form nanodevices having multiple degrees of freedom and ability to apply force and to interact with objects in the nanoscale world. By definition, "Nanorobots are 'smart' structures with the ability to perform actuation, sensing, signaling, information processing, intelligence, manipulation, and swarm behavior at nanoscale (10^{-9} m)" (Khulbe 2014). These devices would be an amalgamation of nature as well as man-made micromachines to generate motion, force, or signal and biological functions, i.e., payload delivery, cellular repair tasks, etc., in response to the specific physiochemical stimuli inside human or animal bodies (Ummat et al. 2005a, b).

Based on the specific need and requirement, nanodevices with individual or combinatorial properties (swarm intelligence and cooperative behavior) can be fabricated for actuation to perform desired tasks at the nanoscale. Basic characteristics for a nanorobot to function may include:

(i) Swarm intelligence: decentralization and distributive intelligence
(ii) Self-assembly and replication
(iii) Information processing and programmability at nano level
(iv) External interface to control and monitor nanorobotic tasks (Ummat et al. 2005b)

High efficiency and consistency is the main advantage of using nature's machine components as proposed by Kinosita et al. (Kinosita et al. 2000). While as in conventional macromachines, forces and motions are generated to accomplish specific tasks, in similar fashion, bio-nanomachines can also be explored to deploy nano-objects to fabricate other nanomachines to perform tasks at cellular levels. Figure 12.1 represents the nature machines and their artificial counterpart and their functions.

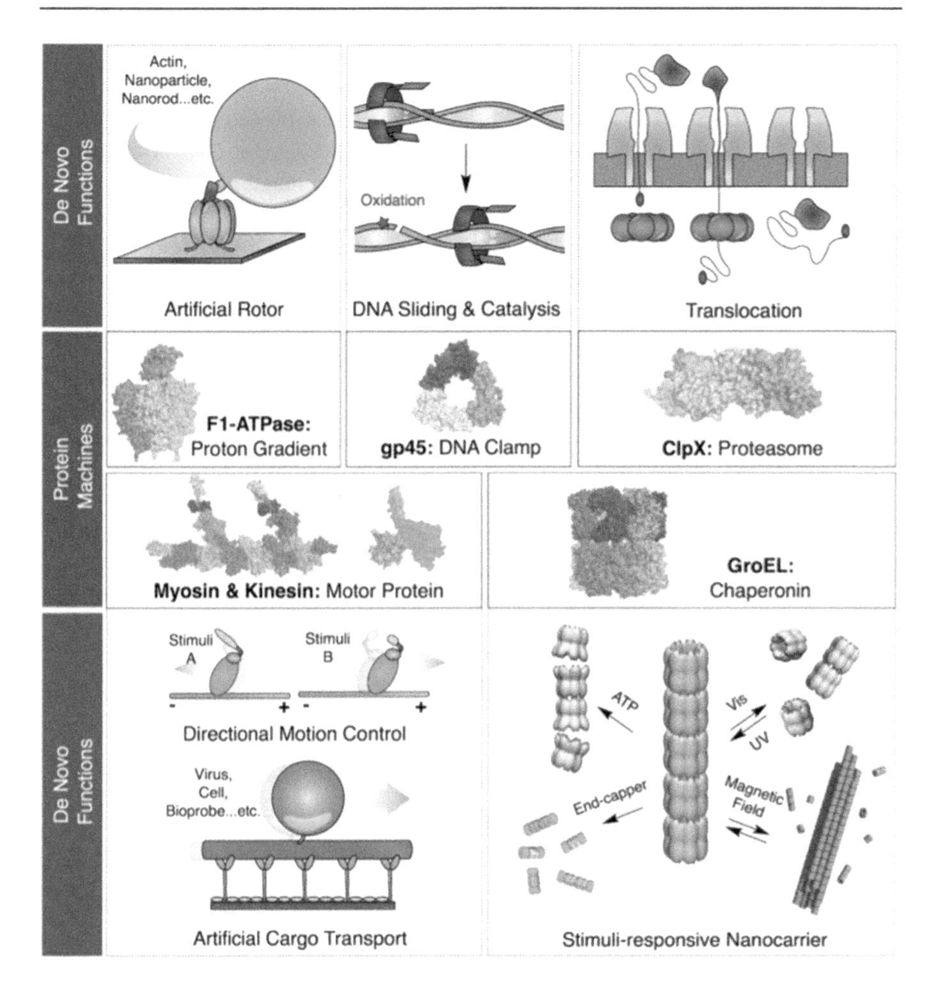

Fig. 12.1 Nature-based nanorobots and their functions

12.1.3 Advantage of Nanorobots over Conventional Medical Techniques

Human civilization has done a remarkable progress in all walks of life, and technological advancements are leading us toward a whole new era of mesmerizing discoveries. We have done a significant progress in medical sciences, and new methods have been developed for providing a better understanding of the vitals as well as to aid diagnosis like CT scan, MRI, endoscopy, etc. But as old order changes yielding place of new, all technology inevitably has to be phased out sometime, and as our current medical advancements overcome the drawbacks of their predecessors, nanorobotics would also surpass many drawbacks of today's medical technology including incisions that relate to painful surgical procedures, with better success rate of sophisticated surgeries with no harm to patients. With the advent of new technological era with robotics, IOT, AI, and machine learning impacting healthcare, conventional techniques of investigation and diagnosis would adapt with the new changes, and we have already seen robotic surgical procedures taking place globally. Basically a surgical robot is a self-powered, computer-controlled device that can be programmed to aid in the positioning and manipulation of surgical instruments, enabling the surgeon to carry out more complex tasks (Gomez 2004). These systems function as *master-slave manipulators* that work as remote extensions completely governed by the surgeon. Till date two master-slave systems have received approval by the US Food and Drug Administration (FDA): (i) *da Vinci* Surgical System (Intuitive Surgical, Mountain View, California) (http://www.intuitivesurgical.com/products/da_vinci.htm.; Ballantyne and Moll 2003) and (ii) the *ZEUS* system (Computer Motion, Goleta, California) (Marescaux and Rubino 2003). Apart from these macro robotic applications, miniaturized nanodevices would defend the body from inside owing to their biocompatibility and nature-inspired architecture. DNA origami-based applications are just an example of how nanorobots would change the healthcare arena. Owing to their tiny architecture and biocompatibility, in vivo applications of nanorobots would provide an upper edge in many aspects of their conventional healthcare practices, i.e., less recovery time, minimal pain, possibility to monitor the treatment regimen, and possibility of rapid self-action in case of emergency. Targeted payload delivery to the specific site for precise therapy and self-excretion and disintegration after completion of task, if required (Bhat 2014).

12.2 Nanorobots: Components and Functions

Nanorobotics is an innovative approach in the development of nature-inspired devices at nanometer scale to perform desired task inside the body and hence known as robots. In its present stage of infancy, nanorobotics is still a theoretical concept. Nanorobots architecture follows the techniques of molecular modeling based on "Energy Minimization" on the hyper surfaces of the bio-modules; "Hybrid Quantum-Mechanical" and "Molecular Mechanical method"; "Empirical Force field" methods; and "Maximum Entropy Production" in least time. Figure 12.2 is showing the circuit diagram of nanorobots. The basic components and design attributes of nanorobots to perform specific task are:

12.2.1 Size and Shape

The basic shape and size that would be less than the size of the blood vessel to traverse into it without damage or blocking the blood flow. A spherical ball like nanorobots can be the simplest structure without self-propulsion mechanisms that could be injected into the bloodstream to navigate their paths through the reservoir. Nano-bearings and nano-gears were the most convenient parts to be fabricated as reported by Drexler et al. He proposed an overlap-repulsion bearing design using

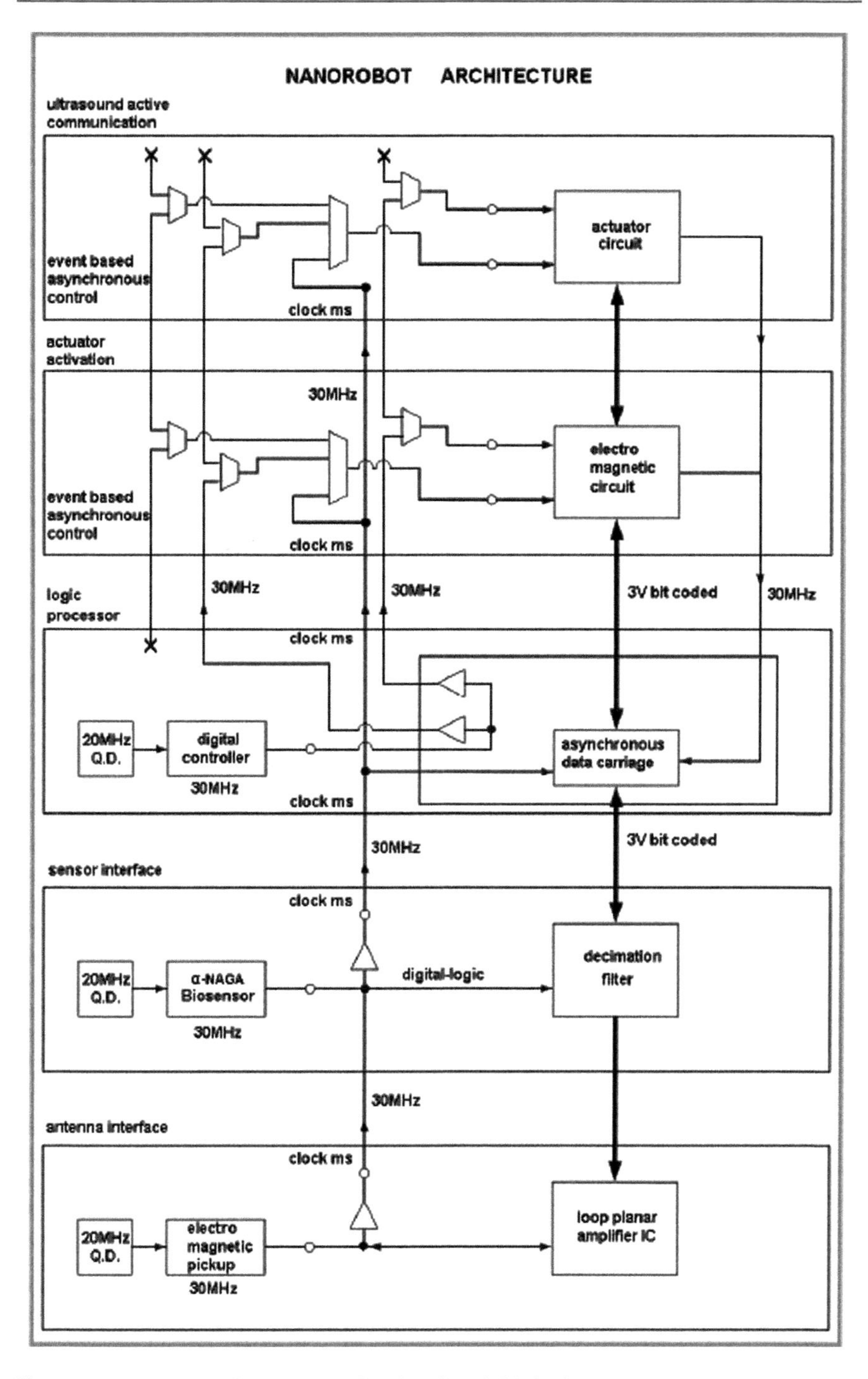

Fig. 12.2 Architecture of nanorobots (Cavalcanti et al. 2008a, b)

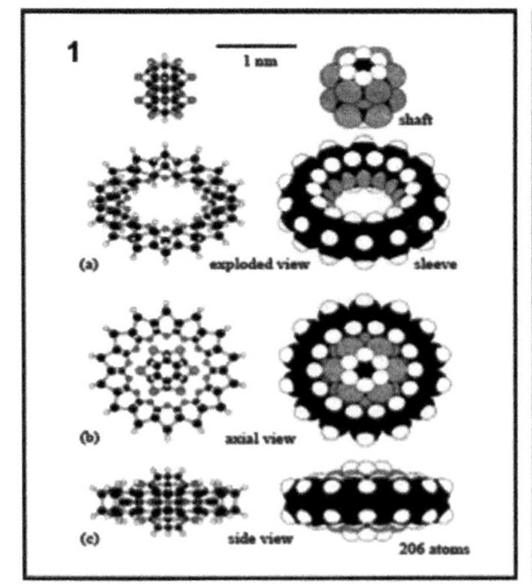

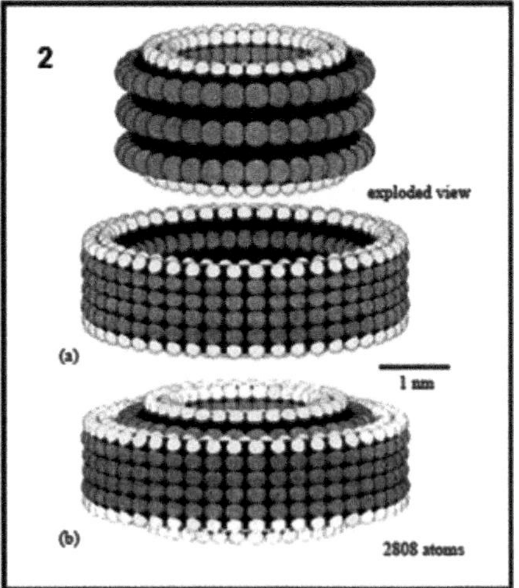

Fig. 12.3 (1) End views and exploded views of overlap-repulsion bearing. (2) Exploded view of strained-shell sleeve bearing. (K. Eric Drexler.© 1992, John Wiley & Sons, Inc)

both ball-and-stick and space-filling representations. His proposed bearing had 206 atoms of carbon, silicon, oxygen, and hydrogen, with a small shaft that rotates within a ring sleeve measuring 2.2 nm in diameter. Following Fig. 12.3 a, b shows design of both ball-and-stick and space-filling representations with end views and exploded views. In this model, atoms of the shaft were arranged in a 6-fold symmetry and the ring had a 14-fold symmetry to provide low energy barriers during shaft rotation (Drexler 1992; Freitas Jr 2005a, b).

12.2.2 Sensors and Actuators

Nanorobots should have the ability to sense temperature, pressure, fluid type and analysis, path, and position while performing the assigned task. In the past decade, scientists have successfully fabricated miniaturized silicon complementary metal oxide semiconductor (CMOS)-based motion sensor, but now to further downscale the process, nanowire-based CMOS devices have been fabricated, and this circuit assembly can achieve maximal efficiency for applications regarding chemical changes (Curtis et al. 2006; Balasubramanian et al. 2005; Zhang et al. 2004), lesser self-heating, and thermal coupling for CMOS functionality (Fung and Li 2004). Risveden et al. developed a region ion-selective bioelectronic nanosensor and observed that low energy consumption and high sensitivity are prime benefits of nano-sensors (Risveden et al. 2007). With the advancement of nanotechnology and nanofabrication, different versions of nano-enabled CMOS sensors are proposed based on carbon nanotubes and other nanomaterials, while researchers have developed new advanced manufacturing techniques, i.e., silicon-on-insulator (SOI) technology to assemble high-performance logic sub-90 nm circuits as described and patented by Park et al. (Park et al. 2005), while Bernstain et al. provided solution to deal with bipolar effect and hysteretic variations based on SOI structures (Bernstein et al. 2003). Although the groundbreaking 10 nm circuits are still far from reality, 45 nm of these nano CMOS ICs represent breakthrough technology devices that are currently being utilized in products (Cavalcanti et al. 2008a, b, Cavalcanti et al. 2007).

12.2.3 Power Generation

Power generation ability is another pillar for a nanorobot device to perform its assigned operations and tasks. At the nanoscale, low power in few pico-watts or micro watts would be needed. Potential means of generating power for the nanobots are:

(a) Power from fluid flow or counter-current motion
(b) Power derived from the reservoir temperature
(c) Power from friction with rock fabrics
(d) Downhole fuel cell generation from in-situation hydrocarbon
(e) Use of downhole recharges station (Freitas 1999; Roundy et al. 2003; Liu et al. 2004)

Few previous reports provide possible applicable solutions to power generation based on CMOS as combination of CMOS with active telemetry and power supply is the most effective and secure way to ensure energy as long as necessary to keep the nanorobot in operation (Cavalcanti et al. 2003). Firstly, Mohseni et al. (2005) described the use of CMOS for active telemetry by developing implantable micro-electrodes arrays for wireless in vivo brain signal recording of spontaneous neural activity at 96.2 MHz from the auditory cortex of a live marmoset monkey at numerous transmission distances ranging from 10 to 50 cm with signal-to-noise ratios in the range of 8.4–9.5 dB. Similarly, Sauer et al. developed a generic chip that can be used to power and interface with an implanted sensor. The chip had dual voltage regulators, used to supply both digital and analog sources. Both voltages were designed to be 3.3 V and nano-circuits with resonant electric properties can operate as a chip providing an electromagnetic energy supplying 1.7 mA, sufficient to operate many tasks with minimal losses during transmission (Sauer et al. 2005). Another possible way for power generation and controlling is radio frequency (RF)-based telemetry procedures as RF-based telemetry procedures have demonstrated good results in patient monitoring and power transmission through inductive coupling reported in many previous reports (Ghovanloo and Najafi 2004; Eggers et al. 2000a, b; Kermani et al. 2006).

RFID-based biomedical applications are already there, i.e., pacemaker, wearable devices and smart fabrics, etc. RF-based power generation would be an extension of this technology for miniaturized nanorobotics devices. In an estimate, ~1 μW of energy can be saved in resting state of nanorobots. Heating is the major drawback of RFID technique, and in vivo environment, only real applications would tell the efficacy of this widely explored technique, although uploading control software in mobile phones for power and data transmission could be an alternative as proposed by Ahuja and Myers (2006).

12.2.4 Propulsion, Control, and Navigation

Nanorobots are designed to perform in vivo operations, and depending on their nanometer dimensions, in the bloodstream their navigation is an important aspect to be considered. As nanorobots are the nature-inspired structures, their control mechanisms are also inspired by nature, and specifically to understand nanorobot motion in vivo, bacterial motions are well-studied for the design and construction of nanorobots. Nature has developed its own innovative mechanisms to overcome the motion-related hurdles in nano-sized environment, i.e., low inertia, high viscous forces, low efficiency, and low convective motion. It's a fact that motion by beating of cilia and flagella at 30 μm/sec with 1% efficiency costs $2x10^{-8}$ erg/s in bacterial movement that shows the energy efficacy of the nature machines (Sharma and Mittal 2008; Purcell 1977). For bio organisms, swimming or flying is preferred over walking or crawling to overcome viscosity in nano-domains, while many cells are able to crawl on a solid substrate to which they stick using adhesive forces. Generally, three prime processes are responsible for microbial motion: (i) formation and protrusion of a thin lamellipod in front of the cell, (ii) adhesion of the lamellipod to the substrate, and (iii) its retraction at the rear, pulling the cell forward (Joanny et al. 2003). As described by Squires and Brady (2005), bacterial motion efficacy depends on the "Peclet number (Pe): ratio of time taken to cover a distance (l) at velocity (v) with a diffusion constant (D) and flagella," and cilia could work only if Pe is greater than unity. While for nanorobots owing to their tiny sizes, it's a difficult task to obtain same efficiency with same parameters and nanorobots are supposed to work in different environments, other propulsion mechanisms would also come into picture and "diffusion" may be a more effective choice in comparison to convective motion. Moreover in nano-domains, the nonrigid nature of nanorobots is also to be taken into account. The nonrigidity of nanomechanisms can be described by "bead-spring models" (Higdon 1979), "slenderness theory," "Kirchhoff's rod theory," and a combination of "resistive force theory" (Kim and Powers 2004; Powers 2002). For nanobots of size less than 600 nm, diffusion (random walk) due to thermal agitation (Brownian motion) is prime mode of propulsion to move small objects in fluid at room temperature as reported first by Purcell in 1977. Later Feringa et al. provided a detailed account of generating controlled movement of objects. They developed a model based on two enantiomers of a bistable molecule that would function as the two distinct states in a molecular information storage system and designed a chiroptical molecular switches and light-driven unidirectional rotary motors. In another report, his group constructed molecular machinery or the incorporation of the light-driven motor into multifunctional systems (Feringa 2001; Feringa et al. 2002). Regarding control of propulsion, two possible mechanisms have been proposed.

12.2.4.1 External Control Mechanisms

As clear from title itself, application of external potential fields control the dynamics of the nanorobot in its work environment. MRI is the preferred choice of researchers as an external control mechanism for guiding the nano particles. Martel et al. have developed a robotic platform that uses magnetic resonance imaging (MRI) for sending information to a controller responsible for the real-time control and navigation of untethered magnetic carriers, nanorobots, and/or magnetotactic bacteria (MTB) having a wireless robotic arm, sensors, and therapeutic agents, toward preset paths in the blood vessels to perform specific remote tasks (Martel et al. 2009).

In another approach, Farahani et al. designed an "adaptive controller" to optimize the motion of nanorobots within the blood vessel. The simulation showed that the proposed control method, by identifying the functional characterization of nanorobots such as transport capability, biocompatibility, planning, receiving, and generating signals, had good performance and high efficiency in controlling the nanorobot motion toward the predetermined reference trajectory path by the signal received from the damaged area, with an error of about 0.01 μm (Farahani and Farahani 2016; Rao et al. 2014).

12.2.4.2 Internal Control Mechanisms

The internal control mechanism can be both passive and active. In passive control, nanorobots could be associated with biomolecules for biochemical sensing and selective binding of various biomolecules. These mechanisms are well explored in designing nano delivery systems. However, a static behavior without any change during the assigned task is the prime limitation of passive control. In this regard, active control mechanisms are better alternatives to control the nanorobots effectively in dynamic environment. Internal control modules can be the best choice in this regard, and fabrication of *inbuilt* molecular computers can solve the problem to program the nanorobots to act smartly based on the situations in vivo by varying their conduct. Molecular computers could be utilized to achieve this target. Leonard Adleman (from the University of Southern California) has introduced DNA computers a decade ago to solve a mathematical problem by utilizing DNA molecules. These DNA structures acted as switches, by changing their position and can be used to perform the logical operations that make computer calculations possible. Researchers at Harvard and Bar-Ilan University in Israel have built different nanoscale robots that can interact with each other, using their DNA switches to react and produce different signals. A bimolecular computer for cancer detection and treatment has recently been developed by Benenson et al. (2001). This device consisted of an input and output module to act together and can be used to disease diagnosis and drug release in response to cure that disease. They explored novel concept of software (made up of DNAs) and hardware (made up of enzymes) molecular elements. It's a kind of generalized device with vast applicability of identifying any disease having a particular pattern of gene expression associated with it (Rao et al. 2014).

12.2.5 Data Storage and Transmission

Data transmission from implantable devices can be very useful for the monitoring of patients. RF is a very promising solution for implanted devices via inductive coupling. In addition to gathering power, RF can also be used to create a two-way link owing to its ability to send data back again to base station (Irazoqui-Pastor et al. 2003; Akin et al. 1998). Sauer et al. have reported a power harvesting chip to provide power, control signals, and a data link for an accompanied sensor to function and can be coupled with other devices. It is a well-known fact that RF energy between 1 and 10 MHz penetrates the body with minimum energy loss (Finkenzeller 2003). So they fabricate their inductive link to operate at a frequency of 4 MHz, where the chip supplies up to 2 mA at 3.3 volts. The chip was designed to let data transmission back to the module that broadcasts the power to eliminate need for a physical connection.

"Acoustic communication" is another alternative to communicate between longer distances. It can detect low energy consumption as compared to light communication approaches. Freitas et al. evaluated the feasibility of in vivo ultrasonic communication for micron-size robots broadcasting into various types of tissues. Frequencies between 10 MHz and 300 MHz gave the best transaction between efficient acoustic generation and attenuation for communication over 100 microns distance. They observed that power available from ambient oxygen and glucose in the bloodstream supported communication rates of about 10,000 bits/second between micron-sized robots. The acoustic pressure fields needed for this communication would not damage nearby tissue, and short bursts at significantly higher power could be explored for therapeutic usage (Hogg and Freitas 2012).

Although optical communication has also been explored with faster rates of data transmission, high-energy demands of this mode make it not the preferred choice for medical nanorobotics (Vasilescu et al. 2005). Chemical signaling is another approach of transmission, reported by Cavalcanti et al. (2005) in nanorobots for some teamwork coordination. To overcome the limitations, CMOS with submicron system on chip (SoC) design uses extremely low-power consumption for nanorobots communicating collectively at longer distances through acoustic sensors. Integrated sensors for data transfer can also be a good alternate to communicate with implanted devices, where a bunch of nanorobots may be equipped with single-chip RFID CMOS-based sensors as reported by Panis et al. (2004).

12.3 Types of Nanorobots

Nanorobots would be the future of medicine and all fields including surgery would be benefitted. As a broad classification, two types of nanorobots are organic and inorganic. The organic nanorobots are bio-nanorobots fabricated using biological entities like viruses and bacterial DNA cells and are more biocompatible in general, while inorganic nanobots are of synthesized proteins and others types of artificial material. These are more toxic and not suitable for direct use without encapsulation. In another way, nanorobots can be classified based on their assembly: "positional assembly" and "self-assembly." In self-assembly, the arm of a miniature robot or a microscopic set is used to pick the molecules and assemble manually. In positional assembly, the billions of molecules are put together and automatically get assembled based on their natural affinities into the desired configuration (Kharwade et al. 2013; Venkatesan and Jolad 2010; Merina 2010).

Apart from these categorizations, there are four types of nanorobots that were conceptualized by *Robert A. Freitas Jr* as artificial blood:

(i) Respirocytes
(ii) Microbivores
(iii) Clottocytes
(iv) Chromallocyte

12.3.1 Respirocyte

Numerous conceptual designs of medical nanorobots have been reported, but in 1998, first theoretical design, describing a hypothetical artificial mechanical red blood cell or "respirocyte" made of 18 billion precisely arranged structural atoms, was given by Freitas (1999). These were proposed replica of red blood cells or erythrocytes which is a blood-borne spherical 1 μm diamondoid 1000-atmosphere pressure vessel with reversible molecule-selective pumps. Respirocyte would mimic the oxygen and carbon dioxide transport functions of erythrocytes and gets their power to function by endogenous serum glucose (Manjunath and Kishore 2014; Freitas 2009).

Each respirocyte had three types of rotors with different functions: one rotor to release the stored oxygen while traveling through the body, second rotor seizes all the carbon dioxide in the bloodstream which are released at the lungs, while the third one utilizes the glucose from the bloodstream as fuel source (Freitas 2009).

This artificial cell was more efficient than the RBCs (red blood cells) in its ability to supply 236 times more oxygen to the tissues per unit volume than RBCs (red blood cells) and would be useful for patients suffering from anemia, in rapid treatment for asphyxia (e.g., monoxide poisoning), as a backup for tissue oxygenation for heart and surgical patients, for site-specific deoxygenation of tumors, and as support for other nanorobots by releasing oxygen as and when needed during in vivo operations (Freitas 2009).

12.3.2 Microbivore Nanorobots

Another conceptual design for the nanorobotic artificial phagocytes described by Freitas et al. was "microbivores" as a surveillance nanorobot to identify and eliminate unwanted pathogens including bacteria, viruses, or fungi by digesting them using a combination of onboard mechanical and artificial enzymatic systems. Microbivore could be considered as an artificial phagocyte having primary function to obliterate pathogens found in human blood and considered as "guardian of bloodstream." Nanorobots recognize a target microbes by contacting its surface antigen markers same as nano delivery systems and uses "digest and discharge" mechanism for removal of pathogens and could be a preferred treatment for sepsis. Basic architecture of microbivore consists of:

(i) An array of reversible binding sites
(ii) An array of telescoping grapples
(iii) A morcellation chamber
(iv) Digestion chamber (Eshaghian-Wilner 2009)

It is an oblate spheroidal device for nanomedical applications with 3.4 μm diameter along its major axis and 2.0 μm diameter along its minor axis, precisely organized by 610 billion atoms in a 12.1 μm^3 geometric volume. These nanobots consume up to 200 pW power to digest trapped microbes. A detailed description of these nanorobots is given elsewhere (Manjunath and Kishore 2014; Freitas Jr 2005a, b, 2009).

12.3.3 Clottocyte Nanorobots

Clottocytes were designed with a unique biological capability of "instant hemostasis" in approximately 1 second (Freitas Jr 2005a, b; Manjunath and Kishore 2014). Clottocyte nanorobots mimic platelet in our blood. When there is a wound, platelet forms a clot to stop the blood flow; in similar way, clottocyte nanorobots form a fiber-like structure around the wound to stop the blood flow. These are spherical nanorobots powered by serum-oxyglucose, approximately 2 μm in diameter containing a fiber mesh specific to the blood group to release it at the site of action to create a clot. The response time of clottocyte is 100–1000 times faster than the natural hemostatic system (Eshaghian-Wilner 2009).

The basic requirement for optimal performance of these nanorobots is consistent communication protocols to control the coordinated mesh release from neighboring clottocytes and to regulate multidevice-activation radius within the local clottocyte population.

12.3.4 Chromallocyte (Mobile Cell Repair Nanorobots)

Chromallocyte nanorobots are hypothetical mobile cell repair nanorobots. These nanorobots were designed to replace entire chromosome. They were the most advanced nanorobots, proposed by Freitas et al. (2005) to perform chromosome replacement therapy (CRT). In CRT, the entire chromatin content of the nucleus in a living cell is extracted and promptly replaced with a new set of prefabricated chromosomes which have been artificially manufactured as defect-free copies of the originals.

A single lozenge-shaped 69 micron^3 chromallocyte has dimensions of 4.18 μm and 3.28 μm along cross-sectional diameters and 5.05 μm length and consumes 50–200 pW in normal operation, and during outmessaging, a maximum of 1000 pW would be needed (Manjunath and Kishore 2014; Freitas Jr 2005a, b).

12.4 Medical Application of Nanorobots

Although nanorobotics is still in its infancy, the advantages associated with proposed designs and ability to perform complex tasks with high precision make them potential candidates and better alternatives for conventional therapies. Some advantages over conventional methods are listed below:

1. Targeted therapy would be possible with no damage to adjacent tissues.
2. Less posttreatment care required.
3. Considerably less recovery time.
4. Continuous monitoring and diagnosis from the inside would be possible.
5. Rapid response to a sudden change.

It is obvious that nanorobots have more applicability in healthcare sector based on their design and size attributes. The development of nanorobots provides remarkable advances in diagnosis and treatment of various complex diseases. Nano-based drug delivery applications have already been reported by many researchers worldwide. These robots deliver specific drug to a target site inside the human body. Although fully mechanized self-sufficient nanorobots are distant dream at present, with the rapid advancements in related fields of electronics, biotechnology, and microfabrication, we would witness their application in treatment of complex medical problems soon.

In this segment, we would focus on in vivo usage of nanorobots in treatment and diagnosis and their potentially entailing benefits in the form of new therapy modules that were otherwise impossible to attain. In few reviews, detailed accounts of nanorobotics in vivo applications are mentioned (Soto and Chrostowski 2018).

12.4.1 Nanorobots in Cancer Treatment

Cancer is defined as a diseased condition where there is an uncontrolled growth of abnormal cells taking place and which may spread to the whole body within short time span. Cancer can be cured with current medical technologies, but many times late diagnosis and adverse side effects of chemotherapy are the limiting factor for an efficient treatment regimen, targeted drug delivery only to cancer cells is still not fully adopted by clinicians, and with conventional therapy, healthy tissue damage can't be prevented.

Nanorobots designed for targeted therapy can work more precisely in releasing the payload at tumor site only after identifying the cancer tissue with no peripheral impact on healthy ones. First described by Freitas Jr., "pharmacytes" are a class of nanorobot enabled to deliver cytocidal agents to tumor site on a cell-by-cell basis. Moreover, unmetabolized cytocidal molecules would be engulfed by pharmacytes after initiating treated cells to be transported out of the patient's body that would mitigate posttreatment collateral damage. There are two mechanisms to deliver the payload at target site; direct or by progressive cytopenetration based on the specific molecules to attach with cancer cell surface receptors to promote cell death by activating death receptors, i.e., CD95L ligand that binds to the extracellular domains of three CD95 death receptors), TNF or lymphotoxin alpha (binds to CD120a), Apo3L ligand aka TWEAK (binds to DR3), or Apo2L ligand aka TRAIL (binds to DR4 and DR5) (Freitas Jr 2005a, b).

In another approach, Douglas et al. devised an autonomous DNA nanorobot to transport molecular payloads to targeted cells after picking signals from cell surface for activation and structure reconfiguring for payload delivery. These devices have been tested for lymphoma and leukemia, consisted of series of DNA strands linked in 2D chains and folded into 3D structures, that could selectively open and close. Their nanobots carried two molecular cargos: a gold nanoshell and an antibody fragment with a DNA hinge at one end and a DNA latch on the other (Douglas et al. 2012). DNA origami-based nanorobots for drug delivery systems were put forth by Harvard University researchers, and till now many applications have been reported based on their biocompatibility and drug delivery efficacy. DNA origami is a technique "to create complex three-dimensional shapes from a single-stranded DNA molecule via self-assembly through Watson–Crick base pairing." With the advancement of computational design tools, we can easily predict the durability of conformational changes in long DNA scaffolds in the presence of short DNA sequences that act as "staples" and "fasteners" to stabilize the final conformation. Till date many applications have been reported, and a detailed account can be obtained from many exciting reviews (Udomprasert and Kangsamaksin 2017; Linko et al. 2015; Baig et al. 2018; Zhang et al. 2014)

Few researchers have investigated the performance of DNA origami nanostructures inside the living insects, i.e., *Blaberus discoidalis*, by hemocoel injection and concluded that these DNA nanorobots can properly function inside living systems (Amir et al. 2014; Arnon et al. 2016). As reported by Perrault and Shih that after i.v. injection into mice, a lipid bilayer-encapsulated nanostructure remained in blood

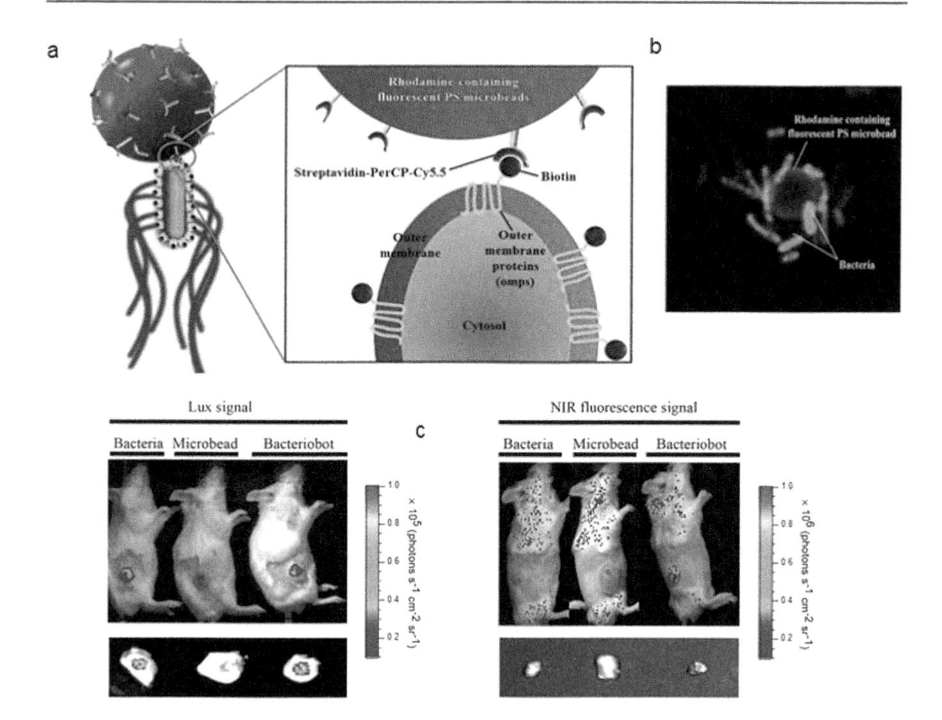

Fig. 12.4 Schematic representation of bacteriobots (**a**, **b**) structure of *S. typhimurium*-attached PS microbeads under confocal laser scanning microscope, (**c**) in vivo application of bacteriobots in mice (Park et al. 2013)

circulation significantly longer than its free form (Perrault and Shih 2014). Li et al. (2018) also reported a DNA "nanorobot" functionalized with target molecules for precise delivery of an active drug only at tumor site. In another approach, Rudchenko et al. (2013) described a molecular automata based on strand displacement cascades directed by antibodies to analyze and sorting of cells based on their surface receptors as inputs while a unique T molecular tag on the cell surface of a specific subpopulation of lymphocytes within human blood cells prepare final output of a molecular automation. This technique was advantageous in identifying the cells with no unique distinguishing feature.

Park et al. came up with a breakthrough idea of "bacteriobot," where nontoxic bacteria, i.e., *Salmonella typhimurium*, after genetic modification were explored to attract chemicals released by cancer cells and delivery of anticancer drugs at target site. The bacteria were engineered to have receptors for higher migration velocity toward tumor cell lysates than normal cells using flagellar motion as shown in Fig. 12.4 a, b.

In this series, few other reports are also there where Akin et al. described biohybrid nanobots for payload delivery based on *'Listeria monocytogenes'* to deliver nanoparticles containing genes and proteins, within a mouse, and gene expression was monitored through differences in the luminescence produced within the

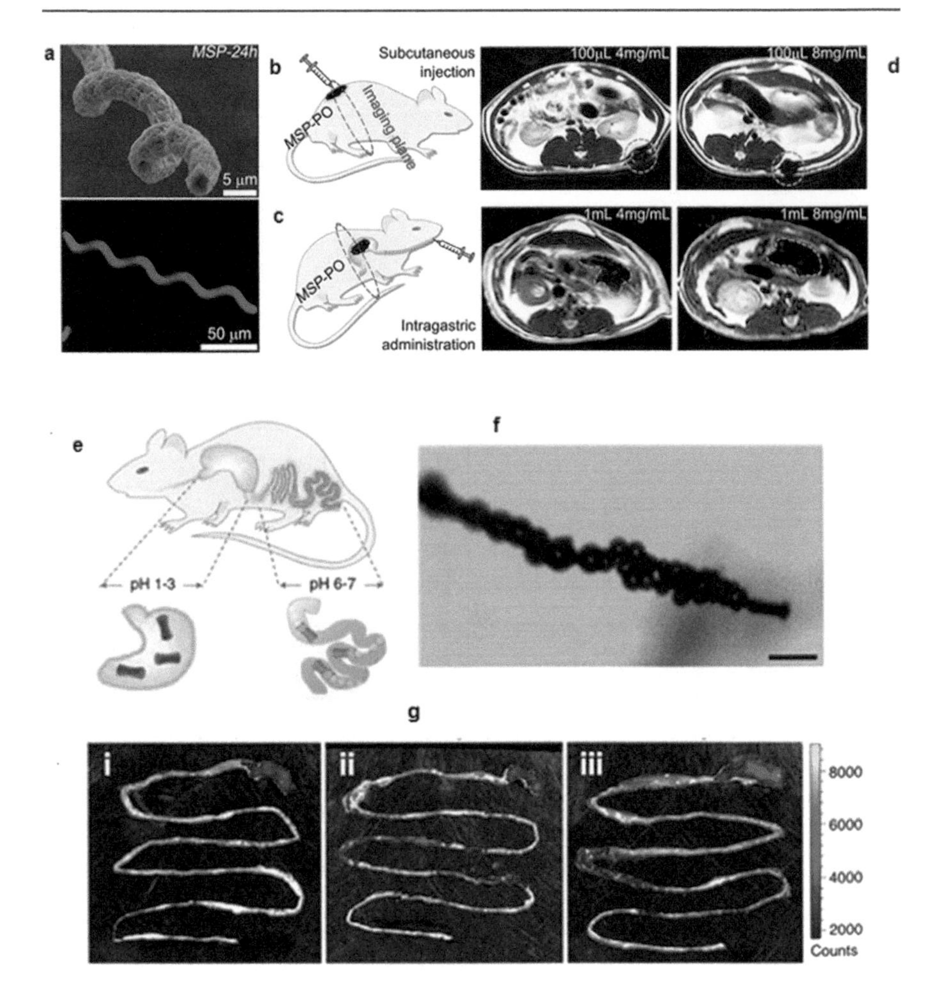

Fig. 12.5 In vivo imaging of magnetically propelled microrobot. (**a**) Scanning electron microscopy (SEM) and fluorescence images of the helical, (**b**) diagrammatic representation of microbots in mouse subcutaneous tissues and (**c**) inside the mouse stomach. (Yan et al. 2017) (**e**) schematic of microrobots for enhanced payloads retention in the gastrointestinal tract. (**f**) Micrograph illustrating the bubble generation at the end of the microrobot responsible for locomotion. (**g**) Ex vivo fluorescent images of the gastrointestinal tract retention of the dye rhodamine (i, control; ii: after 6 h, and iii; after 12 h of administration) (Li et al. 2016)

different mouse organs based on the payload delivery (Akin et al. 2007), where as a step further, Taherkhani et al. (2014) in vitro coupled "magnetotactic" bacteria which produce magnetic iron oxide nanoparticles naturally with liposomes loaded with therapeutic agents. Later these bacteria were guided with an external magnetic field to deliver the drug-loaded liposomes in vivo to a mouse tumor site (Felfoul et al. 2016).

Recently, Hoop et al. explored magnetically guided nanorobots for the delivery of fluorouracil to check its efficacy in tumor reduction in a mice model. This guided method of drug delivery allowed the nanorobotic platform to dispense more amount of the therapeutic agent in a localized area of the tumor as compare to conventional medicine (Hoop et al. 2018). These are few recent but promising applications of in vivo cancer therapy by nanorobots with a bright picture of more efficient future cancer therapies. In very few cases, real-time bioimaging was also achieved via nanorobots. A dual imaging approach was explored by Yan et al. to detect biodegradable magnetic microhelix nanorobots in mice. Fluorescence imaging provided the whereabouts of nanorobots inside the subcutaneous tissue and intraperitoneal cavity of a mouse, while magnetic resonance-based imaging was used to detect the nanorobot's position inside the mouse's stomach, as shown in Fig. 12.5a–d. These approaches can be useful for obtaining good quality images for cancer sites as well as monitoring treatment regimen.

12.4.2 Nanorobots in Diagnosis and Treatment of Diabetes

Diabetes is basically body's inability to process blood glucose, and thus glucose levels in the bloodstream are the prime factors for the diagnosis and treatment of diabetes. Cavalcanti et al. (2008a, b, c) designed novel nanobot for efficient diabetes control based on an alarm system. These nanorobots flew with the RBCs through the bloodstream detecting the glucose levels and whenever glucose attains critical levels it triggers an alarm in nanorobot through the cell phone to alert the patient. At a typical glucose concentration, the nanorobots keep the glucose levels ranging around 130 mg/dl as a target for the blood glucose levels (BGLs). Data storage to cell phone was also possible with these tiny entities by RF signals to keep the records of glucose levels for further clinical investigations of patients.

Recently a nanotech-based approach is described by MIT researchers where nanoparticles not only sensed glucose levels in the body but also secreted the appropriate amount of insulin to get it under control. Gu et al. prepared a nanonetwork loaded with insulin and glucose-metabolizing enzyme. This 3D scaffold dissociated when needed to release insulin in hyperglycemic condition, and later polymeric matrix was degraded. A single nanoparticle injection could maintain the glucose in the blood (200 mg/dL) up to 10 days (Gu et al. 2013). For glucose monitoring the nanorobots use an embedded chemosensor that involves the modulation of human SGLT3 protein glucosensor activity (Wright et al. 2005).

12.4.3 Nanorobots in Retention of Payloads and Wound Healing

Apart from delivery of drug molecules, retention of the drug or cells is also a promising feature of nanobots for sustained drug release in special cases of pain management, bacterial infection treatment, etc. Wang's group is pioneer in proposing biodegradable zinc and magnesium powered microrobots that utilized gastric and intestinal fluids as fuels to withhold payload in the stomach and intestinal tissues

(Gao et al. 2015; Esteban-Fernández de Ávila et al. 2017a, Soto and Chrostowski 2018) and explored these nanorobots for pH neutralization of the gastric fluid (Li et al. 2017) and treatment of bacterial infection (Helicobacter pylori) in the stomach (Esteban-Fernández de Ávila et al. 2017b).

To achieve retention of the microrobot, different mechanisms were applied like direct piercing of the surrounding tissue, or by an improvement in mass transport and nucleation by generation of gas bubbles. Li et al. designed magnesium-based microrobots with built-in "delay activation." They used polymeric coatings to activate the microrobot motion based on their thickness or environmental pH conditions and got dissolved at neutral pH of the intestinal fluids to activate the microrobot (Li et al. 2016) as shown in Fig. 12.5e–g. In a recent approach described by Karshalev et al. nanorobots were incorporated in a pill matrix to streamline their administration with existing pharmaceutical protocols (Karshalev et al. 2018; Soto and Chrostowski 2018).

In wound healing also nanorobots have applications as reported by Baylis et al. where thrombin was delivered at target site to pause the bleeding of wounds in the vasculature of mouse and pig models by chemically propelled calcium carbonate-based microrobots. The distribution of nanorobots at wound region can be explained by "lateral propulsion," "buoyant rise," and "convection" (Baylis et al. 2015). Another laser-based wound sealing approach with locomotive microrobot was explored by He et al. Localized collagen denaturation and melting were initiated by laser generated high temperature followed by temperature decrease that allowed condensation and wound closure (He et al. 2016; Soto and Chrostowski 2018).

12.4.4 Surgery, Biopsy, and Thrombolysis by Nanorobots

Other potential implementation of nanorobots would be in biopsy/surgery. Although few in vitro platforms have been proposed for nano/micro high precision surgery, none have not been converted to in vivo models yet (Kwan et al. 2015; Soto et al. 2015). Gultepe et al. have used their microrobots having star-shaped grippers for easy access to narrow channels in the body and to expunge tissue samples from a pig bile duct (Gultepe et al. 2013). For in vivo surgical application, researchers have developed a module based on magnetic microrobots for controlled navigation inside the eye of a living rabbit (Ullrich et al. 2013; Pokki et al. 2017).

Apart from clot formation, nanorobots have the applicability to remove the clots in narrow arteries reported by Cheng et al. Magnetically actuated nanorobots loaded with tissue plasminogen activator were intravenously injected and sent to the blood clot location. With externally controlled rotations of nanorobots, an interaction of the tissue plasminogen activator molecule with the blood clot interface resulted in thrombolysis (Cheng et al. 2014). More recently, Hu et al. developed a new strategy by incorporating plasminogen activator into porous magnetic iron oxide (Fe_3O_4)-microrods for targeted thrombolytic therapy in ischemic stroke for mechanical destruction of clot induced by distal middle cerebral artery occlusion (Hu et al. 2018). This approach was a major breakthrough not only for the treatment of ischemic stroke but also a vast implication on other fatal thrombotic diseases such as myocardial infarction and pulmonary embolism. Soto and Chrostowski (2018) have covered most of the recent nanorobotics applications in detail.

12.5 Future and Commercialization Aspects of Nanorobots

The field of nanorobotics is gradually achieving maturity with the advancement of supporting fields of microfabrication, electronics, nanotechnology, and medicine. Despite promising medical applications and new therapy modules with many exciting new findings based on nanorobots, few key issues have to be addressed before their full-fledged real-world applications. In current scenario, major challenges with nanorobots in vivo applications can be summarized as:

(a) Lack of sensing and actuation
(b) Integration and control mechanism inefficiency
(c) Lack of understanding of chemistry and biological principles at nanoscale
(d) Clinical risk associated with in vivo application of nanorobots
(e) Toxicity on long-term retention and proper clearance mechanisms
(f) Hurdles in commercialization

Nanorobotics is a highly interdisciplinary platform and its inputs and collaborations from many fields are desired to take this from infancy to maturity. For toxicity and retention issues can be handled by manufacturing of the micro/nanostructure with biocompatible materials but yet a long way to get the necessary approvals for their medicinal usage and applications in vivo. Apart from these pros and cons of nanorobot, this field has a bright future with great impact on lives of millions of people that makes it mandatory to thoroughly consider the commercialization prospects of nanorobotics (Soto and Chrostowski 2018).

Lab to market strategy for this new stream needs enormous efforts to prove its clinical efficacy, validations, and regulatory approvals and to design a strong commercialization plan. Pharmaceutical companies and insurance providers are the target customers for nanorobots with matching areas of interests of reduced costs with more productivity and considering the immense potential of nanorobots this is very much possible. Till date, very few active companies are working toward the commercialization of nanorobots for use in medical applications. Major market player in nanorobotics are Ginkgo Bioworks (USA), EV Group (Austria), Bruker (USA), Imina Technologies (Switzerland), Oxford Instruments (UK), JEOL (Japan), Xidex (USA), Klocke Nanotechnik (Germany), Toronto Nano Instrumentation (Canada), Park Systems (South Korea), and few more.

There are few technical hurdles to be overcome before commercial production and large-scale application of nanorobots. First is the mass fabrication of nonentities with desired efficacy. To overcome this researchers have suggested special nano modules for inspecting, testing, and transporting of nanorobotic designs (Wang and Zhang 2017; Lu et al. 2017; Zhang et al. 2018). Production costs and market penetration can be achieved once sorting out technical difficulties. To keep intellectual property rights of the new techniques developed by key players in the market is one way to keep the initial hefty investment cost under control that would in turn open a gateway to launch a strong business model with recognizable revenue generation streams. In recent years, many patents have been filed ranging from fabrication methods commonly used for parallel mass fabrication, power generation, and tracking of nanorobots to their in vivo intercellular applications (Natan and Mallouk 2007; Odell et al. 2017; Martel and Felfoul 2018).

There may be many possible approaches to explore this technology for wide clinical applications by integrating with conventional medical procedures. Despite everything, there is no doubt about the benefits of nanorobots in medicine, and once established in near future, this technology would be a boon for human civilization.

Acknowledgment Authors are thankful to DBT-BIRAC for the grant and NCL Innovation Park, Pune, for the support to carry out the research work.

References

Adleman LM (1995) On constructing a molecular computer. In: Lipton RJ, Baum E (eds) DNA based computers, DIMACS 27. American Mathematical Society, Providence, RI, pp 1–21
Ahuja SP, Myers JR (2006) A survey on wireless grid computing. J Supercomput 37:3–21
Akin T, Najafi K, Bradley RM (1998) A wireless implantable multichannel digital neural recording system for a micromachined sieve electrode. IEEE J Solid State Circuits 33:109–118
Akin D, Sturgis D, Ragheb K, Sherman D, Burkholder K, Robinson JP et al (2007) Bacteria-mediated delivery of nanoparticles and cargo into cells. Nat Nanotechnol 2:441–449
Amir Y, Ben-Ishay E, Levner D, Ittah S, Abu-Horowitz A, BacheleT I (2014) Universal computing by DNA origami robots in a living animal. Nat Nanotechnol 9:353–357
Arnon S, Dahan N, Koren A, Radiano O, Ronen M, Yannay T et al (2016) Thought-controlled nanoscale robots in a living host. PLoS One 11:e0161227
Baig MS, Babar U, Arshad U, Ullah MA (2018) DNA origami and bionanotechnology: an efficacious tool for modern therapeutics and drug delivery. Int J Dev Res 8:824660–824669
Balasubramanian A, Bhuva B, Mernaugh R, Haselton FR (2005) Si-based sensor for virus detection. IEEE Sensors J 5:340–344
Ballantyne GH, Moll F (2003) The da Vinci telerobotic surgical system: the virtual operative field and telepresence surgery. Surg Clin 83:1293–1304
Baylis JR, Yeon JH, Thomson MH, Kazerooni A, Wang X, John AES, Piret JM (2015) Self-propelled particles that transport cargo through flowing blood and halt hemorrhage. Sci Adv 1:e1500379
Benenson Y, Paz-Elizur T, Adar R, Keinan E, Livneh Z, Shapiro E (2001) Programmable and autonomous computing machine made of biomolecules. Nature 414:430–434.
Bernstein K, Chuang CT, Joshi R, Puri R (2003) Design and CAD challenges in sub-90nm CMOS technologies. International conference on computer aided design, 129–136
Bhat AS (2014) Nanobots: the future of medicine. Int J Manag Eng Sci 5:44–49
Cavalcanti A (2003) Assembly automation with evolutionary nanorobots and sensor-based control applied to nanomedicine. IEEE Trans Nanotechnol 2:82–87
Cavalcanti A, Rosen L, Kretly LC, Rosenfeld M, Einav S (2004) Nanorobotic challenges in biomedical applications, design and control. International conference on electronics, circuits and systems, 447–450. IEEE

Cavalcanti A, Hogg T, Kretly LC (2005) Transducers development for nanorobotic applications in biomedical engineering. IEEE NDSI Conference on nanoscale devices and system integration, Houston TX, USA
Cavalcanti A, Shirinzadeh B, Freitas RA Jr, Hogg T (2007) Nanorobot architecture for medical target identification. Nanotechnology 19:015103. (15pp)
Cavalcanti A, Shirinzadeh B, Kretly LC (2008a) Medical nanorobotics for diabetes control. Nanomedicine 4:127–138
Cavalcanti A, Shirinzadeh B, Zhang M, Kretly L (2008b) Nanorobot hardware architecture for medical defense. Sensors 8:2932–2958
Cheng R, Huang W, Huang L, Yang B, Mao L, Jin K et al (2014) Acceleration of tissue plasminogen activator-mediated thrombolysis by magnetically powered nanomotors. ACS Nano 8:7746–7754
Curtis AS, Dalby M, Gadegaard N (2006) Cell signaling arising from nanotopography: implications for nanomedical devices. Nanomedicine (Lond) 1:67–72
Drexler KE (1992) Nanosystems: molecular machinery, manufacturing, and computation. Wiley, New York, pp 21–26
Douglas SM, Bachelet I, Church GM (2012) A logic-gated nanorobot for targeted transport of molecular payloads. Science 335:831–834
Eggers T, Marscher C, Marschner U, Clasbrummel B, Laur R, Binder J (2000a) Advanced hybrid integrated low-power telemetric pressure monitoring system for biomedical application. In International conference on micro electro mechanical systems, 23–37
Eggers T, Marschner C, Marschner U, Clasbrummel B, Laur R, Binder J (2000b) Advanced hybrid integrated low-power telemetric pressure monitoring system for biomedical applications. In Proceedings IEEE thirteenth annual international conference on micro electro mechanical system, 329–334. IEEE
Eshaghian-Wilner MM (ed) (2009) Bio-inspired and nanoscale integrated computing, vol 1. Wiley, New York
Esteban-Fernández de Ávila B, Angsantikul P, Li J, Lopez-Ramirez MA, Ramírez-Herrera DE, Thamphiwatana S et al (2017a) Micromotor-enabled active drug delivery for in vivo treatment of stomach infection. Nat Commun 8:272
Esteban-Fernández de Ávila B, Angsantikul P, Li J, Gao W, Zhang L, Wang J (2017b) Micromotors go *in vivo*: from test tubes to live animals. Adv Funct Mater 28:1705640
Fann JI, St. Goar FG, Komtebedde J, Oz MC, Block PC, Foster E et al (2004) Beating heart catheter-based edge-to-edge mitral valve procedure in a porcine model: efficacy and healing response. Circulation 110:988–993
Farahani A, Farahani A (2016) An adaptive controller for motion control of nanorobots inside human blood vessels. Biosci Biotechnol Res Commun 9:546–552
Felfoul O, Mohammadi M, Taherkhani S, De Lanauze D, Xu YZ, Loghin D et al (2016) Magneto-aerotactic bacteria deliver drug-containing nanoliposomes to tumour hypoxic regions. Nat Nanotechnol 11:941–947
Feringa BL (2001) In control of motion: from molecular switches to molecular motors. Acc Chem Res 34:504–513
Feringa BL, Koumura N, Van Delden RA, MKJ TW (2002) Light-driven molecular switches and motors. Appl Phys A 75:301–308
Feynman RP (1996) There is plenty of room at the bottom. Eng Sci 23:22–26
Finkenzeller K, RFID handbook, Wiley, Oxford, 2003
Freitas RA Jr (1999) Nanomedicine vol I Basic Capabilities Landes Bioscience. https://www.nanomedicine.com
Freitas RA Jr (2005a) Current status of nanomedicine and medical nanorobotics. J Comput Theor Nanosci 2:1–25
Freitas RA Jr (2005b) Microbivores: artificial mechanical phagocytes using digest and discharge protocol. J Evol Technol 14:1–52

Freitas RA Jr (2009) Medical nanorobotics: the long term goal for nanomedicine. In: Schulz MJ, Vesselin N, Shanov (eds) Nanomedicine design of particles, sensors, motors, implants, robots and devices. Artech House, Norwood Ma, pp 367–392

Fung CK, Li WJ (2004) Ultra-low-power polymer thin film encapsulated carbon nanotube thermal sensors. 4th IEEE conference on nanotechnology, 158–160.IEEE

Gao W, Dong R, Thamphiwatana S, Li J, Gao W, Zhang L, Wang J (2015) Artificial micromotors in the mouse's stomach: a step toward *in vivo* use of synthetic motors. ACS Nano 9:117–123

Ghovanloo M, Najafi K (2004) A wideband frequency-shift keying wireless link for inductively powered biomedical implants. IEEE Trans Circuit Syst I Reg Pap 51:2374–2383

Gomez G (2004) Sabiston textbook of surgery, 17th edn. Elsevier Saunders, Philadelphia, PA. Emerging technology in surgery: informatics, Electronics, Robotics

Gu Z, Aimetti AA, Wang Q, Dang TT, Zhang Y, Veiseh O et al (2013) Injectable nano-network for glucose-mediated insulin delivery. ACS Nano 7:4194–4201

Gultepe E, Randhawa JS, Kadam S, Yamanaka S, Selaru FM, Shin EJ et al (2013) Biopsy with thermally-responsive untethered microtools. Adv Mater 25:514–519

Hagiya M (2000) From molecular computing to molecular programming. In International workshop on DNA-based computers, 89–102. Springer

He W, Frueh J, Hu N, Liu L Gai M, He Q (2016) Guidable thermophoretic Janus micromotors containing gold nanocolorifiers for infrared laser assisted tissue welding. Advanced Science 3(12):1600206

Higdon JJL (1979) A hydrodynamic analysis of flagellar propulsion. J Fluid Mech 90:685–711

Hogg T, Freitas RA Jr (2012) Acoustic communication for medical nanorobots. Nano Communication Networks 3:83–102

Hoop M, Ribeiro AS, Rösch D, Weinand P, Mendes N, Mushtaq F et al (2018) Mobile magnetic nanocatalysts for bioorthogonal targeted cancer therapy. Adv Funct Mater 28:1705920

Hu J, Huang S, Zhu L, Huang W, Zhao Y, Jin K, ZhuGe Q (2018) Tissue plasminogen activator-porous magnetic microrods for targeted thrombolytic therapy after ischemic stroke. ACS Appl Mater Inter 10:32988–32997

Ikeda S, Arai F, Fukuda T, Kim EH, Negoro M, Irie K, Takahashi I (2005) *In vitro* patient-tailored anatomical model of cerebral artery for evaluating medical robots and systems for intravascular neurosurgery. In International conference on intelligent robots and systems, 1558–1563

Irazoqui-Pastor P, Mody I, Judy JW (2003) In-vivo EEG recording using a wireless implantable neural transceiver. In First international IEEE EMBS conference on neural engineering, 2003. Conference proceedings, 622–625. IEEE

Joanny JF, Jülicher F, Prost J (2003) Motion of an adhesive gel in a swelling gradient: a mechanism for cell locomotion. Phys Rev Lett 90:168102

Karshalev E, Esteban-Fernández de Ávila B, Beltrán-Gastélum M, Angsantikul P, Tang S, Mundaca-Uribe R, Zhang F, Zhao J, Zhang L, Wang J (2018) Micromotor pills as a dynamic Oral delivery platform. ACS Nano 12(8):8397–8405

Kermani BG, Mueller J, Hall LC, Nagle HT, Scarantino CW (2006) Methods, systems, and associated implantable devices for dynamic monitoring of physiological and biological properties of tumors US Patent Specification 7010340

Kharwade M, Nijhawan M, Modani S (2013) Nano robots: a future medical device in diagnosis and treatment. Res J Pharm, Biol Chem Sci 4:1299–1307

Khulbe P (2014) Nanorobots: a review. IJPSR 5(6):2164–2173

Kim M, Powers TR (2004) Hydrodynamic interactions between rotating helices. Phys Rev E 69:061910

Kinosita K, Yasuda R, Noji H, Adachi K (2000) A rotary molecular motor that can work at near 100% efficiency. Philos Trans R Soc Lond B Biol Sci 355:473–489

Kwan JJ, Myers R, Coviello CM, Graham SM, Shah AR, Stride E et al (2015) Ultrasound-propelled nanocups for drug delivery. Small 11:5305–5314

Li J, Thamphiwatana S, Liu W, Esteban-Fernández de Ávila B, Angsantikul P, Sandraz E et al (2016) Enteric micromotor can selectively position and spontaneously propel in the gastrointestinal tract. ACS Nano 10:9536–9542

Li J, Angsantikul P, Liu W, Esteban-Fernández de Ávila B, Thamphiwatana S, Xu M et al (2017) Micromotors spontaneously neutralize gastric acid for pH-responsive payload release. Angew Chem Int Ed 56:2156–2161
Li S, Jiang Q, Liu S, Zhang Y, Tian Y, Song C et al (2018) A DNA nanorobot functions as a cancer therapeutic in response to a molecular trigger *in vivo*. Nat Biotechnol 36:258
Linko V, Ora A, Kostiainen MA (2015) DNA nanostructures as smart drug-delivery vehicles and molecular devices. Trends Biotechnol 33:586–594
Liu W, Van Wyk JD, Odendaal WG (2004) Design and evaluation of integrated electromagnetic power passives with vertical surface interconnections. In Nineteenth annual IEEE applied power electronics conference and exposition, 2:958–963. IEEE
Lu X, Soto F, Li J, Li T, Liang Y, Wang J (2017) Topographical manipulation of microparticles and cells with acoustic microstreaming. ACS Appl Mater Interfaces 9(44):38870–38876
Manjunath A, Kishore V (2014) The promising future in medicine: nanorobots. Biomed Sci Eng 2:42–47
Marescaux J, Rubino F (2003) The ZEUS robotic system: experimental and clinical applications. Surg Clin 83:1305–1315
Martel S, Felfoul O (2018) U.S. Patent No. 9,905,347. U.S. Patent and Trademark Office, Washington, DC
Martel S, Felfoul O, Mathieu JB, Chanu A, Tamaz S, Mohammadi M et al (2009) MRI-based medical nanorobotic platform for the control of magnetic nanoparticles and flagellated bacteria for target interventions in human capillaries. Int J Robot Res 28:1169–1182
Merina RM (2010) Use of nanorobots in heart transplantation. INTERACT, 265–268. IEEE
Mohseni P, Najafi K, Eliades SJ, Wang X (2005) Wireless multichannel biopotential recording using an integrated FM telemetry circuit. IEEE Trans Neural Syst Rehabil Eng 13:263–271
Murphy D, Challacombe B, Nedas T, Elhage O, Althoefer K, Seneviratne L, Dasgupta P (2007) Equipment and technology in robotics. Arch Esp Urol 60:349–354
Natan MJ, Mallouk TE (2007) U.S. Patent No. 7,225,082. U.S. Patent and Trademark Office, Washington, DC
Odell L, Nacev AN, Weinberg IN (2017) U.S. Patent No. 9,833,170. U.S. Patent and Trademark Office, Washington, DC
Panis C, Hirnschrott, U, Farfeleder S, Krall A, Laure G, Lazian W, Nurmi J (2004) A scalable embedded DSP core for SoC applications. International symposium on system-on-chip proceedings, 85–88. IEEE
Park JG, Lee GS, Lee SH (2005) U.S. Patent No. 6,884,694. U.S. Patent and Trademark Office, Washington, DC
Park SJ, Park SH, Cho S, Kim DM, Lee Y, Ko SY, Park S (2013) New paradigm for tumor theranostic methodology using bacteria-based microrobot. Sci Rep 3:3394
Perrault SD, Shih WM (2014) Virus-inspired membrane encapsulation of DNA nanostructures to achieve *in vivo* stability. ACS Nano 8:5132–5140
Pokki J, Ergeneman O, Chatzipirpiridis G, Lühmann T, Sort J, Pellicer E et al (2017) Protective coatings for intraocular wirelessly controlled microrobots for implantation: Corrosion, cell culture, and *in vivo* animal tests. J Biomed Mater Res B Appl Biomater 105:836–845
Powers TR (2002) Role of body rotation in bacterial flagellar bundling. Phys Rev E 65:040903
Purcell EM (1977) Life at low Reynolds number. Am J Phys 45:3–11
Rao TVN, Saini HS, Prasad PB (2014) Nanorobots in Medicine-A New Dimension in Bio Nanotechnology. Int J Sci Eng Comp Technol 4:74–79
Reppesgaard L (2002) Nanobiotechnologie: Die Feinmechaniker der Zukunftnutzen Biomaterial alsWerkstoff. Computer Zeitung 36:22
Risveden K, Pontén JF, Calande N, Willander M, Danielsson B (2007) The region ion sensitive field effect transistor, a novel bioelectronics nanosensor. Biosens Bioelectron 22:3105–3112
Roue CC (2002) Aneurysm liner. 6350270US, Feb.
Roundy S, Wright PK, Rabaey JM (2003) Energy scavenging for wireless sensor networks. Norwel:45–47

Rudchenko M, Taylor S, Pallavi P, Dechkovskaia A, Khan S, Butler VP Jr et al (2013) Autonomous molecular cascades for evaluation of cell surfaces. Nat Nanotechnol 8:580–586
Sauer C, Stanacevic M, Cauwenberghs G, Thakor N (2005) Power harvesting and telemetry in CMOS for implanted devices. IEEE Trans Circuit Syst I Reg Pap 52:2605–2613
Saxena S, Pramod BJ, Dayananda BC, Nagaraju K (2015) Design, architecture and application of nanorobotics in oncology. Indian J Cancer 52:236
Sharma NN, Mittal RK (2008) Nanorobot movement: Challenges and biologically inspired solutions. Int J Smart Sens Intell Syst 1:87–109
Soto F, Martin A, Ibsen S, Vaidyanathan M, Garcia-Gradilla V, Levin Y, Wang J (2015) Acoustic microcannons: toward advanced microballistics. ACS Nano 10(1):1522–1528
Soto F, Chrostowski R (2018) Frontiers of medical micro/nanorobotics: *in vivo* applications and commercialization perspectives towards clinical uses. Front Bioeng Biotechnol 6:1–12
Stracke R, Böhm KJ, Burgold J, Schacht HJ, Unger E (2000) Physical and technical parameters determining the functioning of a kinesin-based cell-free motor system. Nanotechnology 11(2):52–56
Sun J, Gao M, Feldmann J (2001) Electric field directed layer-by-layer assembly of highly fluorescent CdTe nanoparticles. J Nanosci Nanotechnol 1:133–136
Squires TM, Brady JF (2005) A simple paradigm for active and nonlinear microrheology. Phys Fluids 17:073101
Taherkhani S, Mohammadi M, Daoud J, Martel S, Tabrizian M (2014) Covalent binding of nanoliposomes to the surface of magnetotactic bacteria for the synthesis of self-propelled therapeutic agents. ACS Nano 8:5049–5060
Udomprasert A, Kangsamaksin T (2017) DNA origami applications in cancer therapy. Cancer Sci 108:1535–1543
Ullrich F, Bergeles C, Pokki J, Ergeneman O, Erni S, Chatzipirpiridis G et al (2013) Mobility experiments with microrobots for minimally invasive intraocular surgery. Invest Ophthalmol Vis Sci 54:2853–2863
Ummat A, Dubey A, Sharma G, Mavroidis C (2005a) Nanorobotics – fractal navigator
Ummat A, Dubey A, Mavroidis C (2005b) Bio-nanorobotics: a field inspired by nature. Biomimetics:219–246
Vasilescu I, Kotay K, Rus D, Dunbabin M, Corke P (2005) Data collection, storage, and retrieval with an underwater sensor network. In Proceedings of the 3rd international conference on Embedded networked sensor systems 154–165. ACM
Venkatesan M, Jolad B (2010) Nanorobots in cancer treatment. INTERACT, 258–264. IEEE
Wang J, Zhang L (2017) U.S. Patent Application No. 15/356, 977
Wieland III CF (2004) Is the US nanotechnology investment paying off? Small Times Magazine, 4(1)
Wright EM, Sampedro AD, Hirayama BA, Koepsell H, Gorboulev V, Osswald C (2005) Novel glucose sensor. United States patent US 0267154
Yan X, Zhou Q, Vincent M, Deng Y, Yu J, Xu J, Xu T, Tang T, Bian L, Wang Y-XJ, Kostarelos K, Zhang L (2017) Multifunctional biohybrid magnetite microrobots for imaging-guided therapy. Science Robotics 2(12):eaaq1155
Zhang M, Sabharwal CL, Tao W, Tarn TJ, Xi N, Li G (2004) Interactive DNA sequence and structure design for DNA nanoapplications. IEEE Trans Nanobioscience 3:286–292
Zhang Q, Jiang Q, Li N, Dai L, Liu Q, Song L et al (2014) DNA origami as an *in vivo* drug delivery vehicle for cancer therapy. ACS Nano 8:6633–6643
Zhang Z, Dai C, Huang JY, Wang X, Liu J, Ru C et al (2018) Robotic immobilization of motile sperm for clinical intracytoplasmic sperm injection. IEEE Trans Biomed Eng 62:2620–2628

13 Nanobiomaterials in Drug Delivery: Designing Strategies and Critical Concepts for Their Potential Clinical Applications

Chang Liu, Zhixiang Cui, Xin Zhang, and Shirui Mao

Abstract

Nowadays nanotechnology has found extensive application in drug delivery. To design efficient nano-based systems as drug vehicles, the selection of appropriate materials as the carrier is of special importance. Owing to their biodegradability, biocompatibility, being renewable and presenting low toxicity, a myriad of biomaterials have been extensively used in the fields of biomedicine and tissue regeneration. Moreover, the use of biomaterials into nano-based drug delivery shows tremendous attraction. Regarding the design of ideal nanotechnology based drug delivery system, the selection of nanocarrier depends not only on the physicochemical features of drugs and materials, but also the administration route. Thus, in this chapter, first of all, commonly used biomaterials for nanocarrier design, including both natural and synthesized polymers, were introduced and their physicochemical properties were summarized. Thereafter the latest advances in drug delivery by using varied biomaterials as the nanocarriers for different administration routes, including parenteral drug delivery by preparing liposomes, micelles, nanoparticles, and mucosal drug delivery with either mucus bioadhesion or mucus penetration nanoparticles, were presented with related designing strategies covered. Finally, challenges and prospective in applying nanobiomaterials based drug delivery systems were discussed.

Keywords

Nanotechnology · Nanocarrier-based drug delivery systems · Nanobiomaterials · Nanoparticles · Route of drug administration

---

C. Liu · Z. Cui · X. Zhang · S. Mao (✉)
School of Pharmacy, Shenyang Pharmaceutical University, Shenyang, China
e-mail: maoshirui@syphu.edu.cn

P. Chandra, R. Prakash (eds.), *Nanobiomaterial Engineering*,
https://doi.org/10.1007/978-981-32-9840-8_13

13.1 Introduction

Nowadays nanotechnology is an emerging and rapidly evolving field, which has found extensive application in drug delivery. Compared with pure molecular therapeutics, nanocarrier-based drug delivery systems have numerous advantages, for example, it can protect drugs against enzymatic and hydrolytic degradation, providing the possibility of targeting for site-specific drug delivery, with controlled release of drugs leading to immense success. Since the 1990s, FDA-approved nanotechnology-based drug products and clinical trials have galloped ahead (Fig. 13.1a). Among them, nanoparticles (NPs) based on various materials as

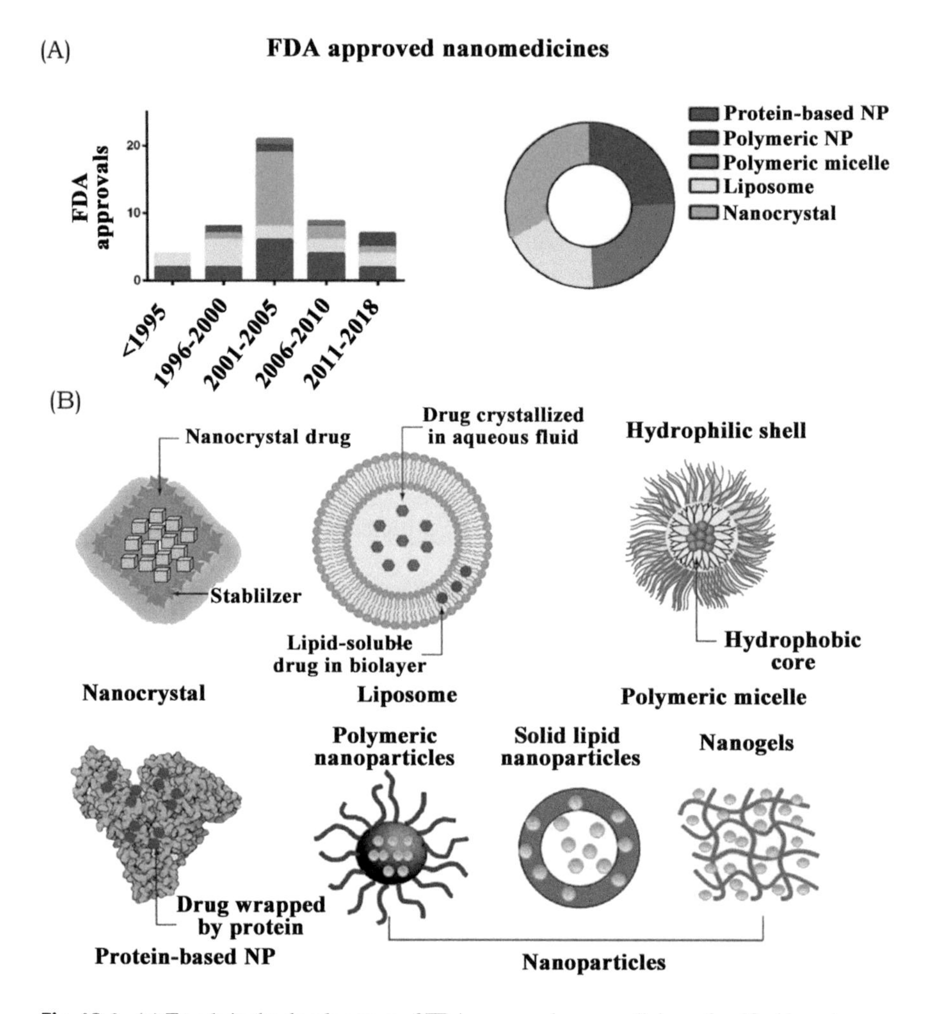

Fig. 13.1 (**a**) Trends in the development of FDA-approved nanomedicines classified by category; (**b**) diagrammatic representation of various types of nanocarriers. (Reproduced from Bobo et al. (2016) with copyright permission)

carriers have been widely explored, with the unique advantages such as the ease of synthesis, biocompatibility, and customizability. Nanocarriers, defined as a carrier with sizes ranging between 1 and 1000 nm, are excellent candidates for drug delivery based on their sub-micrometer size and high surface area to volume ratio. Quite frequently, Doxil® and Abraxane® have been used as examples of nanocarrier-based drug delivery systems. To design efficient nano-based systems for drug delivery, for example, the preparation of nanocrystals, liposomes, polymeric micelles, protein-based NPs, polymeric or lipid-based NPs, nanogels, or any other self-assembled nanosized system for drug delivery (Fig. 13.1b), the selection of appropriate material as the carrier is of special importance.

Owing to their biodegradability, biocompatibility, being renewable, and presenting low toxicity, a myriad of biomaterials have been extensively used in the fields of biomedicine and tissue regeneration. Moreover, the use of biomaterials into nano-based drug delivery shows tremendous attraction. So far, among these nanomaterials that are in phase study, 18 are directed to chemotherapeutics, 15 are intended for antimicrobial agents, 28 are for psychological diseases and autoimmune conditions, and 30 are aimed for nucleic acid-based therapies (Bobo et al. 2016; Bosselmann and Williams 2012).

Regarding the design of ideal nanotechnology-based drug delivery system, the selection of the nanocarrier depends not only on the physicochemical features of drugs and materials but also on the administration route. Thus, in this chapter, first of all, polymeric materials were introduced, and the physicochemical properties of these materials from natural and chemical sources were elaborately introduced. Meanwhile, commonly used materials for nanocarrier design were presented, with a focus on biomaterials. Thereafter the latest advances in drug delivery by using varied nanobiomaterials via different administration routes and related designing strategies were covered. Finally, challenges and prospective in applying nanobiomaterial-based drug delivery systems were discussed.

13.2 Commonly Used Biomaterials for Nanocarrier Design

Nanomaterials can be classified into different types based on their shape, composition, and dimension. Here, the utility of biomaterials for drug delivery was specifically highlighted due to their long history of safe use in humans and unique advantages for fabrication of nanocarriers for drug delivery (Hallan et al. 2016). In terms of polymeric nanomedicine, it consists of two categories: (a) polymer-drug conjugates for prolonged drug half-life and enhanced bioavailability and (b) preparations of NPs for drug delivery based on degradable polymers (Song et al. 2018). Based on their sources, polymeric materials can be classified into natural materials and synthesized ones.

13.2.1 Natural Polymer-Based Biomaterials

Natural polymers, classified as environmentally friendly materials, are a renewable resource considered to be safe in vivo. Commonly used natural polymers include chitosan (CS), alginate, cellulose, hyaluronic acid (HA), carrageenan, chondroitin sulfate, albumin, phospholipid, etc., which are being widely investigated as drug delivery carriers (Han et al. 2018). Among them, CS, HA, albumin, and phospholipid are extensively used.

As the most widely employed natural polysaccharide, CS is a cationic polysaccharide of copolymers glucosamine and N-acetyl glucosamine linked in a β(1–4) manner, prepared by the partial N-deacetylation of crustacean derived from natural biopolymer chitin (Mao et al. 2010). Also, it is naturally found in the fungal cell walls. Deacetylation of chitin renders CS some unique properties, such as bearing positive charge and consequently possesses the capacity to form polyelectrolyte complexes with negatively charged compounds. CS is soluble at acidic pH (pH <5) but precipitates as the physiological pH (pH 7.4) is restored. Besides, due to protonation of the $-NH_2$ group of the D-glucosamine repeating unit, CS is soluble at acetic acid media. The molecular weight and degree of deacetylation of CS can influence its solubility. When the degree of dissociation (α) in solution increases, the role of the cationicity of the amine groups, which depends on the degree of acetylation, plays a more important role in enhancing solubility. Also, the solubility of CS can be increased by decreasing its molecular weight or introducing some hydrophilic groups to the structure of CS. Moreover, a deacetylation of 85% or higher of CS is preferred due to its stronger mucoadhesive properties and biocompatibility. It is also found that CS can increase trans- and paracellular permeability in a reversible, dose-dependent manner (Elgadir et al. 2015). These properties make CS-based materials as an ideal candidate for drug delivery with enhanced mucoadhesion and permeation enhancing properties. CS derivatives, obtained via modification of amino and hydroxyl groups on the CS side chain through acylation, sulfation, hydroxylation, and quaternization, also show immense potential application in biomedical and drug delivery field (Kausar 2017; Wu et al. 2017).

Next to CS, another natural polymeric material HA is a biocompatible, linear glycosaminoglycan, composed of alternating units of N-acetyl-d-glucosamine and glucuronic acid linked together through alternating β-1,3 and β-1,4 glycosidic bonds (Rao et al. 2016; Yadav et al. 2008). Since the pKa value of the carboxyl groups of HA is 3–4, these functional groups are predominantly ionized at pH 7.4, and, therefore, under physiological conditions, HA bear negative charge. Naturally, HA could be found in extracellular matrix, vitreous humor, and synovial fluid of vertebrates, ranging in molecular weight from 5000 to 20,000,000 daltons. Traditionally extracted from rooster combs, HA is now increasingly produced through microbial fermentation. In solution, the chains of HA is highly hydrophilic and surrounded by water molecules linked through hydrogen bonds. Under these conditions, HA adopts random-coil conformation, resulting in forming a very viscous and elastic solution. Thus, HA has been expansively scrutinized for its potential use in biomedical field for visco-supplementation, drug delivery, eye surgery, tissue regeneration, and embryo protection (Jiao et al. 2016; Ossipov 2010). Most notably, due to its ability to specifically bind to various cancer cells which overexpress the CD44 receptor, HA-based NPs have attracted extensive attention in tumor-targeted delivery and imaging (Yu et al. 2013).

With a molecular weight of 66.5 kDa and a diameter of ~10 nm, albumin is regard as the most abundant plasma protein naturally found in blood (Larsen et al. 2016). Under physiological conditions, about 10–15 g of albumin were produced in liver by hepatocytes and released into the vascular space daily. Usually, the circulation time of albumin in the blood proceeds for approximately 20 days (Mariam et al. 2016). This long half-life is thought mainly facilitated by neonatal Fc receptor (FcRn)-mediated recycling and the megalin-cubilin receptor-mediated renal rescue. Possessing multiple ligand binding sites with cellular receptors is in favor of albumin's recycling and cellular transcytosis. Furthermore, the surface of albumin is negatively charged making it highly water-soluble. Benefiting from its physiological transport mechanisms, charge, and solubility, it is regarded as a highly attractive drug carrier for both half-life extension and targeted intracellular delivery (Mariam et al. 2016). Albumin appears as brownish amorphous lumps, scales, or a powder, consisting of a single polypeptide chain of 585 amino acids. Structurally, albumin consists of three repeated homologue domains (sites I, II, and III). Each domain comprises two separate sub-domains (A and B), each of which contains four and six α-helices, respectively (Elzoghby et al. 2012). Besides, 35 cysteine residues were found in albumin domains, of which 34 form disulfide bridges internally in the structure contributed to high stability of albumin. A free cysteine residue at position 34 located on the outer surface of albumin provides a free thiol group (−SH) accounting for 80% of thiol in the plasma (An and Zhang 2017). Therefore, the major role of albumin in serum is mainly focus on covalent conjugation with drugs. Being one of the multifunctional abundant proteins in plasma, it can also play crucial physiological roles in free radical scavenging and maintaining osmotic pressure (Sleep 2015). Besides, being a nonimmunogenic and nontoxic protein, it is readily available and highly soluble and can be modified and manipulated depending on the proposed application.

Phospholipids are also well-established excipients for various applications, such as function as emulsifier, wetting agent, solubilizer, and liposome former. All lipids that contain phosphorus are called phospholipids, which comprise a polar head group and a lipophilic tail. Phospholipids are functional components of all cell membranes and can also be isolated from natural sources such as soybean, rapeseed, and sunflower seed. Moreover, properties of the phospholipids show some differences depending on their natural sources. For example, phosphatidylcholine obtained from egg yolk has a lower content of polyunsaturated fatty acids compared to phosphatidylcholine from soybean (Otto et al. 2018). Structurally, the phospholipid molecule consists of a glycerol backbone, which is esterified in position 1 and 2 with fatty acids and in position 3 with phosphate. Moreover, the phosphate group can be further esterified with an additional alcohol, for instance, in phosphatidylcholine (PC) with choline, in phosphatidylethanolamine (PE) with ethanolamine, and in phosphatidylglycerol (PG) with glycerol (van Hoogevest 2017). In typical membrane phospholipids, most of them have neutral (PC, PE) or negative charge (PG, PS, PI, PA). Positively charged phospholipids rarely exist in nature. An example of a positively charged phospholipid is lysyl-phosphatidylglycerol. Until now, the most common nature phospholipid is PC, which is also the main component of lecithin. On the other hand, the fatty acids esterified to the glycerol backbone of the phospholipid molecule could be saturated (e.g., palmitic acid) or monounsaturated (e.g., oleic acid) or polyunsaturated (e.g., arachidonic acid). The resulting phospholipids are called DPPC, where two palmitic acids are esterified.

13.2.2 Synthetic Polymer-Based Biomaterials

Despite the advantages of naturally available biodegradable materials, limitations and concerns still remain with regard to the use of nature polymers, for example, quality inconsistency, the difficulty to control the mechanical properties and degradation rates, and the potential to elicit an immune response or carry microbes or viruses. In contrast, synthetic polymers are more homogenous in composition and therefore have a higher purity than natural polymers, making the preparation of NPs more reproducible. Furthermore, taking the biocompatibility and immunogenicity into consideration, in the current stage of nanomedicine development, only biodegradable polymers such as polylactide (PLA), polyglycolide (PGA), polylactide-co-glycolide (PLGA), and poly (glutamic acid) have been approved by the US FDA for parenteral use (Guo and Ma 2014).

As a widely reported biodegradable polymer, PLGA are made from a copolymer of PLA and PGA polymers. PLGA can be synthesized by random melting copolymerization of lactic and glycolic acid or their cyclic diesters, lactide, and glycolide, respectively. PLGA copolymers are amorphous in nature with glass transition temperature between 45 and 55 °C. The copolymer is more stable against hydrolytic cleavage than each of the polymers alone and the hydrolysis of PLGA results in lactic and glycolic acids which are natural metabolites found in the body. Depending on their molecular weight, inherent viscosity and ratio of lactic acid to glycolic acid, PLGA is commercially available from different companies and with different composition. The physical-chemical properties of PLGA depend mainly on LA:GA ratio. For example, the solubility of PLGA is closely correlated with its LA:GA proportion. Unlike LA and GA, PLGA is soluble in a wide range of solvents, including dichloromethane, tetrahydrofuran, chloroform, and acetone, ethyl acetate, and benzyl alcohol. The solubility decreases with increasing GA content (Makadia and Siegel 2011; Pandita et al. 2015). Likewise, the degradation rate of PLGA is also LA:GA ratio dependent, and this is mainly due to the different hydrophilic profile of each monomer (Xu et al. 2017). Owing to the absence of methyl side groups, GA is more hydrophilic than LA, and in vivo resorption period of PGA (100% GA) is only 6–12 months, whereas it is between 12 and 24 months for PLA (100% LA). Consequently, PLGA with higher proportion of GA is more hydrophilic and can be degraded faster in vivo (Zhang et al. 2014). At 50:50 LA:GA ratio, PLGA copolymers have high degradation rate, which slows down as the proportion of LA increases from 50 to 100, with GA ratio reducing from 50 to 0. Benefiting from the versatile degradation profile of PLGA, it can be used in biomedical applications, such as surgical implants for controlled drug release. Besides, PLGA is also widely used in the preparation of microspheres, microcapsules, NPs, pellets, implants, and films (Kapoor et al. 2015; Makadia and Siegel 2011).

Poly (glutamic acid) can be achieved by microbial fermentation, which is water-soluble, nontoxic, and completely biodegradable. The molecular weight of poly (glutamic acid) ranges from 100,000 to over 1,000,000, which are largely dependent on the fermentation time. Owing to the presence of a polyglutamyl hydrolase enzyme which can catalyze the hydrolytic breakdown of poly (glutamic acid), their molecular weight decreases as the fermentation time increases (Jeon et al. 2016). Different from traditional proteins structure, poly (glutamic acid), made up of repeating units of L-glutamic acid, D-glutamic acid, or both, is defined as a pseudo-poly (amino acid) linked between the α-amino and γ-carboxylic acid functional groups (Ogunleye et al. 2015). Based on the attachment of the carboxyl group (α and γ, respectively), poly (glutamic acid) can be differentiated into two isoforms, α-poly (glutamic acid) and γ- poly (glutamic acid). α-Poly (glutamic acid) is synthesized chemically by nucleophile-initiated polymerization of the γ-protected N-carboxyanhydride of L-glutamic acid. Microbial production of α-poly (glutamic acid) is difficult, and the polymer can only be produced by recombinant technology. γ-Poly (glutamic acid) has been produced extensively using bacteria, especially those of *Bacillus* species (Ogunleye et al. 2015). So far, it is well known that utility of γ-poly (glutamic acid) plays several advantages over α-poly (glutamic acid) utility. Pure γ-poly (glutamic acid) can be readily obtained in large quantities without any chemical modification step. It is not susceptible to proteases and hence could provide better sustained delivery of conjugated drugs in the body (Shi et al. 2016b).

13.3 Biomaterial-Based Nanocarriers for Drug Delivery

13.3.1 Parenteral Drug Delivery

Compared to conventional injectable solution, injection of nanocarrier-based drug delivery systems offers several advantages, such as controlled drug release and/or selective cell targeting, enhanced cellular uptake, or prolonged circulation time, leading to the potential to maximize the therapeutic effect. However, adverse side effects associated with nanocarrier injection are also noticed, such as undesirable protein adsorption, cell adhesion, as well as inflammation and cytotoxicity. Thus, during the design of nanocarriers for parenteral administration, selection of biodegradable polymers with desirable surface properties is preferred. As shown in Table 13.1, currently marketed nanocarriers used for parenteral drug delivery mainly include liposomes, polymeric micelles, NPs, and polymeric materials stabilized nanosuspension. The materials used for specific nanocarrier design is NPs' type dependent, as described in the following parts.

Table 13.1 List of nanotherapeutics approved by FDA that utilized nanotechnologies for parenteral drug delivery

Formulation	Product	Drug loaded	Carrier materials used	Advantage	Year approved
Liposomes	Doxil	Doxorubicin	Cholesterol, fully hydrogenated soy phosphatidylcholine (HSPC); N-(carbonyl-methoxypolyethylene glycol 2000)-1, 2-distearoyl-sn-glycero-3-phosphoethanolamine sodium salt (mPEG-DSPE) coating	Improved delivery to the site of disease; decreased systemic toxicity of free drugs	1995
	Abelcet	Amphotericin B	l-α-dimyristoylphosphatidylcholine (DMPC) and l-α-dimyristoylphosphatidylglycerol (DMPG)	Reduced toxicity	1995
	DaunoXome	Daunorubicin	Distearoylphosphatidylcholine and cholesterol	Increased delivery to tumor site; lower systemic toxicity	1996
	AmBisome	Amphotericin B	Egg phosphatidylcholine and cholesterol	Reduced nephrotoxicity	1997
	DepoCyt	Cytarabine	Cholesterol; dioleoylphosphatidylcholine (DOPC); and dipalmitoylphosphatidylglycerol (DPPG)	Increased delivery to tumor site; lower systemic toxicity	1997
	Visudyne	Verteporfin	Lactose, egg phosphatidylglycerol, dimyristoyl phosphatidylcholine	Increased delivery to diseased vessels; photosensitive release	2000
	DepoDur	Morphine sulfate	1,2-dioleoyl-sn-glycero-3-phosphocholine (DOPC); cholesterol; 1,2-dipalmitoyl-sn-glycero-3-phospho-rac-(1-glycerol) (DPPG); tricaprylin; and triolein	Extended release	2004
	Marqibo	Vincristine	Sphingomyelin/cholesterol	Increased delivery to tumor site; lower systemic toxicity	2012
	Onivyde	Irinotecan	1,2-distearoyl-sn-glycero-3-phosphocholine (DSPC), cholesterol, and methoxy-terminated polyethylene glycol (MW 2000)-distearoylphosphatidyl ethanolamine (mPEG-2000-DSPE)	Increased delivery to tumor site; lower systemic toxicity	2015
	Vyxeos	Combination of daunorubicin and cytarabine	Distearoylphosphatidylcholine (DSPC), distearoylphosphatidylglycerol (DSPG), and cholesterol	Sustained release of the molecules and co-loading two molecules with synergistic antitumor activity	2017

Nanoparticles	Feridex/ Endorem	Ferumoxides	Superparamagnetic iron oxide associated with dextran	Superparamagnetic character	1996/2008
	Abraxane	Paclitaxel	Human albumin (containing sodium caprylate and sodium acetyltryptophanate)	Improved solubility; improved delivery to tumor	2007
Nano suspension	Invega Sustenna	Paliperidone palmitate	Polysorbate 20, polyethylene glycol 4000	Allows slow release of injectable low-solubility drugs	2009/2014
	Ryanodex	Dantrolene sodium	Mannitol, polysorbate 80, povidone K12	Faster administration at higher doses	2014
Polymeric micelle	Genexol-PM	Paclitaxel	Methoxy poly(ethylene glycol)-poly (lactide) (mPEG-PLA)	The ability to freely circulate throughout the vasculature to avoid being taken up by reticuloendothelial system (RES)	Phase II: IIa IIb approved (South Korea)

13.3.1.1 Nanomaterials as the Carrier of Liposomes

Liposomes, composed of a lipid or phospholipid, have spherical bilayer nanostructures with both hydrophilic and hydrophobic region. With good biocompatibility, low toxicity, and high safety, many hundreds of drugs, such as anticancer and antimicrobial agents, peptide hormones, vaccines, genetic materials, enzymes, and proteins, have been incorporated into the aqueous or lipid phases of liposomes aimed to deliver therapeutic drug at a sufficient concentration to the target tissues for in vivo absorption. In the development of nanomedicine, liposomes are the first nanomedicine transited from concept to clinical application. Early starting with the approval of liposomal formulations Doxil® in 1965, there have been a growing number of trials and approvals using liposome as the nanocarrier for drug delivery. And other liposomal drugs by intravenous administration are in various stages of clinical development.

The ability of liposomes to deliver a range of therapeutic drugs depends on a diverse toolbox of lipids with well-characterized biophysical behavior. Lipids in this toolbox can be naturally occurring or rationally designed using a variety of hydrophilic head groups, linkers, and hydrophobic moieties. The selection of each lipid is closely related to its phase transition temperature, chemical structure, and whether the lipid is unsaturated or not, as well as its charge. Characteristics of lipid affect the capacity of the membrane to accommodate the drug. For instance, unsaturated phosphatidylcholine species from natural sources (i.e., egg or soybean phosphatidylcholine) form much more permeable and less stable bilayer, whereas the saturated phospholipids, with long acryl chains (i.e., dipalmitoylphosphatidylcholine), form a rigid, rather impermeable bilayer structure (Sakai-Kato et al. 2015).

To create liposomes for intravenous administration, naturally occurring lipids are preferred. For example, fully hydrogenated soy phosphatidylcholine (HSPC) and 2-distearoyl-sn-glycero-3-phosphoethanolamine sodium salt (mPEG-DSPE) coating are firstly used in Doxil® approved by FDA. Besides, AmBisome® and DepoCyt®, which used naturally occurring lipids, such as dioleoylphosphatidylcholine (DOPC), dipalmitoylphosphatidylglycerol (DPPG), and egg phosphatidylcholine, have also been approved by FDA. Further, to fulfill multiple functions of liposomes in vivo, synthetic phospholipids are widely used. An example of such a lipid, with extended circulation time, is PEGylated phospholipid such as mPEG-DSPE (methoxy-polyethylene glycol-distearoyl phosphatidylethanolamine) (Doxil®). By forming a steric barrier around the liposome, PEG-modified lipids embedded in a lipid bilayer decrease interaction with serum opsonins, cellular ligands, and other pre-existing serum factors meanwhile reducing adhesion to other membrane surfaces (Jia et al. 2017). Also, in order to prolong circulation half-life, other synthetic polymer-modified lipids are also designed, such as HPMA (poly (N-(2-hydroxypropyl) methacrylamide)), PVP (poly (vinylpyrrolidone)), PMOX (poly (2-methyl-2-oxazoline)), and PVA (poly (vinylalcohol) (Kocisova et al. 2013; Zhang et al. 2016; Zylberberg and Matosevic 2017).

Properties of liposomes can also be modified via changing surface charge of materials used. For example, cationic liposomal formulation has been designed to selectively target tumor vasculature. DOTAP (1, 2 dioleoyl-3-trimethylammonium-propane) is a cationic synthetic lipid, which comprises one positive charge at the head group. EndoTAG-1, the first formulation of cationic liposomes carrying paclitaxel in clinical trial, is prepared by DOTAP, DOPC, and paclitaxel in 50:47:3 molar ratio (Chang and Yeh 2012). Table 13.2 lists some liposomes in clinical trial and their lipid composition.

13.3.1.2 Nanomaterials as the Carrier of Micelles

In addition to liposomes, polymer micelles are another promising nanocarrier for delivering various drugs, such as cytostatic agents, nucleic acids via parenteral delivery, with better stability and stronger mechanical strength. Polymeric micelles can be formed through self-assembly of amphiphilic block copolymers with sizes ranging between 20 and 200 nm (Ohya et al. 2011). The inner hydrophobic core of polymeric micelles acts as a suitable microenvironment for hydrophobic bioactives, while the outer hydrophilic shell provides required colloidal stability. Table 13.3 lists the micellar-based injectable formulations under different phases of clinical study, and there are a large amount of polymeric micelles based formulations still under preclinical investigation. Genexol®-PM injection (paclitaxel), the only polymeric micellar NP-based formulation that has been approved in Bulgaria, Hungary, and South Korea, is being evaluated in Phase II trials in the USA. The formulation consists of 20–50 nm PEG-PLA micelles loaded with PTX, exhibiting superior cytotoxic activity against various human cancer cells, e.g., breast, colon, ovarian, and non-small cell lung cancer compared to Taxol® (Park et al. 2010).

Physical and biological properties of polymeric micelles are of special importance, which depend on the characters of materials used for micelle preparation. The selection of polymeric materials influences many important properties of micelles such as toxicity, bio-distribution, pharmacokinetics, and clinical compatibility. Many amphiphilic copolymers can be used as the carrier of polymeric micelles, which is mainly composed of hydrophilic part and hydrophobic part (Pepic et al. 2013; Qiu et al. 2007). A wealth number of hydrophilic polymers with a flexible nature can be selected as the hydrophilic part of amphiphilic copolymers, such as PEG, poly (ethylene oxide) (PEO), poly (acryloylmorpholine), and poly (vinylpyrrolidone). Sometimes the hydrophilic part is made up of a mixture of polymers like PEO and polyelectrolyte. Especially, PEG has been widely used in the outer layer of the polymeric micelles to block interparticle aggregation (Shiraishi et al. 2013). As for hydrophobic blocks, polyesters, poly (amino acid)s (PAAs), and polyether derivatives are often used as the hydrophobic segments of the copolymers. Polyesters, such as PLA and PCL (Poly (ε-caprolactone)), are biocompatible and biodegradable and have been approved by the FDA for biomedical applications in humans (Janas et al. 2016). PAAs, such as poly (aspartic acid) (P (Asp)) and poly (glutamic acid) (P (Glu)), are biodegradable, and their multiple carboxyl/amine functional groups enable in combination with drugs and formation of complexes with various metals or can be modified to optimize the core-drug compatibility, thus increasing drug loading and formulation stability (Jones 2015; Qiu et al. 2007). Polyethers of pharmaceutical interest are copolymers of PEG-block-poly-(propylene oxide)-block-PEG (PEG-b-PPO-bPEG), known as poloxamers (Pluronic), F68, has

Table 13.2 Liposome-based drugs in clinical trials via injectable route

Product	Drug	Lipid composition	Approved indication	Trial phase
LEP-ETU (powder/12 months) (Zhang et al. 2005)	Paclitaxel	1,2-dioleoyl-sn-glycero-3-phosphocholine (DOPC), cholesterol, and cardiolipin (90:5:5 molar ratio)	Ovarian, breast, and lung cancers	Phase I/II
EndoTAG-1 (powder/24 months) (Dandamudi and Campbell 2007)	Paclitaxel	1,2 dioleoyl-3-trimethylammonium-propane (DOTAP), DOPC, and paclitaxel (50:47:3 molar ratio)	Breast cancer, pancreatic cancer	Phase II
Marqibo (Immordino et al. 2006)	Vincristine	Cholesterol and egg sphingomyelin (45:55 molar ratio)	Metastatic malignant uveal melanoma	Phase III
ThermoDox (Yarmolenko et al. 2010)	Doxorubicin	Dipalmitoylphosphatidylcholine (DPPC), monostearoylphosphatidylcholine (MSPC), and polyethylene glycol-distearoylphosphatidylethanolamine (PEG 2000-DSPE) (90:10:4 molar ratio)	Non-resectable hepatocellular carcinoma	Phase III
S-CKD602 (Immordino et al. 2006)	Camptothecin analog	Distearoylphosphatidylcholine (DSPC) and DSPE-PEG (95:5 molar ratio)	Recurrent or progressive carcinoma of the uterine cervix	Phase I/II
OSI-211 (Tomkinson et al. 2003)	Lurtotecan	Fully hydrogenated soy phosphatidylcholine (HSPC) and cholesterol (2:1 molar ratio)	Ovarian cancer, head and neck cancer	Phase II
INX-0125 (Immordino et al. 2006)	Vinorelbine	Cholesterol and egg sphingomyelin (45:55 molar ratio)	Advanced solid tumors	Phase I

Table 13.3 Current clinical status of polymeric micelle-based formulations for injection

Type of micelles	Product name	Copolymer composition	Drug	Purpose	Treatment	Clinical status
	Genexol®-PM (Cabral and Kataoka 2014)	mPEG-PDLLA	Paclitaxel	Solubilization	Breast and lung cancer	Phase IV/ approved in Korea
	NK105 (Aziz et al. 2017)	PEG-b-poly(α,β-aspartic acid)	Paclitaxel	Solubilization	Breast and gastric cancer	Phase III
	NK911 (Tagami and Ozeki 2017)	PEG-P (Asp)-DOX	Doxorubicin	Targeting	Solid tumors	Phase II
	NC-6300 (Cabral and Kataoka 2014)	PEG-b-poly(aspartatehydrazone)	Epirubicin	Targeting	Breast and liver tumors	Phase I
Single	NC-4016 (Aziz et al. 2017)	PEG-b-poly(L-glutamic acid)	Oxaliplatin	Targeting	Solid tumors	Phase I
	NK012 (Tagami and Ozeki 2017)	PEG-b-poly(L-glutamic acid) (SN 38)	SN-38	Solubilization	Triple negative breast cancer and small lung cancer	Phase II
	NC-6004 (Cabral and Kataoka 2014)	PEG-b-poly(L-glutamic acid) (cisplatin)	Cisplatin	Targeting	Pancreatic cancer	Phase III
	siRNA micelles (Tagami and Ozeki 2017)	cRGD-PEG-b-PAsp(TEP)-Chol/siRNA	siRNA	Targeting	Lung tumor	Preclinical/phase I
Mixed	SP1049C (Valle et al. 2011)	Pluronic L61 and F127	Doxorubicin	Anti-MDR	Adenocarcinoma of the esophagus and gastro esophageal junction	Phase III

been approved by the FDA for parenteral use. F68 is nonbiodegradable, but the individual polymer chains with a size of <50Kda can be excreted by the kidneys (Akbar et al. 2018).

However, one of the main shortcomings for micelles based on single amphiphilic copolymer is their inherent instability upon dilution after their administration, leading to premature release of encapsulated drug before reaching the targeted tissues. Aimed at efficiently improve micellar stability, tremendous researches have focused on mixed micelles by combining two or more dissimilar block copolymers to form micelles. At present, a mixed micelle-based formulation (SP1049C) composed of Pluronic L61 and F127 has reached clinical phase III studies (Valle et al. 2011). Pluronic® L61 copolymer was selected, because it induced a 7.2-fold higher drug uptake in Chinese hamster ovary CHRC5 (resistant) cells, while F127 granted physicochemical stability to the formulation, as it prevented liquid phase separation and preserved the effective size of the micelles below 30 nm, without significantly affecting cytotoxicity of the micelle system.

13.3.1.3 Nanomaterials as the Carrier of Nanoparticles

Similarly, to overcome several inherent problems of liposomes, such as low encapsulation efficiency, rapid leakage of water-soluble drug in the presence of blood components, and poor storage stability, biodegradable polymeric NPs have attracted considerable attention in view of their ability to target particular organs/tissues, as carriers of DNA/siRNA in gene therapy with improved stability. FDA approved the first "nano" particle-based delivery system, a 130 nm albumin-bound paclitaxel NP (Abraxane®), in January 2005 for the treatment of breast cancer. Augmented albumin uptake in tumors is attributed to the interaction of albumin with albondin, a 60-kda glycoprotein (gp60) receptor and SPARC (Secreted Protein, Acidic and Rich in Cysteine), an extracellular matrix glycoprotein which is over expressed in cancer cells (Socinski 2006). Besides, BIND-014, a tumor prostate-specific membrane antigen (PSMA)-targeted NPs (containing docetaxel) formulation, has also garnered attention in the field of cancer therapy. The results of phase I trial of BIND-014 support its further investigation in phase II studies, which are currently ongoing (Autio et al. 2016).

For the design of polymeric NPs, both natural and synthetic polymeric materials can be used, where polymer-drug compatibility is the main criteria for determining drug loading amount in the NPs (Shi et al. 2016a). Among the naturally available materials, CS, HA, gelatin, sodium alginate, and some other biodegradable polymers have gained a lot of attention as nanocarriers (Dong et al. 2018; Guan et al. 2018; Wang et al. 2018). Especially, it was found that CS-based NPs were widely applied not only in the delivery of anticancer agents, proteins, and peptides but also as the nanocarrier for gene therapy (Li et al. 2015; Zhang et al. 2018). Moreover, several CS derivatives, such as trimethyled chitosan (TMC) and hydrophobic groups modified CS, have also been widely evaluated as nanocarriers with additional functions, such as improved mucoadhesive and intestinal permeation capabilities (Liu et al. 2019). Commonly used synthesized polymeric nanomaterials include PLA (poly (D, L-lactide)), PLG (poly (lactide-co-glycolide)), PLGA, and poly (cyanoacrylate) (PCA). It has been reported that the distribution of drugs encapsulated into PLGA NPs increased in the tumor site, with reduced systemic adverse reaction (Xu et al. 2017). However, it should be noted that the acidic nature of PLGA monomers is not suitable for acid sensitive bioactives. Thus, to overcome these problems, PLGA-based NPs can be prepared by blending with other materials, such as alginate, CS, pectin, poly (propylenefumarate), and polyvinylacohol (Bose et al. 2016).

13.3.2 Mucosal Drug Delivery

In addition to the outstanding performance of biomaterial-based nanocarriers in parenteral drug delivery, these nanocarriers have also been found to have extensive applications in mucosal drug delivery, including oral, pulmonary, intranasal, ocular, vaginal, or rectal drug delivery, with improved dissolution of hydrophobic drugs, enhanced cellular uptake, or site-directed drug targeting. However, the existence of the vicious and elastic mucus layer, which covers the gastrointestinal (GI) tract, lung airways, female reproductive tract, nose, and eye, performs an important role as a diffusion barrier for various nutrients, foreign particles, and hydrophobic drugs (das Neves and Sarmento 2018). Typically, the limited permeability of conventional particle-based drug delivery systems leads to their clearance from the mucosal tissue within seconds to a few hours, thereby limiting the duration of sustained drug delivery locally. In order to overcome these challenges, mucoadhesive nanoparticulate delivery systems can be designed, which are expected to remain at mucosal membranes for a longer period of time with prolonged and enhanced drug absorption.

Date back to 1947, mucoadhesive polymers have been developed in order to prolong drug residence time on mucosal surfaces (Wu et al. 2018; Zhang et al. 2018). With the rapid development of nanotechnology, mucoadhesive biomaterials from both natural and synthesized sources are extensively explored (Table 13.4) and used to prepare NPs in order to increase residence time of the particles at the mucus layer, leading to enhanced drug uptake at the site of absorption. Mucus can potentially bind to NPs via various physicochemical mechanisms, such as hydrophobic interaction, electrostatic interaction, and hydrogen binding (Zahir-Jouzdani et al. 2018). Figure 13.2 schemically presented the polymers commonly used for mucoadhesive NPs design and the related mechanism of binding. Anionic nanocarriers are characterized by the presence of carboxyl and sulfate functional groups that give rise to a net overall negative charge at pH values exceeding the pKa of the polymer. Among them, polycarbophil and carbomer, PAA derivatives have been studied extensively as mucoadhesive nanomaterials for drug delivery (Andrews et al. 2009; Makhlof et al. 2008). Notably, PAA, generally recognized as safe (GRAS) for oral use by the FDA, are widely used as mucoadhesive nanomaterials due to their nonirritant, nontoxic properties. Of the cationic polymers, undoubtedly CS is the most extensively investigated one. Whereas PAAs bind to mucus via hydrogen bonds, CS has been reported to bind via ionic interactions between primary amino functional

Table 13.4 Different mucoadhesive nanoparticle systems and their applications

Mucosa targeting	Carrier	Drug	Properties
Buccal mucosa	Thiolated CS/polyvinyl alcohol (Samprasit et al. 2015)	Garcinia mangostana extract	To maintain oral hygiene by reducing the bacterial growth that causes the dental caries
	Polyvinyl alcohol (Singh et al. 2015)	Docetaxel	Enhanced local absorption of anticancer drugs
	Gelatin and photoreactive polyethylene glycol diacrylate (Inoo et al. 2018)	Insulin	Significant reduction of blood glucose level
GI mucosa	Alginate/ wheat germ agglutinin (Dodov et al. 2009)	Insulin	Significant reduction of blood glucose level
	PAA-cysteine/PVP (Dodov et al. 2009)	Insulin	Significant reduction of blood glucose level
	P(MAA-EG) (poly(methacrylic acid-graft-ethylene glycol) (Schoener and Peppas 2013)	Insulin	Significant reduction of blood glucose level
Nasal mucosa	N-trimethyl-CS (Sayin et al. 2008)	Albumin	Significantly enhanced uptake of the model protein by the nasal mucosa
	PAA-cysteine/glutathione (Jespersen et al. 2014)	Human growth hormone (HGH)	Threefold improvement in the relative bioavailability of HGH
	Glycol CS (Lee et al. 2016)	DNA vaccine	Higher mucosal and cellular immune response

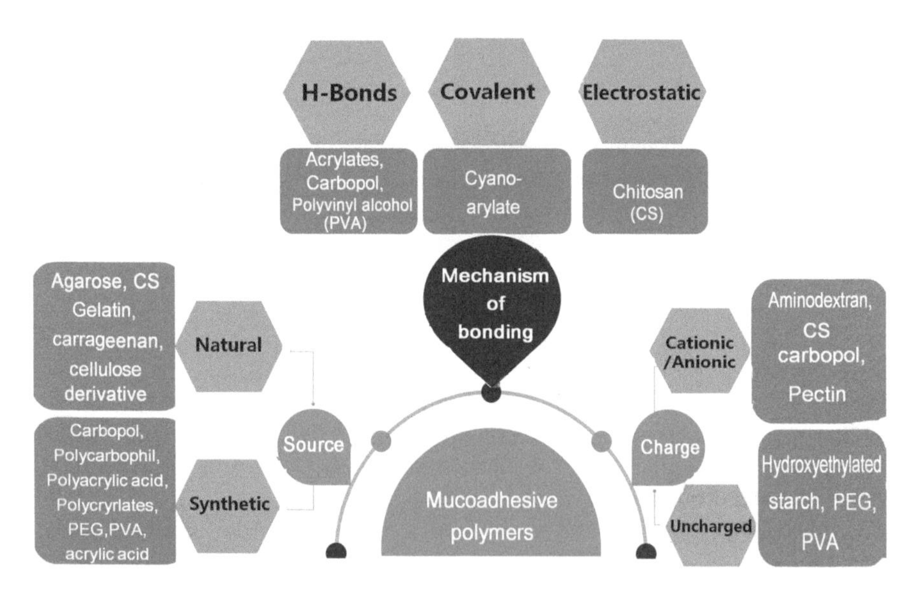

Fig. 13.2 An overview of polymeric materials used for mucoadhesive nanoparticles design and related mechanism of binding

groups and the sialic acid and sulfonic acid substructure of the mucus (Vanic and Skalko-Basnet 2014). Besides, enhancement of mucosal delivery may be obtained through the use of appropriate cytoadhesives nanobiomaterials that can bind to mucosal surfaces. Such widely used nanobiomaterials are lectins, which belong to a group of structurally diverse proteins and glycoproteins, and show the potential to bind reversibly to specific carbohydrate residues (Duennhaupt et al. 2012).

Despite of the advantages of mucoadhesive NPs, its limitation is also quite apparent. Mucoadhesive nanomaterials can efficiently adhere to the mucus layer before reaching the mucosa; however, this might prevent the particle from penetrating across the mucus layer and entering the underlying epithelia (Liu et al. 2018). Thus, for a more efficient transport of drugs to the tissue, the design of NPs with mucus penetration capability is crucially important, which can be achieved by using muco-inert NPs, virus-mimicking NPs, and enzyme-conjugate NPs as schemically described in (Fig. 13.3). Considering the lipophilic nature of mucus, muco-inert NPs, which were coated with "stealth" excipients, can be prepared aiming at decreasing mucous interaction and making the nanomaterials more slippery (Fig. 13.3a). Several polymers including poloxamers and PEG are widely used for surface modification of NPs for this objective (Netsomboon and Bernkop-Schnurch

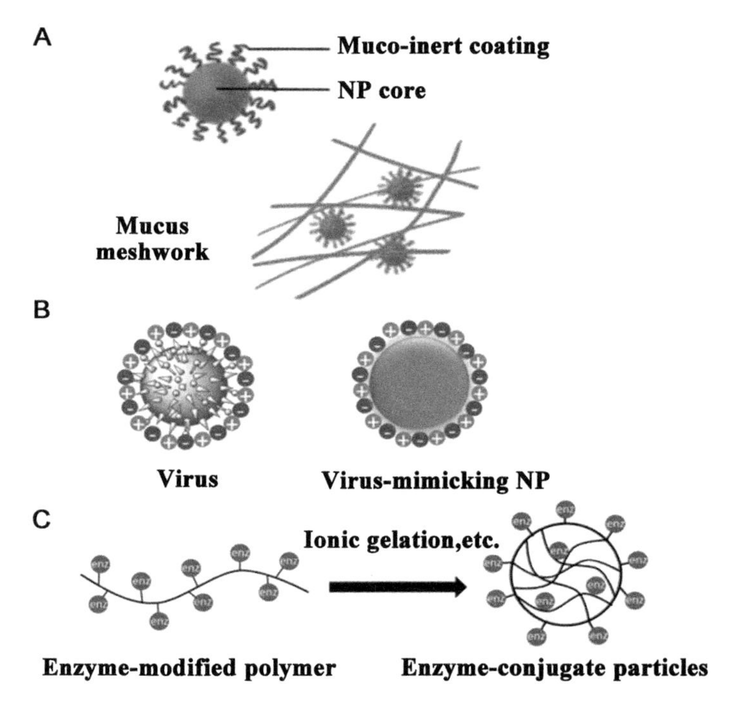

Fig. 13.3 Schematic representatives of the designed muco-permeation nanoparticles including muco-inert nanoparticles (**a**), virus-mimicking nanoparticles (**b**), and enzyme-conjugate nanoparticles (**c**). (Reproduced from (Netsomboon and Bernkop-Schnurch 2016) with copyright permission)

2016). Pluronic F-127, a type of poloxamer, is extensively used for mucopenetrating NPs preparation. Moreover, inspired by natural virus, NPs, which present a highly densely charged surface bearing both positive and negative charges, are formulated with the purpose to increase their mucus permeation ability (Fig. 13.3b). De Sousa et al. prepared NPs by combining chitosan with chondroitin sulfate with a slightly positive (4.02 mV) charge. Not surprising, positively charged NPs (high chitosan content) was found not permeating across the mucus due to the electrostatic interaction with the negatively charged components in the mucus, while negative (high chondroitin sulfate content) and near neutral particles revealed a higher permeation (de Sousa et al. 2015). Another strategy that can be used to create mucous-penetrating NPs is to coat particles with enzymes. Enzyme-conjugate NPs (Fig. 13.3c) are capable of cleaving certain substructures within the three-dimensional network of the mucus, without destroying the whole mucus gel layer. So far, mucolytic enzymes such as bromelain, papain, pronase, and trypsin are immobilized on the surface of nanomaterials (Shan et al. 2017). These enzymes are capable of cleaving amide bonds within mucin glycoproteins in a very efficient manner, which makes coated nanomaterials much easier to penetrate the mucus.

However, the variability of mucosa and their properties challenges the design of polymeric NP-based systems. Fortunately, the rapid development of natural, synthetic, and semisynthetic polymers commercially available and the fact that many of them have been approved by regulatory agencies enable the design of polymeric material-based NPs based on the intrinsic feature of a specific mucosa.

13.4 Challenges and Future Perspectives on Nanobiomaterials

So far, a series of natural or synthetic material-based nanodrugs are playing an important role in clinic for disease therapy or in clinical trials. A number of examples of both FDA-approved products and those under clinical trials are designed by utilization of nanomaterials as a modifying agent for drug delivery. The versatility and diversity of potential biomaterials allows for a flexible design of nanocarriers with tailor-made properties based on the requirement in clinical application. The fact that some nanosystem, such as liposomes, albumin NPs, and polymeric micelles, are on the market and several biomaterial-based nanodrug delivery systems are in clinical trials indicated that potentially more nanobiomaterials can be used as a drug carrier in the near future. However, the key to transform nanotechnology from basic research into clinical products involves further understanding of the surface chemistry of nanomaterials and the interaction of these nanomaterials with drug and in vivo environment. To design smarter, functional nanomaterials with maximized therapeutic efficacy and good safety in drug delivery, several impediments must be addressed right now, such as ambiguous structure-function relationship of these nanomaterials in drug delivery system design and varied material characteristics such as molecular weight, shape, charge, composition, and complex architectures of nanocarriers. Another challenge posing as a major hindrance for nanomedicine design is the large-scale production of nanomaterials for commercialization purpose under Good Manufacturing Practices (GMP) conditions. And the safety of their administration must be carefully considered. With recent draft guidelines published by the FDA on the importance of nanomaterials characterization for different regulated environments, biomaterial-based nanocarriers will show superior capacity in drug delivery with tunable properties in the near future.

Acknowledgments This project is financially supported by the Natural Science Foundation of China (Grant No. 31870987).

References

Akbar MU, Zia KM, Nazir A, Iqbal J, Ejaz SA, Akash MSH (2018) Pluronic-based mixed polymeric micelles enhance the therapeutic potential of curcumin. AAPS PharmSciTech 19:2719–2739

An FF, Zhang XH (2017) Strategies for preparing albumin-based nanoparticles for multifunctional bioimaging and drug delivery. Theranostics 7:3667–3689

Andrews GP, Laverty TP, Jones DS (2009) Mucoadhesive polymeric platforms for controlled drug delivery. Eur J Pharm Biopharm 71:505–518

Autio KA, Garcia JA, Alva AS, Hart LL, Milowsky MI, Posadas EM, Ryan CJ, Summa JM, Youssoufian H, Scher HI, Dreicer R (2016) A phase 2 study of BIND-014 (PSMA-targeted docetaxel nanoparticle) administered to patients with chemotherapy-naive metastatic castration-resistant prostate cancer (mCRPC). J Clin Oncol 34

Aziz ZABA, Ahmad A, Mohd-Setapar SH, Hassan H, Lokhat D, Kamal MA, Ashraf GM (2017) Recent advances in drug delivery of polymeric Nano-micelles. Curr Drug Metab 18:16–29

Bobo D, Robinson KJ, Islam J, Thurecht KJ, Corrie SR (2016) Nanoparticle-based medicines: a review of FDA-approved materials and clinical trials to date. Pharm Res 33:2373–2387

Bose RJ, Lee S-H, Park H (2016) Lipid-based surface engineering of PLGA nanoparticles for drug and gene delivery applications. Biomater Res 20:34–34

Bosselmann S, Williams RO (2012) Has nanotechnology led to improved therapeutic outcomes? Drug Dev Ind Pharm 38:158–170

Cabral H, Kataoka K (2014) Progress of drug-loaded polymeric micelles into clinical studies. J Control Release 190:465–476

Chang HI, Yeh MK (2012) Clinical development of liposome-based drugs: formulation, characterization, and therapeutic efficacy. Int J Nanomedicine 7:49–60

Dandamudi S, Campbell RB (2007) Development and characterization of magnetic cationic liposomes for targeting tumor microvasculature. BBA-Biomembranes 1768:427–438

das Neves J, Sarmento B (2018) Technological strategies to overcome the mucus barrier in mucosal drug delivery preface. Adv Drug Deliver Rev 124:1–2

de Sousa IP, Steiner C, Schmutzler M, Wilcox MD, Veldhuis GJ, Pearson JP, Huck CW, Salvenmoser W, Bernkop-Schnuerch A (2015) Mucus permeating carriers: formulation and characterization of highly densely charged nanoparticles. Eur J Pharm Biopharm 97:273–279

Dodov MG, Calis S, Crcarevska MS, Geskovski N, Petrovska V, Goracinova K (2009) Wheat germ agglutinin-conjugated chitosan-ca-alginate microparticles for local colon delivery of 5-FU: development and in vitro characterization. Int J Pharm 381:166–175

Dong W, Wang X, Liu C, Zhang X, Chen X, Kou Y, Mao S (2018) Chitosan based polymer-lipid hybrid nanoparticles for oral delivery of enoxaparin. Int J Pharm 547:499–505

Duennhaupt S, Barthelmes J, Thurner CC, Waldner C, Sakloetsakun D, Bernkop-Schnuerch A (2012) S-protected thiolated chitosan: synthesis and in vitro characterization. Carbohyd Polym 90:765–772

Elgadir MA, Uddin MS, Ferdosh S, Adam A, Chowdhury AJK, Sarker MZI (2015) Impact of chitosan composites and chitosan nanoparticle composites on various drug delivery systems: a review. J Food Drug Anal 23:619–629
Elzoghby AO, Samy WM, Elgindy NA (2012) Albumin-based nanoparticles as potential controlled release drug delivery systems. J Control Release 157:168–182
Guan J, Liu Q, Zhang X, Zhang Y, Chokshi R, Wu H, Mao S (2018) Alginate as a potential diphase solid dispersion carrier with enhanced drug dissolution and improved storage stability. Eur J Pharm Sci 114:346–355
Guo B, Ma PX (2014) Synthetic biodegradable functional polymers for tissue engineering: a brief review. Sci China Chem 57:490–500
Hallan SS, Kaur P, Kaur V, Mishra N, Vaidya B (2016) Lipid polymer hybrid as emerging tool in nanocarriers for oral drug delivery. Artif Cells Nanomed Biotechnol 44:334–349
Han J, Zhao D, Li D, Wang X, Jin Z, Zhao K (2018) Polymer-based nanomaterials and applications for vaccines and drugs. Polymers 10
Immordino ML, Dosio F, Cattel L (2006) Stealth liposomes: review of the basic science, rationale, and clinical applications, existing and potential. Int J Nanomedicine 1:297–315
Inoo K, Bando H, Tabata Y (2018) Enhanced survival and insulin secretion of insulinoma cell aggregates by incorporating gelatin hydrogel microspheres. Regen Ther 8:29–37
Janas C, Mostaphaoui Z, Schmiederer L, Bauer J, Wacker MG (2016) Novel polymeric micelles for drug delivery: material characterization and formulation screening. Int J Pharm 509:197–207
Jeon YO, Lee JS, Lee HG (2016) Improving solubility, stability, and cellular uptake of resveratrol by nanoencapsulation with chitosan and gamma-poly (glutamic acid). Colloids Surf B Biointerfaces 147:224–233
Jespersen GR, Matthiesen F, Pedersen AK, Andersen HS, Kirsebom H, Nielsen AL (2014) A thiol functionalized cryogel as a solid phase for selective reduction of a cysteine residue in a recombinant human growth hormone variant. J Biotechnol 173:76–85
Jia H-J, Jia F-Y, Zhu B-J, Zhang W-P (2017) Preparation and characterization of glycyrrhetinic-acid loaded PEG-modified liposome based on PEG-7 glyceryl cocoate. Eur J Lipid Sci Tech 119
Jiao Y, Pang X, Zhai G (2016) Advances in hyaluronic acid-based drug delivery systems. Curr Drug Targets 17:720–730
Jones M-C (2015) Thinking outside the 'Block': alternative polymer compositions for micellar drug delivery. Curr Top Med Chem 15:2254–2266
Kapoor DN, Bhatia A, Kaur R, Sharma R, Kaur G, Dhawan S (2015) PLGA: a unique polymer for drug delivery. Ther Deliv 6:41–58
Kausar A (2017) Scientific potential of chitosan blending with different polymeric materials: a review. J Plast Film Sheet 33:384–412
Kocisova E, Antalik A, Prochazka M (2013) Drop coating deposition Raman spectroscopy of liposomes: role of cholesterol. Chem Phys Lipids 172:1–5
Larsen MT, Kuhlmann M, Hvam ML, Howard KA (2016) Albumin-based drug delivery: harnessing nature to cure disease. Mol Cell Ther 4:3
Lee YH, Park HI, Choi JS (2016) Novel glycol chitosan-based polymeric gene carrier synthesized by a Michael addition reaction with low molecular weight polyethylenimine. Carbohyd Polym 137:669–677
Li L, Zhang X, Gu X, Mao S (2015) Applications of natural polymeric materials in solid Oral modified-release dosage forms. Curr Pharm Des 21:5854–5867
Liu C, Kou Y, Zhang X, Cheng H, Chen X, Mao S (2018) Strategies and industrial perspectives to improve oral absorption of biological macromolecules. Expert Opin Drug Deliv 15:223–233
Liu C, Kou Y, Zhang X, Dong W, Cheng H, Mao S (2019) Enhanced oral insulin delivery via surface hydrophilic modification of chitosan copolymer based self-assembly polyelectrolyte nanocomplex. Int J Pharm 554:36–47
Makadia HK, Siegel SJ (2011) Poly lactic-co-glycolic acid (PLGA) as biodegradable controlled drug delivery carrier. Polymers 3:1377–1397

Makhlof A, Werle M, Takeuchi H (2008) Mucoadhesive drug carriers and polymers for effective drug delivery. J Drug Delivery Sci Technol 18:375–386
Mao S, Sun W, Kissel T (2010) Chitosan-based formulations for delivery of DNA and siRNA. Adv Drug Deliv Rev 62:12–27
Mariam J, Sivakami S, Dongre PM (2016) Albumin corona on nanoparticles - a strategic approach in drug delivery. Drug Deliv 23:2668–2676
Netsomboon K, Bernkop-Schnurch A (2016) Mucoadhesive vs. mucopenetrating particulate drug delivery. Eur J Pharm Biopharm 98:76–89
Ogunleye A, Bhat A, Irorere VU, Hill D, Williams C, Radecka I (2015) Poly-gamma-glutamic acid: production, properties and applications. Microbiology 161:1–17
Ohya Y, Takeda S, Shibata Y, Ouchi T, Kano A, Iwata T, Mochizuki S, Taniwaki Y, Maruyama A (2011) Evaluation of polyanion-coated biodegradable polymeric micelles as drug delivery vehicles. J Control Release 155:104–110
Ossipov DA (2010) Nanostructured hyaluronic acid-based materials for active delivery to cancer. Expert Opin Drug Deliv 7:681–703
Otto F, Brezesinski G, van Hoogevest P, Neubert RHH (2018) Physicochemical characterization of natural phospholipid excipients with varying PC content. Colloid Surface A 558:291–296
Pandita D, Kumar S, Lather V (2015) Hybrid poly(lactic-co-glycolic acid) nanoparticles: design and delivery prospectives. Drug Discov Today 20:95–104
Park C-K, Kang H-W, Kim T-O, Kim E-Y, Ban H-J, Oh I-J, Kwon Y-S, Lim S-C, Yoon B-K, Choi Y-D, Kim Y-I, Ki H-S, Kim Y-C, Kim K-S (2010) A case of ischemic colitis associated with paclitaxel loaded polymeric micelle (Genexol-PM®) chemotherapy. Tuberc Respir Dis 69:115–118
Pepic I, Lovric J, Filipovic-Grcic J (2013) How do polymeric micelles cross epithelial barriers? Eur J Pharm Sci 50:42–55
Qiu L, Zheng C, Jin Y, Zhu K (2007) Polymeric micelles as nanocarriers for drug delivery. Expert Opin Ther Pat 17:819–830
Rao NV, Yoon HY, Han HS, Ko H, Son S, Lee M, Lee H, Jo D-G, Kang YM, Park JH (2016) Recent developments in hyaluronic acid-based nanomedicine for targeted cancer treatment. Expert Opin Drug Deliv 13:239–252
Sakai-Kato K, Nanjo K, Kawanishi T, Okuda H, Goda Y (2015) Effects of lipid composition on the properties of doxorubicin-loaded liposomes. Ther Deliv 6:785–794
Samprasit W, Kaomongkolgit R, Sukma M, Rojanarata T, Ngawhirunpat T, Opanasopit P (2015) Mucoadhesive electrospun chitosan-based nanofibre mats for dental caries prevention. Carbohyd Polym 117:933–940
Sayin B, Somavarapu S, Li XW, Thanou M, Sesardic D, Alpar HO, Senel S (2008) Mono-N-carboxymethyl chitosan (MCC) and N-trimethyl chitosan (TMC) nanoparticles for non-invasive vaccine delivery. Int J Pharm 363:139–148
Schoener CA, Peppas NA (2013) pH-responsive hydrogels containing PMMA nanoparticles: an analysis of controlled release of a chemotherapeutic conjugate and transport properties. J Biomat Sci Polym E 24:1027–1040
Shan W, Cui Y, Liu M, Wu L, Xiang Y, Guo Q, Zhang Z, Huang Y (2017) Systematic evaluation of the toxicity and biodistribution of virus mimicking mucus-penetrating DLPC-NPs as oral drug delivery system. Int J Pharm 530:89–98
Shi C, Sun Y, Wu H, Zhu C, Wei G, Li J, Chan T, Ouyang D, Mao S (2016a) Exploring the effect of hydrophilic and hydrophobic structure of grafted polymeric micelles on drug loading. Int J Pharm 512:282–291
Shi D, Ran M, Zhang L, Huang H, Li X, Chen M, Akashi M (2016b) Fabrication of biobased polyelectrolyte capsules and their application for glucose-triggered insulin delivery. ACS Appl Mater Interfaces 8:13688–13697
Shiraishi K, Hamano M, Ma H, Kawano K, Maitani Y, Aoshi T, Ishii KJ, Yokoyama M (2013) Hydrophobic blocks of PEG-conjugates play a significant role in the accelerated blood clearance (ABC) phenomenon. J Control Release 165:183–190

Singh H, Sharma R, Joshi M, Garg T, Goyal AK, Rath G (2015) Transmucosal delivery of docetaxel by mucoadhesive polymeric nanofibers. Arti Cells Nanomed Biotechnol 43:263–269
Sleep D (2015) Albumin and its application in drug delivery. Expert Opin Drug Deliv 12:793–812
Socinski M (2006) Update on nanoparticle albumin-bound paclitaxel. Clin Adv Hematol Oncol 4:745–746
Song R, Murphy M, Li C, Ting K, Soo C, Zheng Z (2018) Current development of biodegradable polymeric materials for biomedical applications. Drug Des Devel Ther 12:3117–3145
Tagami T, Ozeki T (2017) Recent trends in clinical trials related to carrier-based drugs. J Pharm Sci 106:2219–2226
Tomkinson B, Bendele R, Giles FJ, Brown E, Gray A, Hart K, LeRay JD, Meyer D, Pelanne M, Emerson DL (2003) OSI-211, a novel liposomal topoisomerase I inhibitor, is active in SCID mouse models of human AML and ALL. Leukemia Res 27:1039–1050
Valle JW, Armstrong A, Newman C, Alakhov V, Pietrzynski G, Brewer J, Campbell S, Corrie P, Rowinsky EK, Ranson M (2011) A phase 2 study of SP1049C, doxorubicin in P-glycoprotein-targeting pluronics, in patients with advanced adenocarcinoma of the esophagus and gastro-esophageal junction. Invest New Drug 29:1029–1037
van Hoogevest P (2017) Review - An update on the use of oral phospholipid excipients. Eur J Pharm Sci 108:1–12
Vanic Z, Skalko-Basnet N (2014) Mucosal nanosystems for improved topical drug delivery: vaginal route of administration. J Drug Deliv Sci Technol 24:435–444
Wang X, Gu X, Wang H, Yang J, Mao S (2018) Enhanced delivery of doxorubicin to the liver through self-assembled nanoparticles formed via conjugation of glycyrrhetinic acid to the hydroxyl group of hyaluronic acid. Carbohydr Polym 195:170–179
Wu T, Li Y, Lee DS (2017) Chitosan-based composite hydrogels for biomedical applications. Macromol Res 25:480–488
Wu L, Shan W, Zhang Z, Huang Y (2018) Engineering nanomaterials to overcome the mucosal barrier by modulating surface properties. Adv Drug Deliver Rev 124:150–163
Xu Y, Kim C-S, Saylor DM, Koo D (2017) Polymer degradation and drug delivery in PLGA-based drug-polymer applications: a review of experiments and theories. J Biomed Mater Res Part B Appl Biomater 105:1692–1716
Yadav AK, Mishra P, Agrawal GP (2008) An insight on hyaluronic acid in drug targeting and drug delivery. J Drug Target 16:91–107
Yarmolenko PS, Zhao Y, Landon C, Spasojevic I, Yuan F, Needham D, Viglianti BL, Dewhirst MW (2010) Comparative effects of thermosensitive doxorubicin-containing liposomes and hyperthermia in human and murine tumours. Int J Hyperth 26:485–498
Yu M, Jambhrunkar S, Thorn P, Chen J, Gu W, Yu C (2013) Hyaluronic acid modified mesoporous silica nanoparticles for targeted drug delivery to CD44-overexpressing cancer cells. Nanoscale 5:178–183
Zahir-Jouzdani F, Wolf JD, Atyabi F, Bernkop-Schnuerch A (2018) In situ gelling and mucoadhesive polymers: why do they need each other? Expert Opin Drug Deliv 15:1007–1019
Zhang JA, Anyarambhatla G, Ma L, Ugwu S, Xuan T, Sardone T, Ahmad I (2005) Development and characterization of a novel Cremophor (R) EL free liposome-based paclitaxel (LEP-ETU) formulation. Eur J Pharm Biopharm 59:177–187
Zhang K, Tang X, Zhang J, Lu W, Lin X, Zhang Y, Tian B, Yang H, He H (2014) PEG-PLGA copolymers: their structure and structure-influenced drug delivery applications. J Control Release 183:77–86
Zhang Y, Mintzer E, Uhrich KE (2016) Synthesis and characterization of PEGylated bolaamphiphiles with enhanced retention in liposomes. J Colloid Interf Sci 482:19–26
Zhang X, Cheng H, Dong W, Zhang M, Liu Q, Wang X, Guan J, Wu H, Mao S (2018) Design and intestinal mucus penetration mechanism of core-shell nanocomplex. J Control Release 272:29–38
Zylberberg C, Matosevic S (2017) Bioengineered liposome-scaffold composites as therapeutic delivery systems. Ther Deliv 8:425–445

Future Prospected of Engineered Nanobiomaterials in Human Health Care

14

Guilherme Barroso L. de Freitas and Durinézio J. de Almeida

Abstract

In recent years we have witnessed the beginning of the third medical revolution with the introduction of personalized treatments and the use of biotechnology to obtain antibodies and peptides with high selectivity for the therapeutic target. However, although this paradigm is still very recent, it is possible that we are simultaneously experiencing, even if still discreet, the fourth revolution of medicine. It is characterized by the development and miniaturization of devices. This allows remote and real-time monitoring of patient conditions, the introduction of smart dressings and implants, and the use of matrices made with organic and synthetic materials for varied purposes, such as serving as a biodegradable matrix for tissue growth or even as substituents for parts of the human body (cyborgs). These biotechnological products became possible because nanomaterials with greater flexibility, electrical conductivity, biocompatibility, resilience, and extremely sensitive were developed. In this chapter some of these materials and their practical applications are presented.

Keywords

Tattoo-Based wearable electrochemical devices · Smart dressing · Nanosensors · Cyborgs · Nano implant microchip devices

G. B. L. de Freitas (✉)
Department of Biochemistry and Pharmacology, Federal University of Piauí (UFPI), Teresina, Brazil
e-mail: guilhermebarroso@ufpi.edu.br

D. J. de Almeida
Department of Biology, Midwestern State University (UNICENTRO), Guarapuava, Brazil

P. Chandra, R. Prakash (eds.), *Nanobiomaterial Engineering*,
https://doi.org/10.1007/978-981-32-9840-8_14

14.1 Introduction

The association of nanotechnology with the treatment, medical diagnosis, monitoring, implants, tissue-engineered constructs, and even the design of cyborgs has brought a new concept in science, namely, nanomedicine. These devices have emerged to leverage the advances of the applications of nanobiomaterials in traditional medicine. In this interdisciplinary field of science, different areas of expertise, particularly pharmacists, chemists, engineers, physicians, and computer scientists, ought to work together.

Many advances in nanomaterial devices have recently been introduced or are currently seeking a development method and a further validation to be employed in clinical routine. The range of applicability of new products is huge due to the level of the scientific advancements. They have enabled devices to monitor the health status in real time, in a micronized way, integrated to the body to overcome some physiological limitations and even to generate superhuman responses as in the senses of hearing and sight.

In this chapter, we will highlight the recent progress in these emerging areas and briefly outline the possible future directions and advances. The most innovative topics within these fields, such as biomonitors (wearable or implants) and the development of cyborgs (implants and cyborganics), will be addressed. The monitoring of physiological parameters by the devices allows the patient's treatment to be faster, more precise, efficient, and at a lower cost. This characteristic was obtained after the introduction of nanomaterials such as coated carbon nanotubes, polyaniline (PANI), copper nanowires, and poly(N-isopropylacrylamide) (pNIPAM), which are reported during this chapter. They allow the measurement of temperature, pH, humidity, and even electrolytes or organic markers such as glycemia. The development and applicability of cyborgs are two of the most prominent areas within the field of nanobiomaterials. There is a clear need for biocompatible nanomaterials with high flexibility. We can highlight the silicon nanowire field-effect transistor (FET)-based nanoelectronic biomaterials and the use of 3D printer techniques using the raw material Ag nanoparticles and cartilage cells embedded within an alginate hydrogel matrix for cyborg implementation. Throughout the chapter, other nanomaterials will be reported.

14.2 Biomonitors

Many of the advances in bioelectronics occurred by the emergence of biocompatible materials, flexible, resilient, and smarter and less expensive that can be integrated with the dynamics of the human body. Nanobiomaterials are able to improve the detection systems and parameters such as reproducibility, precision, and reduced cost of experiments. That is possible due to their favorable properties such as high surface-to-volume ratio and sensibility, desired electrochemical properties and stabilities, and the demand for a very small amount of reagents and samples. Some nanomaterials exhibit conductivity and peculiar quantum effects that can be used to amplify the signal stemming from the detection. However, the appropriate choice of nanomaterials is essential to successfully miniaturize the tests.

The desire to develop devices that can monitor real-time health status of a patient and evaluate the prognosis or therapeutic response of an intervention is directly dependent on technological advances and practical application of miniaturized sensors and the manufacture of "smart" materials. The development of flexible, efficient, and biocompatible materials allows commercial viability and great reliability of health monitors. Currently, the challenges are to be able to adapt and incorporate these materials into smart wearable systems (SWSs) (smart wound bandages, e-skin, and tattoo-based sensors), implantable diagnostic devices (IDDs), and miniaturized diagnostic devices (lab-on-a-chip and organ-on-a-chip) (Fig. 14.1).

Some of these new technologies enable us to perform personalized diagnosis, even in real time or a comprehensive outlook of a disease genetic makeup. Therefore, such information can guide decisions for patient-specific diagnosis and a smart therapy (pharmacogenetics, clinician-oriented or patient-oriented point of care [POC], and the best therapeutic regime). Possibly, one of the main advancements allowed by this association lies in the implementation of the POC. It will allow easier access from the primary care physician and the patient to clinical diagnostic testing. The perception of physiological changes allows the maintenance of homeostasis through feedback by the release of active substances. That can be considered

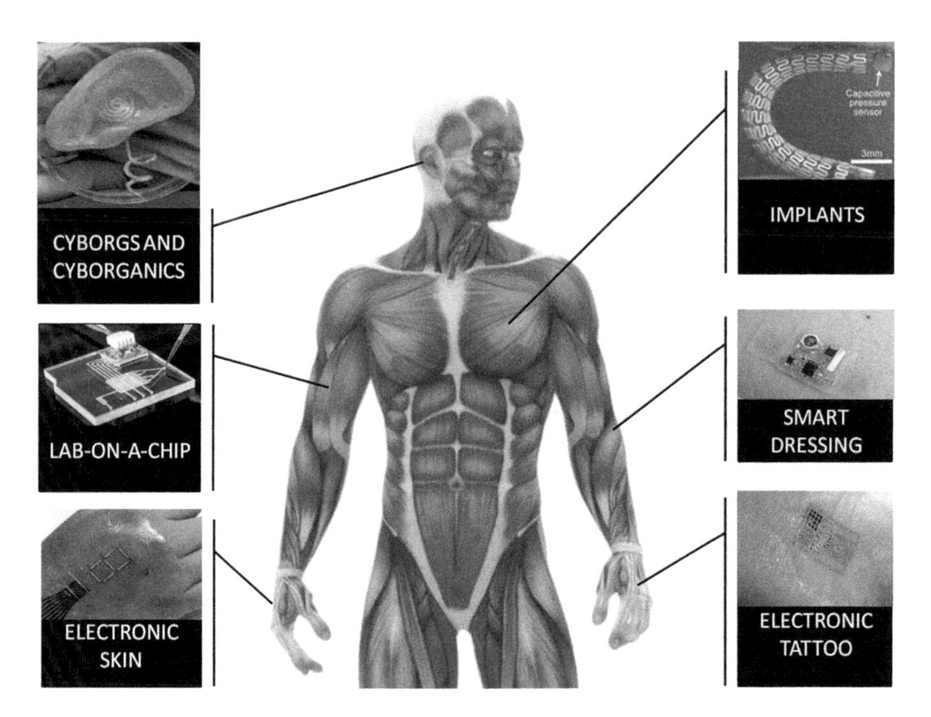

Fig. 14.1 The applicability of nanomaterials in medicine

as an autonomous and intelligent first-aid care with great utility in cancer, epilepsy, and cardiovascular diseases, among other clinical conditions.

The reduced size of these devices makes manipulation difficult by hand; therefore, they are automated or autonomous. Even though seductive, the miniaturization process of biosensors or cyborgs has a physical limitation because of three main reasons. The first is the size of blood cells, measured in micrometers. Therefore, the reduction below this diameter would prevent the flow of these cells by the device generating altered results or, in the case of cyborganics, nutritional deficiency and hypoxia. Another impediment is the volume of the aliquot, since the larger the device reduction, the smaller the sample volume. However, depending on what one wants to analyze, a larger volume of the aliquot is needed to find the analyte in the medium; for example, there is on average only one tumor cell per milliliter of blood; therefore, there is recommendation for the collection of at least 7 mL for a more reliable analysis. The third point is the increase in the sensitivity of excessive form, which generates false-positive results due to the contamination of substances present in the surface of the skin or in the ambient air, which can alter parameters like sanguine gases, pH, and even the presence of microorganisms.

In addition, another challenge to the large-scale introduction of these technologies is to obtain a high barrier property to avoid dimensional instability of the polymer material because of moisture and gas permeation. However, there are some materials that should not be completely impermeable such as smart bandages because gaseous exchange and moisture control are necessary. In these cases, semi-impermeable surfaces are desired. Therefore, the choice of nanomaterial is fundamental for the proper functioning of the devices. They must present flexibility, biocompatibility, and semi-conductivity as key characteristics needed in these applications. For instance, nanomaterials built with polyimide and parylene have been extensively used as components for bionanosensors as IDDs and SWSs (Mostafalu and Sonkusale 2014).

14.3 Wearable Devices

14.3.1 Smart Bandages

Wound healing is a dynamic and interactive biological process involving delivery of several inflammatory mediators, matrix molecules, and nutrients to the wound site. The wound healing stages are made up of three main phases, inflammatory, proliferative, and maturation, also known as remodeling. In normal healthy subjects, the healing process will occur properly. However, there is a possibility of errors occurring in numerous steps along this pathway, chiefly if the patient has certain pathological conditions (e.g., diabetes, infections, severe burns, and low nutrition content). In these cases, the prognosis worsens, and it can create chronic and honorable wounds with a high chance of becoming larger morbidities. Tissue healing in elderly patients is usually a slower and more costly process. Therefore, it is expected that chronic wounds stand out as one of the major global health challenges in the coming years. Part of this clinical inefficiency is in the management of the wound site because it is neither monitored nor attended properly.

There is great difficulty in treating chronic wounds, which means that costs can exceed $ 90 billion. These costs serve as an incentive for the development of new technologies. It is represented by a billionaire commercial industry divided into four main curative groups: passive, interactive, advance, and bioactive (Han and Ceilley 2017; Nussbaum et al. 2018). *Passive dressings* are considered as the elementary ones and include bandages and cloth gauze as basic physical barriers of protection. The adherence of these dressings to the wound bed is habitual, stable, and with difficult displacement. They are suitable for dry shallow wounds because they are not porous, nor do they retain local moisture (Han and Ceilley 2017).

On the other side, the introduction of polymeric films, hydrogels, and foams make interactive dressings permeable to gases and fluids. When the patient has exudate-producing wounds, for example, burns, deep wounds, and diabetic ulcers, the foam or sponge dressings can absorb this fluid better than gauze. The hydrogels present in this type of dressing exhibit several positive features, as they are able to reshape the wound site, maintain local humidity, and facilitate dressing removal, unlike what is observed with passive dressings (Fonder et al. 2008).

Advanced dressings maintain wound moisture within appropriate levels to improve the kinetics of tissue regeneration and avoid bacterial proliferation. They are usually made using alginate or hydrocolloids. Alginate is indicated when the wound contains heavy exudate levels because it can form a gel with the ion exchange between the sodium secreted from the wound and the calcium released from the alginate. Platelets are activated by the released calcium salts, which contributes to hemostasis.

Finally, the bioactive dressings are those that include biological scaffolds or drug delivery systems and stimulate healing through tissue regeneration, called tissue-engineered dressings, and stimulating angiogenesis (Brown et al. 2018). This type of dressing showed clinical efficacy in the tissue regeneration of patients with ulcers and burns, and, therefore, despite the higher cost, they reduce hospitalization time and shorten treatment duration. Some commercially available examples are Dermagraft, AlloDerm, Apligraf, and Biobrane.

The next generation of dressings has led us to think totally different about wound care, mainly because of the inclusion of wearable sensors. These sensors have been created in order to precisely identify real-time biomarkers, such as inflammatory markers, pH, and ionic concentration values and temperature at the wound site. The drug delivery system can be integrated with the sensors. With this, the kinetics of drug release will follow instantaneous responses and autonomously to varied situations, for example, to potential presence of pathogenic microorganisms, inflammation, or local allergy. These integrated systems are called multifactorial smart dressings because they allow therapeutic direction, responses to the clinical prognosis, reduction of health care costs, and availability of information on the patient's clinical situation. This information cannot be provided with traditional (not sensory) wound dressings. This information is obtained and transmitted, locally or remotely (Wi-Fi transmission), by sensors in the wound site coupled to the dressings.

The introduction of porous nanomaterials has increased the robustness of these devices, making them more resistant to mechanical changes and electroconductors. The most widely used nanomaterials are nanofibrous substrates (e.g., polyaniline [PANI]), poly(styrene-b-(ethylene-co-butylene)-b-styrene), and multiwall carbon nanotubes and their combination.

Sensors intended for use in wound care products should be able to measure potential biomarkers under abnormal conditions such as changes in pH, humidity, and ion concentrations, which occur, for example, during inflammatory or infectious processes. The sensors must also have the ability to adhere properly to the contours of the body without causing antigenic or toxic responses to the tissues. The pH of a wound can provide important information for interpreting what is occurring in a given tissue, for example, microorganism infection, angiogenic process, and protease activity. While an acidic environment enables healing to occur more quickly, extremely acidic pH values can also indicate bacterial infection. Various types of *pH sensors* have been manufactured for wound care applications, and all of them are flexible, biocompatible, and permeable enough to maintain conformal contact with curved surfaces and to yield smart wound bandages. Basically, they are divided into colorimetric and electrochemical. *While electrochemical pH sensors use, for example, potentiometric measurement and ion-selective field-effect transistors, the colorimetric pH sensors are simpler, because they work without the need for coupled electronic components* (Guinovart et al. 2014). The readings are generally performed by image processing. However, identification even to the naked eye can estimate pH at the wound site. Nevertheless, leakage of dye into the skin is still a recurring problem in the development of many colorimetric sensors. In order to solve this problem, the pH-responsive dyes can be incorporated into mesoporous silica particles and then encompassed by hydrogel fibers to prevent leakage into the skin during the analysis (Sridhar and Takahata 2009).

An example of the potentiometric approach is the thread-based pH sensor consisting of conductive threads and a microfluidic splitter with three channels for the delivery of a sample to the sensing chambers (Fig. 14.2a–d). In this method, the open circuit potential of a working electrode, carbon nanotubes (CNTs) coated with doped PANI, was measured. The dynamics of readout electronic circuitry starts with the detection of the parameters from the sensors, followed by wireless transmission of the results to an electronic device (smartphone or computer) near or in another location (Mostafalu et al. 2016).

A similar study also applied PANI as pH-sensitive nanomaterial in the development of a flexible smart bandage coupled to a Bluetooth module (Fig. 14.2e) (Punjiya et al. 2017). However, this device still needs to be tested in vivo to confirm the good sensitivity and real-time communication. It is noteworthy that the board, yet rigid, is small and inserted externally so as not to influence the performance of the sensor or to cause any discomfort to the prospective patient. In addition, it is possible to insert the electronic components onto the flexible printed circuit board on a polyimide substrate (Siegel et al. 2010).

Electrochemical and colorimetric *temperature sensors* are also usually employed in smart dressings. One of the strategies employed to reduce the cost of the devices

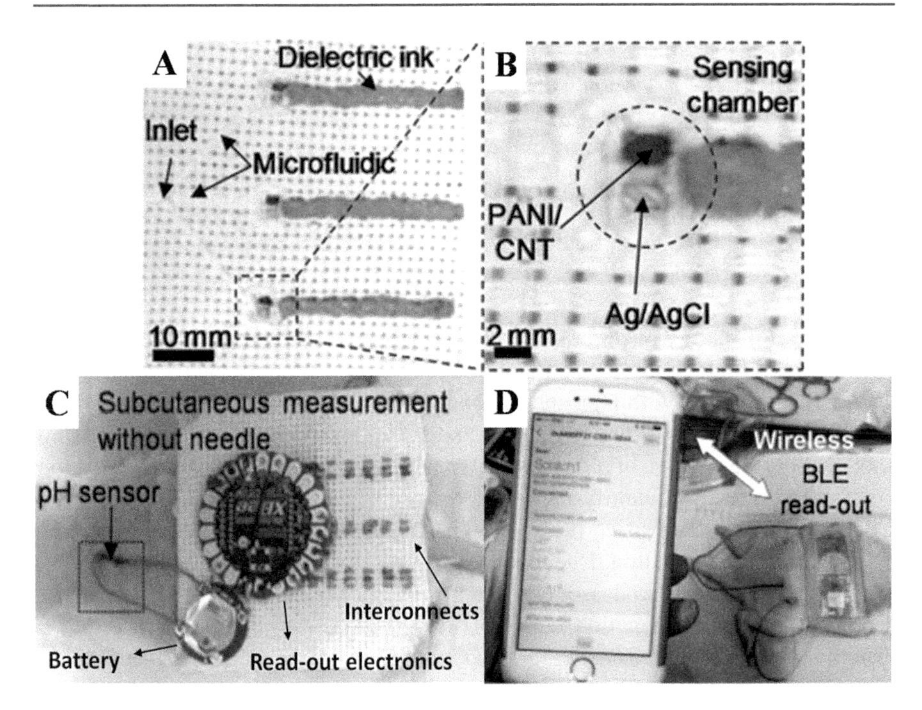

Fig. 14.2 Example of pH sensor. (**a** and **b**) Optical image of a multiplexed microfluidic pH sensor. (**c**) Implanted sensors connected to the patch. (**d**) Sensing system communicating with an external device. (Adapted from Mostafalu et al. 2016)

is the use of material-based inks. However, despite facilitating the flexibility of the final product, conductive paints suffer from the change in mechanical resistance induced by mechanical stresses and skin-shaped contact.

The introduction of the surgical suture made from silk with nanobioelectronic sensor was a great advance. It allows wound closure and monitoring of clinical response. This suture system makes it possible to obtain more than one information at a time, that is, temperature changes and temperature conductivity, which are related to the inflammation progress and the moisture content of the wound. The highest sensitivity in temperature variation (≈ 0.2 °C) was obtained with the incorporation of the silicon-based temperature sensor within this silk suture (Mehrali et al. 2018). This system consists of a polyimide substrate containing a sensor array that is connected with copper nanowires (CuNWs).

Other nanomaterials have enabled several utilities for suture. Nano-silver particles coated on the surface of conventional silk suture may improve inflammation process, promote wound healing, and have antibacterial action (Zakeri et al. 2016). There are studies demonstrating advantages in nano-coating the conductive materials. For example, CuNWs are highly conductive and fundamental for data transfer and heat conduction. Nevertheless, the degree of miniaturization is inversely proportional to electrical and thermal conductivity. The use of graphene to form graphene-encapsulated CuNWs increased electrical and thermal conductivity compared to uncoated CuNWs, which allows a greater sensitivity in the detection of data even with miniaturization (Mehta et al. 2015).

During a tissue injury, there will be instantaneous damage of local blood vessels, edema, bleeding, and interruption of the nutritional oxygen supply to the cells previously nourished by these vessels. Local hypoxia activates angiogenic factors to restore vascularization and to avoid tissue necrosis. However, in certain situations, this process cannot occur correctly, impairing healing, and may even lead to possible amputations. Therefore, it is evident that monitoring the local tissue oxygenation allows us to evaluate the evolution of the healing process.

The great difficulty of introducing *oxygen sensors* in the intelligent dressings is to selectively analyze the oxygen concentration of the tissue without being influenced by the contact with atmospheric oxygen. There are available dressings with coatings impermeable to oxygen; however, as mentioned above, the ideal dressing should allow the passage of gases not to impair healing (Li et al. 2014). Actually, the gold standard test to quantify the tissue oxygen concentration is the needle-type Clark electrodes, an invasive procedure that can lead to painful, local injury, and changes in microcirculation. Due to these possible adverse effects of monitoring by needle-type Clark electrodes, many products have been developed to allow the detection of tissue oxygenation, e.g., laborious methods with hypoxia markers such as pimonidazole and those using radioactively labeled indicators. Another mechanism developed is a thin film with a flourishing feature dependent on the presence of oxygen, which is painted on the skin. It can be used to identify the oxygen concentration of the underlying skin tissue. The results of in vivo studies have been shown to be exciting and promising (Li et al. 2014).

During the healing process, there is the formation of inflammatory exudate. It is expelled by the wound and helps to maintain moisture. It is well documented and accepted that moisture balance is important in achieving optimum wound healing conditions. However, when the site is protected by some types of dressing, it prevents the maintenance of this moisture and may cause unbalance of this parameter. Excessive moisture wounds facilitate the proliferation of microorganisms, leading to an increased risk of infections (Farrow et al. 2012). Therefore, changing the bandage within the appropriate frequency is recommended. Nevertheless, clinicians have been unable to observe the moisture status of the dressing without disturbing it. Consequently, selecting the ideal time for this change has been a task with high failure rate. It is estimated that almost half of the changes occur before the ideal time, which results in increased patient costs and private or public health policies and unnecessary physical stress. Therefore, the development of *moisture sensors* is considered as a key tool in patient follow-up (McColl et al. 2007).

Many materials are used in moisture sensors and most of them can be classified based on organic and inorganic sensitive materials. Inorganic-based sensing materials, for example, Al_2O_3, In_2O_3, $MnWO_4$, SiO_2, SnO_2, TiO_2, WO_3, and $ZrTiO_3$, can detect the slightest variations of humidity in the wound environment and have limited utility in the microchip manufacturing. Manufacturing becomes more viable with products processed at low temperatures, such as hydrosoluble organic or polymer-based sensing materials polyvinyl alcohol, poly(sodium 4-styrenesulfonate), poly(vinylpyrrolidone) (PVP), PANI, and its congeners. However, two disadvantages were noted: the high-water solubility of these materials in environments with high humidity and their relatively slow response time, which varies from 10 s until impractical 2 min. Therefore, these sensors are not considered adequate to integrate the sensorial devices, because they have low sensitivity and energy efficiency, in addition to an extended response time (Bruinink 2018).

Despite advances in the area of intelligent sensors, the limited and low number of wound care devices available in clinical practice is surprising. In 2013, WoundSense ™ sensor came to the market. It can measure the moisture at the site of the wound and be adapted in any dressing (Fig. 14.3). It is based on two parallel electrodes with a defined distance measuring the impedance between them. This parameter is used to classify the local humidity into five rates: wet, wet to moist, moist, moist to dry, and dry (Milne et al. 2016).

Another promising approach is the use of graphene oxide (GO) to bind water molecules. This is due to the GO's ability to be an amphiphilic soft material. It allows excellent response time and recovery time of the humidity sensors (0.2/0.7 s, respectively) when comparing with other forms of graphene or nanomaterials, such as reduced GO (4 s/10 s), graphene quantum dots (10 s/20 s), carbon nanotube (6 s/120 s), or H-doped graphene (3 min/several hours).

The development of a wireless device to continuously and remotely monitor the wound healing process generates benefits to everyone involved, both patients and professionals. This remote communication allows the reduction of the physical inspection of the wounds so that one can know the ideal moment of some clinical intervention, reducing prolonged hospitalizations or repeated visits to clinics. Smart integrated wireless dressings follow a standard system in most devices (Fig. 14.4). The interesting thing to reduce the costs of the devices is always to try to reuse the part of the dressings represented by the sensors. The sensors to detect parameters on

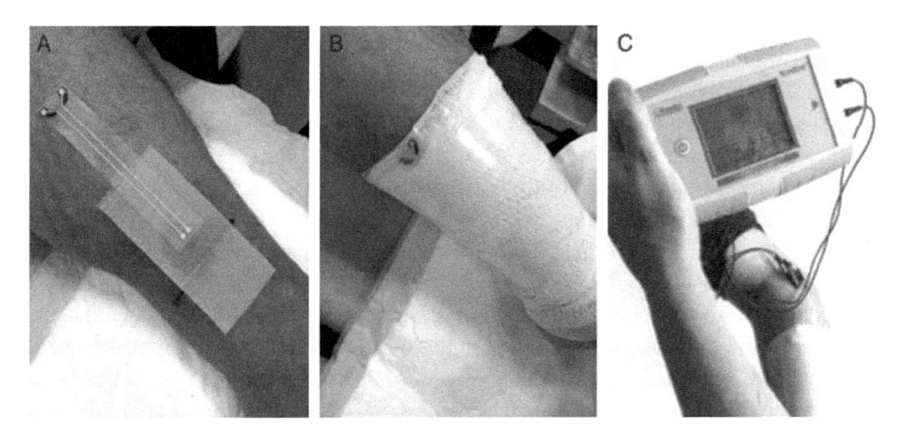

Fig. 14.3 (**a**) WoundSenseTM sensor implanted at the wound site, (**b**) after dressing and (**c**) being measured by a mobile device (WoundSenseTM meter) (Milne et al. 2016)

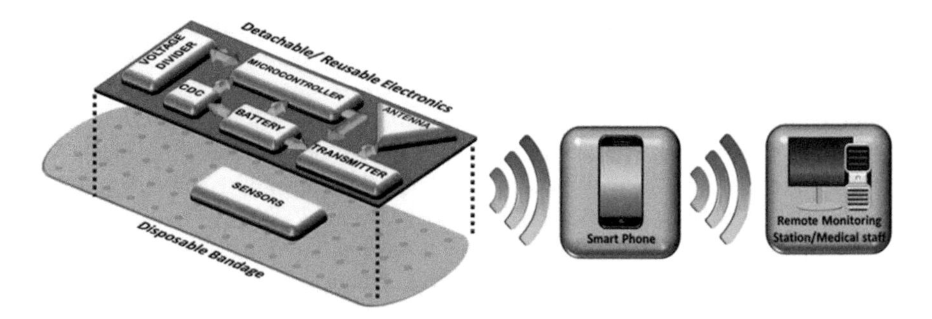

Fig. 14.4 Representation of the system design. Highlight of the two parts of the device, a disposable part and a reusable part, and its components (Farooqui and Shamim 2016)

the wound make up the disposable part, whereas the electronics are integrated and can be detached and reused multiple times. The values obtained by the sensors are processed by the electronics, and the information is sent wirelessly to a personal smartphone and remote monitoring station or medical staff (Farooqui and Shamim 2016).

14.3.1.1 Dressings with Active Delivery of Drugs

Monitoring the characteristics of a wound, as mentioned earlier in the text, is a fundamental part of modern rational treatment. Complementing this information with "smart" drug and healing factor delivery would put the dressings on a plateau above. Basically, we can divide the drug delivery system into two main processes: a passive one and an active one. The passive process is simpler than the active one because the drug embedded is released from dressings regardless of wound needs. However, the wound environment undergoes dynamic changes. Consequently, the need for drugs could be different at singular time points and curative stages, which requires adjustments in the release profiles in real time. On the other hand, active drug delivery systems allow greater control of release profile. These systems are usually designed to release only drugs in response to external stimulus sites such as temperature, pH, pressure, electrolyte, and mechanical concentration (Saghazadeh et al. 2018). Currently, researchers are seeking the ideal combination of local markers that signal the physiological changes of each clinical status, such as skin cancer, burn, psoriasis, and chronic wound. The definition of such desired markers enables the selection and arrangement of the sensors that will be incorporated into devices for active control of the release of drugs. Miniaturized pumps integrated with dressings are considered the most used strategy for delivering solutions containing drugs into the wound site. These pumps can be used integrated with microneedles to inject accurate concentration of drugs. Thermal stimulation is considered a safe, popular, and low-cost method, which can increase transdermal drug permeability and reduce healing time (Prausnitz et al. 2004).

An integrated multilayer bandage was developed in 2015. It features electrochemical pH sensors embedded in a hydrogel matrix composed also by thermoresponsive drug carriers cast on a flexible heater. When the sensor detects pH values outside the previously determined range as acceptable, the heater is triggered to release antibiotics (Saghazadeh et al. 2018; Mostafalu et al. 2015).

Some thermally activated wound dressings use thermoresponsive particles of pNIPAM produced by means of a microfluidic system. Hydrogel-based dressing matrix structure with a flexible heater was used as thermoresponsive drug and growth factors carrier to the injured tissue. This device has the ability to store and release various chemical structures. An example of a complex consisting of different nanomaterials are the chitosan-based nanoparticles mixed with a polymeric solution. This flexible complex forms nanofibrous substrates composed of nanoparticles in its interior. Changes to local temperature stimulate the release of the drug (Huang et al. 2014). Numerous studies about the appropriate drug delivery to the wound site were recently discussed by Saghazadeh and collaborators (2018).

14.3.2 Electronic Skin (e-Skin)

Unlike dressings, e-skin was not designed to protect a wound or to serve as a platform for the release of active substances. Adequate functioning of the e-skin is dependent on many of the parameters also required for the performance of smart dressings, i.e., flexibility, sensitivity, and the ability to provide information on certain pathophysiological parameters. These devices can identify and signal changes in the most superficial layers of the skin even in the deep ones, e.g., invasive melanomas (Dagdeviren et al. 2015).

The viscoelastic changes of the skin are perceived by nanoelectronics and pressure sensors, which transform them into electrical signals for OLED processing into pixelated signals. These changes are detected by elastomeric pressure sensor built by the incorporation of carbon nanotubes (CNTs) with polydimethylsiloxane (PDMS) substrate and an electronic system that enables the conversion of the viscoelastic parameters into a pixelated format (Kim et al. 2015). New multifunctional circuits have been developed and promise to revolutionize the range of responses achieved by e-skin. Such multifunctional and mechanically robust platforms identify variations in thermic, moisture, and chemical components. Among the materials evaluated, CNT- and graphene-based nanoelectronics have been outstanding due to the excellent sensitivity towards temperature and humidity changes (Lu et al. 2019; Kim et al. 2015). However, when comparing some polymer nanocomposite-based strain sensors, it is concluded that 1D nanomaterials (CNTs and AgNWs) are preferable than other nanomaterials such as carbon blacks (CBs) and graphene and silver nanoparticles (AgNPs). This observation is confirmed by the values of gauge factor and stretchability and explained by the high-aspect ratios as well as the capability to sustain the percolation network at large strain presented by the 1D nanomaterials. A deficiency of conductive nanomaterials is that they do not support strain over 5%, so a designed conductive network is required that controls the transfer voltage to elastomeric matrix (Lu et al. 2019).

In order to obtain proper e-skin devices, there are still many obstacles to be overcome, such as materials that meet the practical needs (conductive nanomaterials, high purity, adequate stretchability, and affordable price). Carbon black does not present good electrical conductivity and needs high loading to form the percolation network. Graphene-based strain sensors exhibit inadequate stretchability, which facilitates the appearance of irreversible cracks. Metal nanowires, for example, AgNWs and AuNWs, are expensive, which impede the widespread trade of devices. Low purity is highlighted as the main deficiency of CNTs. Therefore, it is clear that the science of nanomaterials needs to advance a little further before introducing this new technology on a large scale.

14.3.3 Tattoo-Based Electrochemical Sensors

Biosensors and electrochemical sensors have recently found extensive applications in medicine, mainly in diagnostic fields. These sensors are largely used in a number of analytical instruments in clinical laboratories and commercial point-of-care devices, e.g., pH electrodes and glucose biosensors. Electrochemical sensors work when they transform electrochemical information into an analytical signal. It is formed by two main components, a chemical recognition system and a physico-chemical transducer device that converts the chemical response into an electrical signal recognized by a working electrode. Otherwise, biosensors recognize bio-chemical markers when contacting their chemical sensors. The great challenge is building a sensor with high sensitivity, selectivity, accuracy, and low cost (Chandra 2016).

Screen printing, photolithography, and stamping methods are commonly used to introduce electrodes on planar substrates. This allowed the development of a simple manufacturing sensory device for low-cost monitoring, capable of performing non-invasive electrochemical analyzes of biomarkers present in body fluids, mainly body sweat. They are called functional sensing "tattoos," also known as tattoo-based electrochemical or colorimetric sensors (TBECS) (Kabiri et al. 2017). One interesting feature of these wearable sensors is that they can evaluate parameters related to patient health in the form of varied artistic tattoo patterns. These parameters include quantification of electrolytes, lactate production, and even xenobiotics eliminated by sweating, such as heavy metals and drugs. In 2013, Jia et al. demonstrated the first example of a noninvasive enzymatic temporary-transfer tattoo electrochemical biosensor that provided a real-time analysis of sweat lactate during exercise (Fig. 14.5). This tattoo is composed of a surface of the electrodes imprinted with tetrathiafulvalene (TTF) and multiple wall carbon nanotubes (CNT), functionalized by the presence of the LOx enzyme adhered to the outer layer of chitosan. Screen techniques were used obeying the body contours for the synthesis of this sensor. Mixed with the printing ink, carbon fibers were added, which allowed greater resilience and robustness towards extreme mechanical stresses expected from physical activity. Unlike the traditional methods for measuring lactate, the tattoos allow to evaluate the results of anaerobic exercises, which produce lactate in the muscles by the consumption of glycogen stores, in a noninvasive and simpler way (Jia et al. 2013).

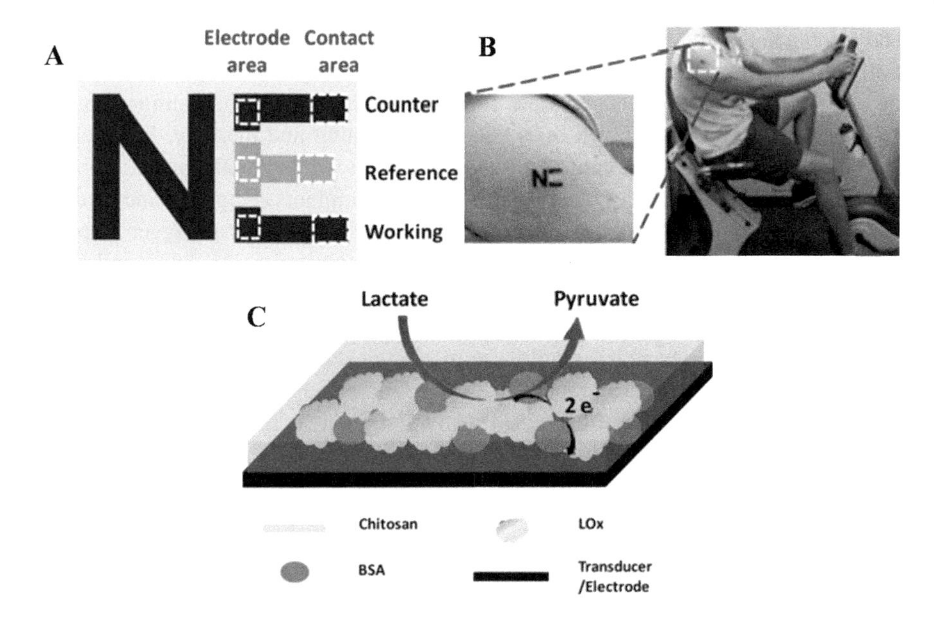

Fig. 14.5 (**a**) Schematic illustration of a three-electrode "NE" tattoo biosensor for electrochemical epidermal monitoring of lactate. (**b**) An "NE" lactate biosensor applied to a deltoid. (**c**) Constituents of the reagent layer of the working electrode that is coated by biocompatible polymer (chitosan). (Adapted from Jia et al. 2013)

14.4 Internal Healthcare Monitors (Implants)

As well as smart dressings, the growing number of breakthroughs in bioelectronics driven by the emergence of nanomaterials and micronization of devices allows real-time internal monitoring of health risk factors. Cardiovascular diseases are the leading causes of deaths and morbidities around the world. Therefore, monitoring the characteristics of blood pressure and cardiac function is vital for the treatment and prevention of these diseases. For this purpose, the electrocardiograms are used: a simple procedure, noninvasive, and capable of real-time monitoring of heart activity. However, it requires the patient to go to a hospital facility. Another questioning of the method is that the electrodes are not directly in contact with the myocardium, which may impair the precision of the stimulation and control of inotropic and chronotropic responses. Electrode implants should be flexible enough to adapt to the sinuous characteristics of organic tissues. Among the materials used, the 1D nanomaterials (silicon nanowires, silver nanowire, and CNT) are incorporated into biocompatible and flexible elastomers with good success rate. An in vivo study obtained signal-to-noise ratios and high spatial resolution from a silicon-based nanocircuit assembled onto a flexible polyimide (PI) substrate (Timko et al. 2009). A second study developed a device that consisted of an Au-based electrode network deposited onto a flexible PI substrate (25 μm). Due to its 288 electrical contact points, it was possible to record electrical signals with high sensitivity for signal-to-noise ratio and temporal resolutions (≈34 decibels and < 2 ms, respectively) in the test in a porcine animal model (Viventi et al. 2010). Neural connections are complex, interdependent, multifunctional, and fundamental for the proper functioning of basic responses to the most complex ones, such as reasoning and memory. In order to understand and monitor the central nervous system, it would be innovative to apply the network of injectable, miniature wireless neural implants (Ferguson and Redish 2011). For example, identifying the exact location of the electrical trigger responsible for an epileptic seizure or previously excited regions prior to a panic syndrome allows researchers to better understand the diseases. A silicon nanomembrane device transistor into the electrode array is able to identify specific responses on the surface of visual cortex, for example, sleep spindles and single-trial visual evoked responses. The array was composed of 720 silicon nanomembrane transistors. Active electrode arrays were produced using a multilayer process, which can be divided into four layers: (i) doped silicon nano-ribbons (~ 260 nm, first layer); (ii) horizontal and vertical metal interconnect layers insulated using layers of PI; (iii) polymeric encapsulation layers (PI and epoxy); (iv) and platinum (~ 50 nm) deposited onto the surface electrodes (fourth layer). The device was either implanted on the cerebral surface or inserted interhemispheric fissure of the brain (Viventi et al. 2010).

14.5 Development of Cyborgs (Prosthetics and Cyborganics)

The idea of merging inanimate materials with living organisms, despite being innovative in several of its purposes, cannot be taken as recent. It derives largely from the cybernetic term, created in the early 1960s. At present, with the advancement of medicine, this term also encompasses technical healthcare monitors, implants, and prosthetic materials integrated to the body to overcome some physiological limitations and even create superhuman responses as in the senses of hearing and sight (Mehrali et al. 2018). With the development of these cybernetic products, another term emerged: cyborgs (short for "cybernetic organisms"). There are also lines of research that associate living tissues within intricate nanowire-based materials to create hybrid synthetic tissues, half-synthetic and half-organic. They are called cyborg organic constructs (cyborganics) (Dai et al. 2018).

Proper integration between the material to be introduced and the human body should ensure stability, durability, and quality of performance. Therefore, the material should have sufficient flexibility and biocompatibility to adapt to body characteristics without resorting to bodily movements (Simon et al. 2013; Zhou et al. 2017; Tamayol et al. 2016). The applications of cybernetic prosthetics become more numerous and rapidly expanding at the same speed as new nanomaterials and bioelectronic products are developed. Some examples of implants will be developed to treat still irreversible problems such as retinal implants that bridge the optic nerves with the brain region that interprets the signs, paraplegia (spinal cord implants), and deafness (ear implants that reestablish the auditory capacity). There is already a product successfully developed in vivo that is characterized by the union of inanimate electronics with living tissues to create cyborganic transplants (Mehrali et al. 2018). In addition to allowing replacement parts for those physically disabled, the prostheses can also be used to enhance human capabilities beyond normality. It is hard to imagine the limit of this improvement in abilities that can range from an increase in hearing to even the creation of cyborganic muscle tissue with high efficiency and endurance.

Cyberorganics represent the evolution of science of materials, medicine, tissue engineering, and electronics. The breaking of a paradigm is evident when imagining the creation of hybrid tissues and hydrogel carriers made from a combination of inanimate and biological matter. These devices can be classified as permanent hybrid organs or biodegradable which may be of nanomaterials or hydrogel carriers embedded with biological matter (Mehrali et al. 2017). In short, the great difference between the two classes is that the permanent hybrid organs, as the name itself says, do not undergo purposive biodegradation of the nonorganic part. On the other hand, the biodegradable class uses hydrogel carriers and nanomaterials as a matrix of organic material until it reorganizes and assumes its function in the local tissue.

An example of the application of this technology can be observed in the study that carried out the maturation of stem cells inside a web of silicon nanowires (Tian et al. 2012). Silicon nanowire field-effect transistor (FET)-based nanoelectronic biomaterials have the capability of recording both extracellular and intracellular signals with subcellular resolution. The synthetic process can be described in three main steps: In the first, silicon nanowires were deposited randomly or in regular patterns for single nanowire FETs (Fig. 14.6 step A). Then, individual nanowire FET devices were lithographically patterned and integrated into free-standing macroporous scaffolds (Fig. 14.6 step B), the nanoelectronic scaffolds, nanoES, which simulate the extracellular matrices with high porosity and biocompatible flexibility. NanoES were then combined with synthetic or natural macroporous extracellular matrices providing (i) electrical sensory function and (ii) nanoES with biochemical environments suitable for tissue culture. Finally, cells were cultured within the nanoES (Fig. 14.6 step C) to yield 3D hybrid nanoelectronic tissue constructs. This technique was extended to the development of tissue embedded in nanoelectronic sensors, for example, human aortic smooth muscle cells (HASMCs) were cultured on 2D mesh vascular nanoES (Fig. 14.6D). The cells of hybrid nanoES/HASMC were able to produce the contractile protein in smooth muscles, α-actin.

The interest in creating/retrieving lost or damaged tissues has grown stronger through the advances of cyborganics (Fig. 14.7). Artificial ear was created, which was capable of auditory sensing in the radio-frequency range. This device was generated by a 3D printer using the raw material Ag nanoparticles and cartilage cells embedded within an alginate hydrogel matrix. The product of this mixture was deposited into a bioelectronic hybrid ear, where 3D structural matrix of the ear underwent bioxidation and the cartilage cells retained the ear-shaped morphology and developed with functionality (Mannoor et al. 2013).

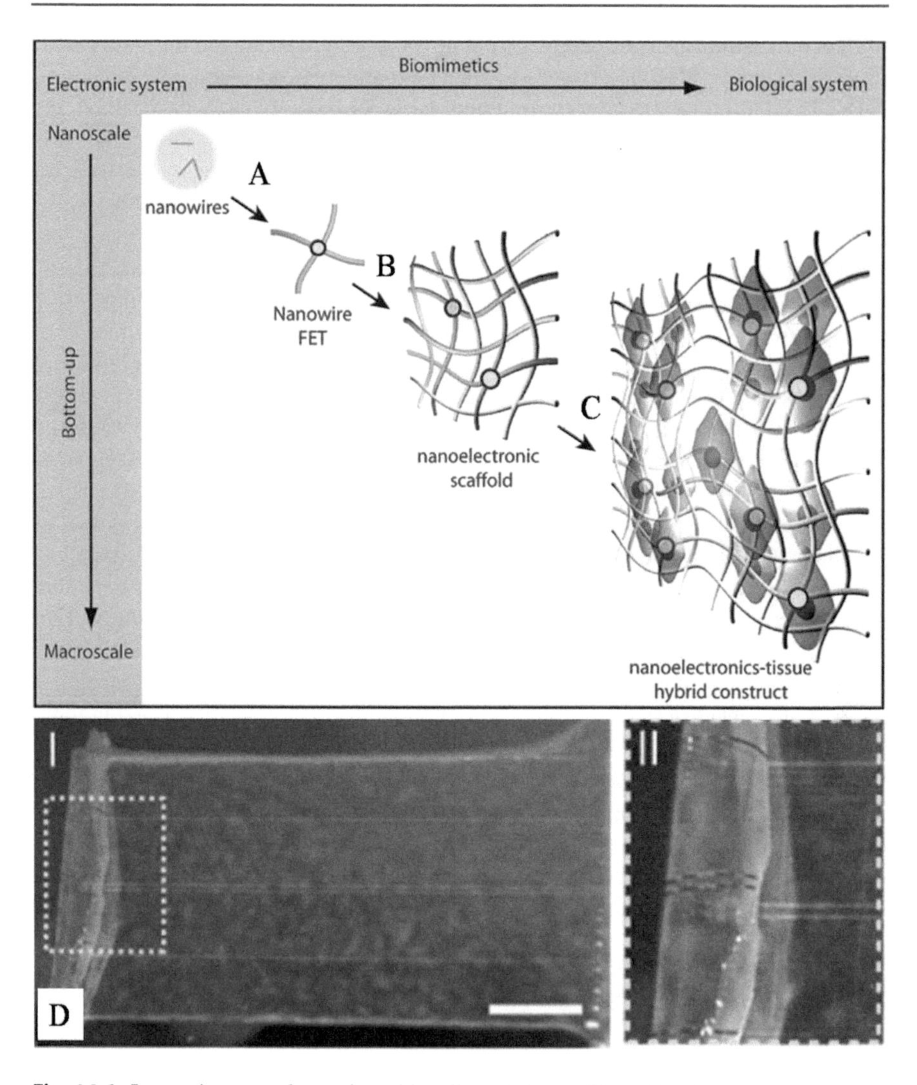

Fig. 14.6 Integrating nanoelectronics with cells and tissue. Three biomimetic and bottom-up steps have been designed: (A) patterning, metallization, and epoxy passivation for single nanowire FETs; (B) forming 3D nanowire FET matrices (nanoelectrical scaffolds); (C) incorporation of cells and growth of synthetic tissue via biological processes. (D) Synthetic vascular construct enabled for sensing. (I) Photograph of a single HASMC sheet cultured with sodium L-ascorbate on a nanoES. (II) Zoomed-in view of the dashed area in (I), showing metallic interconnections macroscopically integrated with cellular sheet. Note Fig. 14.6 A–C. Yellow dots, nanowire components; blue ribbons, metal and epoxy interconnects; green ribbons, traditional extracellular matrices; pink, cells. (Adapted from Tian et al. 2012)

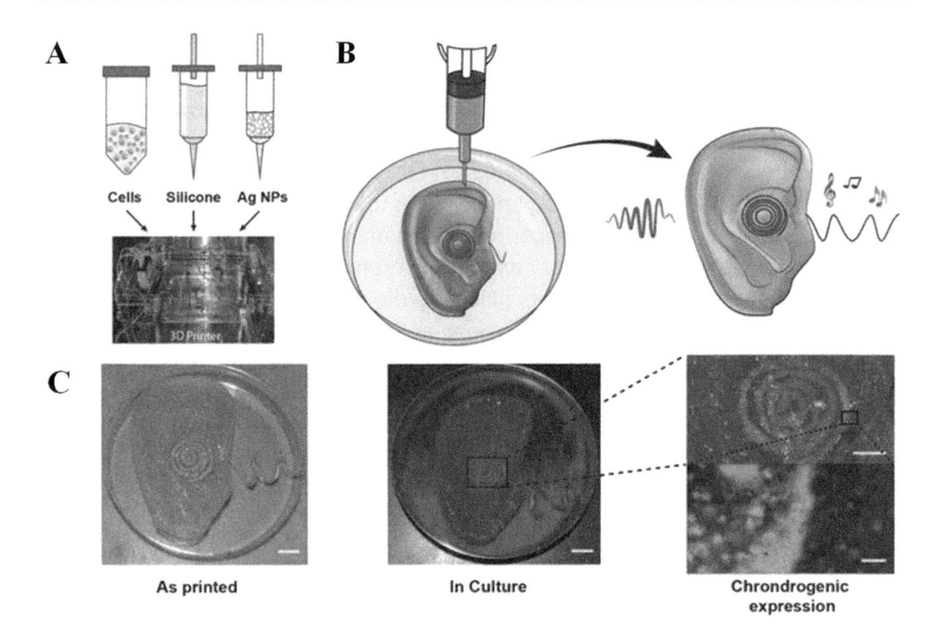

Fig. 14.7 Development of a cyborg ear. (**a**). Organic and synthetic constituents employed (**b**). Demonstration of the technique of 3D printing of the artificial ear (**c**). Images of the printed ear before and after culturing and highlighting the evaluation of chrondogenic cell viability (Mannoor et al. 2013)

14.6 What Are the Next Steps?

We can undoubtedly say that the field of nanomaterials has no limits and that predicting future events is bordered by a utopia. However, some predictions related to devices that are already produced are more feasible. Smart dressings are expected to integrate even more active markers and substances (drugs, growth factors, and even stem cells) into the same device. Measuring the concentration of a circulating drug in plasma or in a specific tissue in which it might accumulate would allow real-time monitoring of therapeutic regimens and toxic events. Sensors to identify the type of microorganism that colonizes a wound or diagnose cancers will be developed in the near future. However, it will be necessary to improve the nanomaterials and miniaturize the sensors so that the lab-on-a-chip devices perform the genetic sequencing. There is still a need to make nanoelectronic devices even smaller, to reduce production costs, and to conduct more clinical studies with surface dressings and implanted sensors. In this way, the products can be popularized and commercially viable. The tendency to universalize remote monitoring is a reality of modern medicine, which will reduce the medical costs of moving patients and mobile health services, face-to-face consultations, and variations in the patient's health status, which has difficulty receiving the adequate presential attendance of the medical team.

Cyborgs will revolutionize various areas of modern medicine, such as transplantation and implanting organs and tissues made from nanomaterials or using them as a matrix for stem cells. The cinematic idea of having a robotic member built of nanomaterials and coated with human tissues, with sensitivity to heat, pain, and touch, is ever closer. These "biotechnological tissues" will allow to reestablish lost functions as in blind individuals, in paraplegic, or in the queue of organ transplants, as well as to create tissues with superhuman responses. These super responses are fantastic: imagine seeing objects at great distances, performing facial identification of people through smart nanosensors placed in the eyes, or creating super athletes with high-resilience muscles. Let us imagine an individual with a family history of lung cancer with poor prognosis; he would clearly be a perfect candidate to replace this natural organ with a cyborg lung, 100% constructed with nanomaterials and with zero risk of tumor growth. Autoimmune diseases such as rheumatoid arthritis, Hashimoto's disease, and type 1 diabetes would also be controlled without major problems by replacing the biological tissue (antibody antigen) with highly functionalized nanomaterials.

So, what are the next steps? Different answers can arise, but one thing is certain, we will certainly be surprised and experienced with this new industrial revolution!

References

Brown MS, Ashley B, Koh A (2018) Wearable technology for chronic wound monitoring: current dressings, advancements, and future prospects. Front Bioeng Biotechnol 6:1–21

Bruinink A (2018) Biosensor-bearing wound dressings for continuous monitoring of hard-to-heal wounds: now and next. Biosens Bioelectron Open Acc: BBOA-117. https://doi.org/10.29011/BBOA-117.100017

Chandra P (2016) Nanobiosensors for personalized and onsite biomedical diagnosis. IET, London

Dagdeviren C, Shi Y, Joe P, Ghaffari R, Balooch G, Usgaonkar K, Gur O, Tran PL, Crosby JR, Meyer M, Su Y, Chad Webb R, Tedesco AS, Slepian MJ, Huang Y, Rogers (2015) Conformal piezoelectric systems for clinical and experimental characterization of soft tissue biomechanics. Nat Mater 14:728–736

Dai X, Hong G, Gao T, Lieber CM (2018) Mesh nanoelectronics: seamless integration of electronics with tissues. Acc Chem Res 51(2):309–318

Farooqui MF, Shamim A (2016) Low cost inkjet printed smart bandage for wireless monitoring of chronic wounds. Sci Rep 6:1–13

Farrow MJ, Hunter IS, Connolly P (2012) Developing a real time sensing system to monitor bacteria in wound dressings. Biosensors 2:171–188

Ferguson JE, Redish AD (2011) Wireless communication with implanted medical devices using the conductive properties of the body. Expert Rev Med Devices 8(4):427–433

Fonder MA, Lazarus GS, Cowan DA, Aronson-Cook B, Kohli AR, Mamelak AJ (2008) Treating the chronic wound: a practical approach to the care of nonhealing wounds and wound care dressings. J Am Acad Dermatol 58:185–206

Guinovart T, Ramirez GV, Windmiller JR, Andrade FJ, Wang J (2014) Bandage-based wearable potentiometric sensor for monitoring wound pH. Electroanalysis 26:1345–1353

Han G, Ceilley R (2017) Chronic wound healing: a review of current management and treatments. Adv Ther 34:599–610

Huang CH, Kuo TY, Lee CF, Chu CH, Hsieh HJ, Chiu WY (2014) Preparation of a thermo- and pH-sensitive nanofibrous scaffold with embedded chitosan-based nanoparticles and its evaluation as a drug carrier. Cellulose 21:2497–2509

Jia W, Bandodkar AJ, Valdés-Ramírez G, Windmiller JR, Yang Z, Ramírez J, Chan G, Wang J (2013) Electrochemical tattoo biosensors for real-time noninvasive lactate monitoring in human perspiration. Anal Chem 85:6553–6560

Kabiri AS, Ho R, Jang H, Tao L, Wang Y, Wang L, Schnyer DM, Akinwande D, Lu N (2017) Graphene Electronic Tattoo Sensors. ACS Nano 11(8):7634–7641

Kim SY, Park S, Park HW, Park DH, Jeong Y, Kim DH (2015) Highly sensitive and multimodal all-carbon skin sensors capable of simultaneously detecting tactile and biological stimuli. Adv Mater 28:4178–4185

Li Z, Roussakis E, Koolen PG, Ibrahim AM, Kim K, Rose LF, Wu J, Nichols AJ, Baek Y, Birngruber R, Apiou-Sbirlea G, Matyal R, Huang T, Chan R, Lin SJ, Evans CL (2014) Non-invasive transdermal two-dimensional mapping of cutaneous oxygenation with a rapid drying liquid bandage. Biomed Opt Express 5:3748–3764

Lu Y, Biswas MC, Guo Z, Jeon JW, Wujcik EK (2019) Recent developments in bio-monitoring via advanced polymer nanocomposite-based wearable strain sensors. Biosens Bioelectron 123:167–177

Mannoor MS, Jiang ZW, James T, Kong YL, Malatesta KA, Soboyejo WO, Verma N, Gracias DH, McAlpine MC (2013) 3D printed bionic ears. Nano Lett 13(6):2634–2639

McColl D, Cartlidge B, Connolly P (2007) Real-time monitoring of moisture levels in wound dressings in vitro: An experimental study. Int J Surg 5:316–322

Mehrali M, Thakur A, Pennisi CP, Talebian S, Arpanaei A, Nikkhah M, Dolatshahi-Pirouz A (2017) Nanoreinforced hydrogels for tissue engineering: biomaterials that are compatible with load-bearing and electroactive tissues. Adv Mater (8):1–26

Mehrali M, Bagherifard S, Akbari M, Thakur A, Mirani B, Mehrali M, Hasany M, Orive G, Das P, Emneus J, Andresen TL, Dolatshahi-Pirouz A (2018) Blending electronics with the human body: a pathway toward a cybernetic future. Adv Sci 5:1700931

Mehta R, Chugh S, Chen Z (2015) Enhanced electrical and thermal conduction in graphene encapsulated copper nanowires. Nano Lett 15:2024–2030

Milne SD, Seoudi I, Al Hamad H, Talal TK, Anoop AA, Allahverdi N, Zakaria Z, Menzies R, Connolly P (2016) A wearable wound moisture sensor as an indicator for wound dressing change: an observational study of wound moisture and status. Int Wound J 13(6):1309–1314

Mostafalu P, Sonkusale S (2014) Flexible and transparent gastric battery: energy harvesting from gastric acid for endoscopy application. Biosens Bioelectron Apr 54:292–296

Mostafalu P, Amugothu S, Tamayol A, Bagherifard S, Akbari M, Dokmeci MR, Khademhosseini A, Sonkusale S (2015) Smart flexible wound dressing with wireless drug delivery. Biomedical Circuits and Systems Conference (BioCAS), IEEE:1–4

Mostafalu P, Akbari M, Alberti KA, Xu Q, Khademhosseini A, Sonkusale SR (2016) A toolkit of thread-based microfluidics, sensors, and electronics for 3D tissue embedding for medical diagnostics. Microsyst Nanoeng 2:1–10

Nussbaum SR, Carter MJ, Fife CE, DaVanzo J, Haught R, Nusgart M, Cartwright D (2018) An economic evaluation of the impact, cost, and medicare policy implications of chronic nonhealing wounds. Value Health 21(1):27–32

Prausnitz MR, Mitragotri S, Langer R (2004) Current status and future potential of transdermal drug delivery. Nat Rev Drug Discov 3:115–124

Punjiya M, Mostafalu P, Sonkusale S (2017) Smart bandages for chronic wound monitoring and on-demand drug delivery. 2017 IEEE 60th International Midwest Symposium on Circuits and Systems (MWSCAS), Aug 6–9:495–498

Saghazadeh S, Rinoldi C, Schot M, Kashaf SS, Sharifi F, Jalilian E, Nuutila K, Giatsidis G, Mostafalu P, Derakhshandeh H, Yue K, Swieszkowski W, Memic A, Tamayol A, Khademhosseini A (2018) Drug delivery systems and materials for wound healing applications. Adv Drug Deliv Rev 127:138–166

Siegel AC, Phillips ST, Dickey MD, Lu N, Suo Z, Whitesides GM (2010) Foldable printed circuit boards on paper substrates. Adv Funct Mat 20:28–35

Simon D, Ware T, Marcotte R, Lund BR, Smith DW Jr, Di Prima M, Rennaker RL, Voit W (2013) A comparison of polymer substrates for photolithographic processing of flexible bioelectronics. Biomed Microdevices 15(6):925–939

Sridhar V, Takahata K (2009) A hydrogel-based passive wireless sensor using a flex-circuit inductive transducer. Sensors Actuators A: Physical 155:58–65

Tamayol A, Akbari M, Zilberman Y, Comotto M, Lesha E, Serex L, Bagherifard S, Chen Y, Fu G, Ameri SK, Ruan W, Miller EL, Dokmeci MR, Sonkusale S, Khademhosseini A (2016) Flexible pH-sensing hydrogel fibers for epidermal applications. Adv Healthc Mater 6:711–719

Tian B, Liu J, Dvir T, Jin L, Tsui JH, Qing Q, Suo Z, Langer R, Kohane DS, Lieber CM (2012) Macroporous nanowire nanoelectronic scaffolds for synthetic tissues. Nat Mater 11:986–994

Timko BP, Cohen-Karni T, Yu G, Qing Q, Tian B, Lieber CM (2009) Electrical recording from hearts with flexible nanowire device arrays. Nano Lett 9(2):914–918

Viventi J, Kim DH, Moss JD, Kim YS, Blanco JA, Annetta N, Hicks A, Xiao J, Huang Y, Callans DJ, Rogers JA, Litt B (2010) A conformal, bio-interfaced class of silicon electronics for mapping cardiac electrophysiology. Sci Transl Med (24):24ra22

Zakeri M, Arjmand N, Forouzanfar A, Zakeri M, Koohestanian N (2016) Nano-silver suture as a new application for healing of periodontal flaps. Int J Dent Oral Health (7):1–5

Zhou J, Xu X, Yu H, Lubineau G (2017) Deformable and wearable carbon nanotube microwire-based sensors for ultrasensitive monitoring of strain, pressure and torsion. Nanoscale 9(2):604–612

9789813298392